Springer Climate

Series editor

John Dodson (iD), Institute of Earth Environment, Chinese Academy of Sciences
Xian, Shaanxi, China

Springer Climate is an interdisciplinary book series dedicated to climate research. This includes climatology, climate change impacts, climate change management, climate change policy, regional climate studies, climate monitoring and modeling, palaeoclimatology etc. The series publishes high quality research for scientists, researchers, students and policy makers. An author/editor questionnaire, instructions for authors and a book proposal form can be obtained from the Publisher, Dr. Michael Leuchner (michael.leuchner@springer.com).

More information about this series at http://www.springer.com/series/11741

Simone Lucatello • Elisabeth Huber-Sannwald
Ileana Espejel • Natalia Martínez-Tagüeña
Editors

Stewardship of Future Drylands and Climate Change in the Global South

Challenges and Opportunities
for the Agenda 2030

 Springer

Editors
Simone Lucatello
Estudios Ambientales y Territoriales
Instituto Mora
Mexico City, Mexico

Ileana Espejel
Facultad de Ciencias
Universidad Autónoma de Baja California
UABC
Ensenada, Baja California, Mexico

Elisabeth Huber-Sannwald
División de Ciencias Ambientales
Instituto Potosino de Investigación
Científica y Tecnológica
San Luis Potosi, San Luis Potosí, Mexico

Natalia Martínez-Tagüeña
Cátedra CONACYT
Consortium for Research, Innovation and
Development of Drylands
Instituto Potosino de Investigación
Científica y Tecnológica
San Luis Potosi, San Luis Potosí, Mexico

ISSN 2352-0698   ISSN 2352-0701   (electronic)
Springer Climate
ISBN 978-3-030-22463-9   ISBN 978-3-030-22464-6   (eBook)
https://doi.org/10.1007/978-3-030-22464-6

This Springer imprint is published by the registered company Springer Nature Switzerland AG
The registered company address is: Gewerbestrasse 11, 6330 Cham, Switzerland

Foreword

The world has entered a period of unprecedented change, as conveyed by the Great Acceleration. Our planetary life support system has the potential to be overwhelmed by the combined forces of climate change, declining biodiversity, pollution and social unrest or more positively to chart a course to a sustainable future that pacifies these forces for coming generations. Equally positively, in 2015, the world came together to endorse Agenda 2030 with its set of Sustainable Development Goals (SDGs), a remarkable agreement among all nations of the world as to the form of the future we want.

Nowhere are the challenges and opportunities more evident than in the drylands of the world. Drylands are the 'canary in the mine' for human disruption due to global change. The dependence of a billion of their inhabitants' livelihoods on ecosystem services means that dryland peoples are among the first to be affected by widespread changes such as land degradation, climate change and undermined water cycles. Not surprisingly, many of the world's refugee movements can be linked to resource pressures in drylands. Yet in the past, their challenging environments also made them a source of many social and technical innovations; and this continues today, with persisting traditional linguistic and cultural diversity. Thus, there is hope.

But this hope will only be realized through systematic efforts to entrain research and local knowledge towards an understanding of how to support the drylands better, as well as the implementation of this understanding. This book is a first major product of a relatively new network, RISZA, focused on collective learning about the sustainability of the drylands of the Global South. It builds on advances in drylands thinking over the past two decades, including the evolving Drylands Development Paradigm, but is the first to frame this effort in conjunction with the SDGs.

The contributors to the book frame the challenges of drylands as guiding complex adaptive social-ecological systems, looking through various sectoral lenses but with the whole framework of the SDGs in view. Parts of the book emphasize the potential for alliances at all scales from local transdisciplinary engagement to regional alliances, like the Agadir Platform, and global networks, like RISZA itself, as well as alliances across disciplines and technologies and even in to the arts and humanities that are so important for framing the cultural norms and messaging through which the world views drylands.

I look forward to see the fruits of the discussions and partnerships which have taken root through the development of this book and which are so important for the future of the drylands and their inhabitants.

CSIRO Land and Water, Canberra, Australia   Mark Stafford Smith
May 2019

Introduction

Current global risks emerging from socio-environmental changes are clearly linked to inappropriate and misleading models adopted for natural and socio-economic development (Sena et al. 2016). Among them, the destruction of ecosystems, loss of biodiversity and ecosystem function, land use, occupation, land use change and deforestation and the increasing expansion of drylands, together with misgovernance and other socio-political factors, constitute clear threats to the provision of natural resources, especially at the local level. These processes affect the environment and its interrelationship with society by modifying local populations' conditions of life, health, education and future development perspectives, among others. At the same time, the recent adoption of the Sustainable Development Goals (SDGs) adopted in 2015 paves the way for a new global framework under which nations worldwide must increase their efforts to stop poverty and improve life conditions of millions of people while conserving nature as life support systems by 2030 (UN 2015). The SDGs are firmly rooted in the sustainable development paradigm, which renders them conceptually appealing.

Ever since the conception of 'sustainability' as a guiding paradigm, it has become evident how difficult it is to integrate social, political, ecological and economic aspects—because of their complex interrelations and trade-offs (Berg 2015). In the specific case of drylands, challenges multiply due to the complex management of the so-called *fragile ecosystems*, like deserts which are constantly under pressure by climatic variations and human-induced activities. Desertification affects as much as one-sixth of the world's population, 70% of all drylands and one-quarter of the total land area of the world (WAD 2018). It results in the widespread poverty as well as in the degradation of billion hectares of rangeland and cropland (UNCED 2015). Understanding the drylands socioecological systems, integrated with stewardship (i.e. planning, management and governance), must be set out in order to fulfill the ambitious agenda of the SDGs.

Cross-sectoral aspects of decision-making for the sustainable use and development of natural and cultural resources as well as a transdisciplinary approach to the study of drylands are essential for the implementation of a robust and integral 2030 agenda.

This collective book is meant to explore cutting-edge views from different scholars about drylands and their interactions with a socio-ecosystemic environment and its projections towards the compliance of the SDGs agenda. The authors will explore from different angles the issue of drylands and will analyse the trade-offs as well as the link of social and economic development with environmental protection and enhancement for reaching the goals set by the 2030 agenda.

Contents

Contributors

E. M. Abraham Instituto Argentino de Investigaciones de las Zonas Aridas (IADIZA-CONICET-Universidad Nacional de Cuyo), Mendoza, Argentina

A. El Alaoui Natural Resources and Environment Research Team (NR & E), Department of Chemistry, Faculty of Science and Technology Errachidia, Moulay Ismail University, Meknes, Morocco

M. T. Alarcón-Herrera Centro de Investigación en Materiales Avanzados, S.C. Durango, Durango, Mexico

S. Alfonso de Nehren Institute for Technology and Resources Management in the Tropics and Subtropics, TH Köln, Köln, Germany

G. Arámburo Facultad de Ciencias Administrativas y Sociales, Universidad Autónoma de Baja California UABC, Ensenada, Mexico

M. C. Arredondo-García Facultad de Ciencias Marinas, Universidad Autónoma de Baja California, Ensenada, Mexico

N. Badan Rancho El Mogor, Ensenada, Mexico

M. Ballardo Grupo Mujeres con Alas, Domicilio conocido Bahía de los Ángeles, Ensenada, Mexico

L. Bouchaou Laboratoire de Géologie Appliquée et Géo-Environnement (LAGAGE) Faculté des Sciences, Université Ibn Zohr, Agadir, Morocco

R. Camacho-López Programa de Maestría en Manejo de Ecosistemas de Zonas Áridas, Universidad Autónoma de Baja California, Ensenada, Mexico

L. Carreño Programa de Doctorado en Medio Ambiente y Desarrollo, Universidad Autónoma de Baja California UABC, Ensenada, Mexico

R. Cázares Reyes Ejidatario de La Soledad, Reserva de la Biosfera de Mapimí, Mapimi, Mexico

D. L. Coppock Department of Environment and Society, Utah State University, Logan, UT, USA

A. Cota Programa de Maestría en Manejo de Ecosistemas de Zonas Áridas, Universidad Autónoma de Baja California UABC, Ensenada, Mexico

C. Delgado-Ramírez Escuela de Antropología e Historia del Norte de México, INAH, Chihuahua, Mexico

N. Dennig Dindum Kulturkommunikation e.V., Pulheim-Manstedten, Germany

I. Espejel Facultad de Ciencias, Universidad Autónoma de Baja California UABC, Ensenada, Baja California, Mexico

E. Estrada Grupo Mujeres con Alas, Domicilio conocido Bahía de los Ángeles, Ensenada, Mexico

B. Ferraz Centro de Gestão e Estudos Estratégicos (CGEE), Brasilia, Brazil

D. Galván-Martínez Programa de Doctorado en Medio Ambiente y Desarrollo, Universidad Autónoma de Baja California, Ensenada, Mexico

S. García Independent Consultant, Mexico City, Mexico

C. Gutiérrez Programa de Doctorado en Medio Ambiente y Desarrollo, Universidad Autónoma de Baja California, Ensenada, Mexico

G. Gutiérrez Secretaria de Protección al Ambiente de Baja California, Ensenada, Mexico

M. Gutiérrez Department of Geography, Geology and Planning, Missouri State University, Springfield, MO, USA

A. Hernández El Colegio de la Frontera Norte, Tijuana, Mexico

E. Huber-Sannwald División de Ciencias Ambientales, Instituto Potosino de Investigación Científica y Tecnológica, A.C. (IPICYT), San Luis Potosí, Mexico

L. Ibarra Programa de Maestría en Manejo de Ecosistemas de Zonas Áridas, Universidad Autónoma de Baja California UABC, Ensenada, Mexico

O. Jiménez-Orocio Facultad de Ciencias Marinas, Universidad Autónoma de Baja California, Ensenada, BC, Mexico

C. Leyva Facultad de Ciencias, Universidad Autónoma de Baja California (UABC), Ensenada, Mexico

S. Lichtenberg Department of Physical Geography, University of Passau, Passau, Germany

D. Núñez López Centro de Investigación en Materiales Avanzados, S.C., Durango, Mexico

J. J. López Pardo Instituto Potosino de Investigación Científica y Tecnológica, A.C., San Luis Potosí, Mexico

S. Lucatello Estudios Ambientales y Territoriales, Instituto Mora, Mexico City, Mexico

D. Martínez Centro de Investigación en Materiales Avanzados, S.C. Durango, Durango, Mexico

N. Martínez-Tagüeña Cátedra CONACYT, Consortium for Research, Innovation and Development of Drylands, Instituto Potosino de Investigación Científica y Tecnológica, San Luis Potosi, San Luis Potosí, Mexico

R. I. Mata Páez Instituto Potosino de Investigación Científica y Tecnológica, A.C., San Luis Potosí, Mexico

H. Mertin Freelance Musician, Sound Performer and Music Ethnologist, Cologne, Germany

M. Millones Department of Geography, University of Mary Washington, Fredericksburg, VA, USA

J. Morán El Colegio de San Luis, San Luis Potosí, Mexico

T. Moreno-Zulueta Programa de Doctorado en Medio Ambiente y Desarrollo, Universidad Autónoma de Baja California UABC, Ensenada, Mexico

D. M. Muñoz-Pizza Programa de Doctorado en Medio Ambiente y Desarrollo, Universidad Autónoma de Baja California, Ensenada, Mexico

U. Nehren Institute for Technology and Resources Management in the Tropics and Subtropics, TH Köln, Köln, Germany

M. Niamir-Fuller Vice-Chair International Support Group, International Year of Rangelands and Pastoralists, Purcellville, VA, USA

E. Nickl University of Delaware, Newark, DE, USA

A. Ocaña Grupo Mujeres con Alas, Domicilio conocido Bahía de los Ángeles, Ensenada, Mexico

L. Ojeda-Revah Colegio de la Frontera Norte, Tijuana, Mexico

L. Ortega El Colegio de San Luis, San Luis Potosí, Mexico

B. Parmentier Department of Geography, University of Mary Washington, Fredericksburg, VA, USA

SESYNC, Annapolis, MD, USA

L. Pedrín Programa de Maestría en Manejo de Ecosistemas de Zonas Áridas, Universidad Autónoma de Baja California UABC, Ensenada, Mexico

D. Pinedo Facultad de Ciencias Marinas, Universidad Autónoma de Baja California, Ensenada, Mexico

C. Raedig Institute for Technology and Resources Management in the Tropics and Subtropics, TH Köln, Köln, Germany

J. C. Ramírez Instituto Municipal de Investigación y Planeación de Ensenada, Ensenada, Mexico

R. F. Rentería-Valencia Anthropology and Museum Studies in Central Washington University, Ellensburg, WA, USA

V. M. Reyes Gómez Red Ambiente y Sustentabilidad, Instituto de Ecología, A.C., Chihuahua, Mexico

M. Reyes-Orta Facultad de Turismo y Mercadotecnia, Universidad Autónoma de Baja California, Ensenada, Mexico

A. Rizzo UMR ESPACE-DEV (IRD, Université de Montpellier, Université de la Réunion, Université de Guyane, Université des Antilles), Montpellier, France

P. Rojas Rancho El Mogor, Ensenada, Mexico

M. A. Cordero Grupo Mujeres con Alas, Domicilio conocido Bahía de los Ángeles, Ensenada, Mexico

J. Sandoval Instituto Municipal de Investigación y Planeación de Ensenada, Ensenada, Mexico

D. S. Savín Grupo Mujeres con Alas, Domicilio conocido Bahía de los Ángeles, Ensenada, Mexico

Y. Savín Grupo Mujeres con Alas, Domicilio conocido Bahía de los Ángeles, Ensenada, Mexico

G. Seingier Facultad de Ciencias Marinas, Universidad Autónoma de Baja California, Ensenada, BC, Mexico

M. Silva Grupo Mujeres con Alas, Domicilio conocido Bahía de los Ángeles, Ensenada, Mexico

A. Sifeddine UMR LOCEAN (IRD, CNRS, MNHN, Sorbonne Université), Departamento de Geoquimica-UFF-Brazil, UNAM-IRD-Mexico, Ciudad de México, Mexico

M. C. Tonche Grupo Mujeres con Alas, Domicilio conocido Bahía de los Ángeles, Ensenada, Mexico

Y. Torres Grupo Mujeres con Alas, Domicilio conocido Bahía de los Ángeles, Ensenada, Mexico

A. Trejo Universidad Nacional Autónoma de México, Mexico City, Mexico

C. Turrent Taller de Arquitectura Contextual ClaCla, Ensenada, Mexico

J. Urquidi Macías Ejidatario de La Soledad, Reserva de la Biosfera de Mapimí, Mapimi, Mexico

C. Uscanga Programa de Maestría en Manejo de Ecosistemas de Zonas Áridas, Universidad Autónoma de Baja California, Ensenada, Mexico

I. Vaillard Rancho Tres Mujeres, Ensenada, Mexico

C. Vázquez-León El Colegio de la Frontera Norte, Tijuana, Mexico

Á. Vela Instituto Municipal de Investigación y Planeación de Ensenada, Ensenada, Mexico

L. Vera Instituto Mora, Mexico City, Mexico

M. Villada-Canela Instituto de Investigaciones Oceanológicas, Universidad Autónoma de Baja California UABC, Ensenada, Mexico

C. Villarreal Wislar Comisión Nacional de Áreas Naturales Protegidas, Torreón, Mexico

H. Zgou Polydisciplinary Faculty of Ouarzazate, Ibn Zohr University, Agadir, Morocco

M. Zortea UNESCO Engineering for Human and Sustainable Development, DICAM – Department of Civil, Environmental and Mechanical Engineering, University of Trento, Trento, Italy

About the Editors

Simone Lucatello is a full-time researcher and professor at the Instituto Mora, a public research centre belonging to the Mexican National Agency for Science and Technology (CONACYT) in Mexico City, Mexico. He is one of the leading authors of the IPCC (Intergovernmental Panel on Climate Change) Working Group II, which deals with the impacts, adaptation and vulnerability to climate change for the next IPCC Sixth Assessment Report (AR6, due in 2021), and coordinating leading author of the AR6 North American Chapter. His research interests include climate change impacts in the Global South and risk assessment and disaster risk reduction in Latin America. He served as a consultant to several international organizations, such as the Inter-American Development Bank (IDB), UNEP, UNIDO, OCHA and European Union (Europe Aid) in the Balkans, Central America and Mexico. He is member of the International Network for Sustainable Drylands (RISZA) and of the Scientific Committee of the Humanitarian Encyclopedia at the Genève Centre for Education and Research in Humanitarian Action (CERAH, Switzerland). He is also actively engaged in national and international academic networks and projects across the Americas, Europe and Africa. He holds a master's degree in International Relations from the London School of Economics (LSE) and a PhD in Governance for Sustainable Development from the Venice International University, Italy.

Elisabeth Huber-Sannwald is research professor at the Division of Environmental Sciences, Instituto Potosino de Investigación Científica y Tecnológica (IPICYT), San Luis Potosi, Mexico. She co-founded the area of Global Environmental Change and Ecology and co-designed the Environmental Sciences Postgraduate Program with the specialty in Global Environmental Change and Ecology. She is an expert in dryland systems considering the influence of global and social changes. She has 15 years of experience in interdisciplinary and 5 years of transdisciplinary research. Her work focuses on the integrity and resilience of dryland socioecological systems and the sustainable development of rural livelihoods. She is founder of and currently coordinating the International Network for Drylands Sustainability (RISZA) and is member of the Executive Committee of the Transatlantic Agadir Platform fostering transdisciplinary dryland research in Latin America, Southern Europe and

Northern Africa. She is member of the Continuing Committee of the International Rangeland Congress since 2011. In addition, she served as international member of the Science Council of the Ecological Society of America from 2010 to 2014 and as vice chair and chair of the Rangeland Section of the Ecological Society of America from 2016 to 2018 and is regional president of the Mexican Scientific Ecological Society since 2012. She forms part of the Executive Committee of the International Network for Desertification ARIDnet and contributed to the formulation of the Dryland Development Paradigm. She has been associate editor of *Rangeland Ecology and Management* from 2008 to 2010 and of *Ecological Applications* since 2012. She has organized national and international scientific and multi-sector workshops, meetings and congresses. Her publications have over 5290 citations (without self-citations). Moreover, she has been PI of 15 national and international research grants; has graduated 1 bachelor, 12 masters and 6 doctorate students; and has taught 15 different graduate-level courses at national and international institutions.

Ileana Espejel is teacher-researcher at the School of Sciences, Universidad Autónoma de Baja California (UABC), and leader of the academic group Manejo de Recursos Costeros y Terrestres. In 2011, she received Ecology and Society Award and an honorific mention by Semarnat-Ecological Award in 2014. She is cofounder of three postgraduate programmes and a bachelor's in environmental issues, member of the Research National System level III and jury member of several academic and meeting committees. She has professional experience of 35 years in interdisciplinary research and teaching in arid and coastal environments and received Academic Award in 2012 and Semarnat-Ecological Award in 2017. She is part of the editorial board of Frontera Norte and Economía Sociedad y Territorio and peer reviewer of journals like *Tourism Management* and *Ocean and Coastal Management*. She has been responsible of 30 projects on environmental and ecosystem management issues. Most projects and papers have a main goal of inter- and transdisciplinary research seeking for the sustainable development of arid and coastal communities in Mexico. As a professor, she has given 40 types of lessons on diverse issues and institutions (ecology ecosystem management, interdisciplinary research methodology and thesis workshops). Moreover, she is advisor of four bachelor's theses, 7 diploma levels, 37 master's theses and 17 PhD theses besides being member of the interdisciplinary thesis committees of several postgraduate students. She completed her bachelor's degree in Biology in 1980 at the Autonomous National University of México; master's at INIREB in Xalapa, Veracruz; and PhD in Vegetation Ecology at Uppsala University, Sweden, in 1986 and received the Leadership for Environment and Development (LEAD) (Fundación Rockefeller) in 1997 at El Colegio de México.

Natalia Martínez-Tagüeña started her career as an anthropologist specialized in archaeology with a degree from the Universidad de las Américas, Puebla. Since then, her work focuses on dryland regions. Her first studies reconstructed a past that is relevant for the future, looking for an understanding of subsistence practices and climate change in the following topics: transitions from hunter-gatherers to agriculture, agave cultivation, coastal adaptations in dryland regions and human impact in ancient environments. She continued her graduate studies at The University of Arizona where she started teaching and participated in several interdisciplinary and transdisciplinary projects. During her graduate work, she transitioned from an ethnoarchaeological approach to a participatory and community-based research. Since 2009, she established a long-term commitment with the Comcaac indigenous community to develop in collaboration the documentation of their cultural landscape, through archaeology, ethnography, oral history and oral tradition. She employs participatory methodologies like community workshops and participatory mapping. She is now a member of the National Research System (SNI) as candidate level and is a researcher in CONACYT, Instituto Potosino de Investigación Científica y Tecnológica (IPICYT), Centro de Investigación, Innovación y Desarrollo para las Zonas Áridas (CIIDZA). She has the opportunity to collaborate in and to develop transdisciplinary research projects to jointly understand humans and nature as a unity whose particularities vary upon each context and temporal trajectories to thus develop management plans, adopt social technologies and sustainably use natural and cultural resources.

Abbreviations

ACI	Adaptation Capacity Index
AI	Aridity Index
AMIMP	Asociación Mexicana de Institutos de Planeación
AMO	Atlantic Multidecadal Oscillation
ANP	Área Natural Protegida
BR	Biosphere Reserve
BRM	Biosphere Reserve of Mapimí
CAP	Community Action Plan
CAS	Complex Adaptive Systems
CC	Climate Change
CGIAR	Consultative Group for International Agricultural Research
CESPE	Comisión Estatal de Servicios Públicos de Ensenada
CHARISMA	Changement et VariabilitéS Climatiques
CLCU	Changes in Land Cover/Use
CMSD	Community Model of Sustainable Development
CNEED	Conseil National de l'Environnement pour un Développement Durable
CONABIO	Comisión Nacional de Biodiversidad
CONACYT	Consejo Nacional de Ciencia y Tecnología
CONANP	Comisión Nacional de Áreas Naturales Protegidas
CONAGUA	Comisión Nacional del Agua
COP	Conference of the Parties
COTAS	Comité Técnico de Aguas Subterráneas
CFSR	Climate Forecast System Reanalysis
CV	Coefficient of Variation
DAC	Development Assistance Committee
DESA	Department of Economic and Social Affairs
DSES	Dryland Socioecological Systems
DWL	Depletion of Aquifers
ECOWAS	Economic Community of West African States
EM	Environmental Mainstreaming

ENSO	El Niño-Southern Oscillation
ERIS	Engaged Research Within an Innovation System
FAO	Food and Agriculture Organization of the United Nations
FSR&E	Farming Systems Research and Extension
FST	Faculté des Sciences et Techniques
FPO	Faculté Polydisciplinaire de Ouarzazate
GAD	Gender and Development
GDI	Global Drylands Imperative
GDP	Gross Domestic Product
GEF	Global Environment Facility
GESP	General Ecological Spatial Plan
GFDL	Geophysical Fluid Dynamics Laboratory
GHG	Greenhouse Gases
GIS	Geographic Information System
GO(s)	Governmental Organization(s)
GPS	Global Positioning System
ICSU	International Council for Science
IEA	International Energy Agency
IGES	Institute for Global Environmental Strategies
IGM	Integrated Groundwater Management
INDC	Intended Nationally Determined Contribution
INIREB	Instituto Nacional de Investigaciones sobre Recursos Bióticos
IPCC	Intergovernmental Panel on Climate Change
IS	Innovation Systems
ITT	Institute for Technology and Resources Management in the Tropics and Subtropics
IWRM	Integrated Water Resources Management
JCDAS	Japanese Climate Data Assimilation System
JMA	Japan Meteorological Agency
JRA	Japanese Reanalysis
LDN	Land Degradation Neutrality
LH	Laguna de Hormigas
LITK	Indigenous knowledge and technology
MAB	Man and the Biosphere Programme
MAP	Mean Annual Precipitation
MASL	Metres Above Sea Level
MDG	Millennium Development Goals
MEA	Millennium Ecosystem Assessment
MEICS	Model for the Estimation of Indigenous Community Sustainability
MPI	Message Passing Interface
MPL	Maximum Permissible Limits
NAO	North Atlantic Oscillation
NAFTA	North American Free Trade Agreement
NAP(s)	National Action Programme
NCAR	National Center for Atmospheric Research

NGO(s)	Non-governmental Organization(s)
NOAA/CDC	National Oceanic and Atmospheric Administration/Climate Diagnostics Center
NPA	Natural Protected Area
NCEP	National Centers for Environmental Prediction
NSSD	National Strategies for Sustainable Development
ODA	Official Development Assistance
OECD	Organization for Economic Co-operation and Development
OUV	Outstanding Universal Value
P	Precipitation
PANDSOC	Program for the Sustainable Development of Oceans and Coasts of Mexico
PAR	Participatory Action Research
PCA	Principal Component Analysis
PDO	Pacific Decadal Oscillation
PET	Potential Evapotranspiration
PROCODES	Programa de Conservación para el Desarrollo Sostenible
PRONATURA	Pronatura México A.C
PRSP	Poverty Reduction Strategy Paper
PVSC	Photovoltaic Solar Cells
PRA	Participatory Rural Appraisal
RAN	Registro Agrario Nacional
RBM	UNESCO Biosphere Reserve of Mapimí
RCP	Representative Concentration Pathways
REPDA	Registro Público de Derechos de Agua
RISZA	International Network for Dryland Sustainability (Red Internacional para la Sostenibilidad de Zonas Áridas
RPC	Rotated Principal Component
RS	Remote Sensing
RW	Reclaimed Water
SAGARPA	Secretaría de Agricultura y Desarrollo Rural
SD	Aldama, San Diego
SD	Sustainable Development
SDG(s)	Sustainable Development Goals(s)
SEDAGRO	Secretaría de Desarrollo Agropecuario
SEDATU	Secretaría Desarrollo Agrario, Territorial y Urbano
SEDESOL	Secretaría de Desarrollo Social
SEDUE	Secretaría de Desarrollo Urbano y Ecología
SEGOB	Secretaría de Gobernación
SES(s)	Social/Socioecological System(s)
SEMARNAT	Secretaría de Medio Ambiente y Recursos Naturales
SI	Sensitivity Index
SLP	San Luis Potosí
SPA	Salud por Agua
SPI	Standard Precipitation Index

SU/TCDC	Special Unit for Technical Cooperation Among Developing Countries
TA	Tabalaopa-Aldama
TEK	Traditional Ecological Knowledge
TDS	Total Dissolved Solids
TUC	Tierra de uso común
UABC	Universidad Autónoma de Baja California
UN	United Nations
UNAM	Universidad Nacional Autónoma de México
UNCBD	United Nations Convention on Biological Diversity
UNCCD	United Nations Convention to Combat Desertification
UNCOD	United Nations Conference on Desertification
UNEP	United Nations Environment Programme
UNESCO	United Nations Educational, Scientific and Cultural Organization
UNFCCC	United Nations Framework Convention on Climate Change
UNRIP	United Nations Rights of Indigenous Peoples
UNDP	United Nations Development Programme
USA	United States of America
UV	Universidad Veracruzana
WAD	World Atlas of Desertification
WHS	World Heritage Sites
WID	Women in Development
WIPO	World Intellectual Property Organization
WL	Water Level
WQI	Water Quality Index
WRI	World Resources Institute

Chapter 1
Introduction: International Network for the Sustainability of Drylands— Transdisciplinary and Participatory Research for Dryland Stewardship and Sustainable Development

E. Huber-Sannwald, N. Martínez-Tagüeña, I. Espejel, S. Lucatello, D. L. Coppock, and V. M. Reyes Gómez

Abstract Drylands are the largest biome complex on Planet Earth and home to over 40% of the human population. Their extraordinary high biotic and cultural richness is endangered by global climate change, land use pressures including coastal/marine systems, and environmental degradation. Understanding and maintaining the functional integrity of dryland socio-ecological systems (DSES) is fundamental for sustainable development. It requires resilience-based dryland stewardship, where land users, managers and decision-makers incorporate change, as understood from the multiple actors' perspective of a SES, into their planning and governance. The linkage of America's drylands with west Africa and Southern Europe is often overseen, however increasing economic activities in these DSES have enormous impacts on their functional integrity. In response to this daunting

E. Huber-Sannwald (✉)
División de Ciencias Ambientales, Instituto Potosino de Investigación Científica y Tecnológica, San Luis Potosi, San Luis Potosí, Mexico
e-mail: ehs@ipicyt.edu.mx

N. Martínez-Tagüeña
Cátedra CONACYT, Consortium for Research, Innovation and Development of Drylands, Instituto Potosino de Investigación Científica y Tecnológica, San Luis Potosí, Mexico

I. Espejel
Facultad de Ciencias, Universidad Autónoma de Baja California UABC, Ensenada, Baja California, Mexico

S. Lucatello
Estudios Ambientales y Territoriales, Instituto Mora, Mexico City, Mexico

D. L. Coppock
Department of Environment and Society, Utah State University, Logan, UT, USA

V. M. Reyes Gómez
Red Ambiente y Sustentabilidad, Instituto de Ecología, A.C., Chihuahua, México

© Springer Nature Switzerland AG 2020   1
S. Lucatello et al. (eds.), *Stewardship of Future Drylands and Climate Change in the Global South*, Springer Climate,
https://doi.org/10.1007/978-3-030-22464-6_1

task, academic and government institutions founded the Agadir Platform as a coordinating instrument for cooperation in the Global South. As focal node of this platform, Mexico established the first international network to co-generate knowledge through transdisciplinary research partnerships. We present the conceptual framework of this network highlighting 1) the socio-ecological system's approach, 2) the transdisciplinary scope of participatory research, 3) the intercultural action scheme, and 4) the repercussions of this integrated approach on polycentric governance. This book includes diverse examples of the application of this framework in DSES ranging from co-designing socio-ecological development projects, to adaptive management, and policy development.

Keywords RISZA · Transdisciplinary networks · Co-designed projects · Arid lands · Participative research · South-South and triangular cooperation

Drylands are the largest biome complex on Planet Earth and home to over 40% of the human population. Their extraordinary high biotic and cultural richness is endangered by global climate change, land use pressures including coastal/marine systems, and environmental degradation. Understanding and maintaining the functional integrity of dryland socio-ecological systems (DSES) is fundamental for sustainable development. It requires resilience-based dryland stewardship, where land users, managers and decision-makers incorporate change, as understood from the multiple actors' perspective of a SES, into their planning and governance. The linkage of America's drylands with west Africa and Southern Europe is often overseen, however increasing economic activities in these DSES have enormous impacts on their functional integrity. In response to this daunting task, academic and government institutions founded the Agadir Platform as a coordinating instrument for cooperation in the Global South. As focal node of this platform, Mexico established the first international network to co-generate knowledge through transdisciplinary research partnerships. We present the conceptual framework of this network highlighting 1) the socio-ecological system's approach, 2) the transdisciplinary scope of participatory research, 3) the intercultural action scheme, and 4) the repercussions of this integrated approach on polycentric governance. This book includes diverse examples of the application of this framework in DSES ranging from co-designing socio-ecological development projects, to adaptive management, and policy development.

Aridity is often characterized by an aridity index (AI) (Thomas and Middleton 1992), calculated as annual precipitation divided by annual potential evapotranspiration, and ranges from a minimum of 0.05 to a maximum of 0.65 (Hulme 1996; Safriel et al. 2005). Based on the AI drylands can be classified as hyperarid, arid, semi-arid, and dry sub-humid (UNCCD 1994). In comparison to other biomes, life in the drylands has evolved under highly variable precipitation, extreme water scarcity, pronounced fluctuations in diurnal temperatures, and extended exposure to high levels of solar radiation (Noy-Meir 1973). These factors continuously exert strong selection pressures on specialized life forms (Whitford 2002). However, there is an exceptionally high species diversity across all categories of biota that contributes to varied ecosystems that span from coastal drylands to intracontinental basins and highland plateaus.

Dryland ecosystems offer a wealth of ecosystem goods and services for human well-being (Safriel et al. 2005; Stafford Smith et al. 2009). Large populations of agriculturalists, pastoralists, and coastal fishermen have enormous cultural wealth and ecological knowledge. Over millennia, humans have adapted to the scarcity and abundance cycles of natural resources, shaping their livelihoods accordingly (Stafford Smith and Cribb 2009; Davis 2016a). The long history of fine-tuning socio-economic and political life among drylands peoples reflects some of the oldest legacies of socio-ecological system (SES) development, and today are characterized by both their ecological significance in sustaining the supply of ecosystem services and their capacity to support millions of people (Safriel et al. 2005; Cherlet et al. 2018). Variability is an inherent structural property of drylands (Stafford Smith et al. 2009) to which local communities have adapted and evolved under, thereby lowering their vulnerability to unpredictable environmental changes (Krätli 2015; Davis 2016b). These adaptive social–ecological interdependencies of human activities and ecosystem services require collective knowledge-based actions supporting dryland stewardship (Chapin III et al. 2009a, b, c).

However, over recent decades, drylands have suffered substantial losses of productivity and biodiversity, increasing the severity and frequency of droughts, food insecurity, poverty, violence, emigration, and social disintegration (Reed and Stringer 2016; Cherlet et al. 2018; Middleton 2018). In addition, some areas have been converted to irrigated lands to expand high-input agriculture and to pastures for intensive livestock production (Jia et al. 2004; Squires 2010) triggering irreversible systemic changes. The processes underlying all these changes are often termed desertification (UNCCD 1994; Reynolds et al. 2007) undermining the sustainable regional development and threatening the global dryland SES (UNCCD 1994; Cherlet et al. 2018), which are mainly situated in the Global South. According to the sustainable development goals, the objectives include thriving lives and livelihoods, sustainable food security, sustainable water security, universal clean energy, healthy and productive ecosystems, and governance for sustainable societies (Griggs et al. 2013).

The scope of this chapter is to elucidate the challenges of understanding current human and environmental conditions in the drylands and identify emerging research needs that can help forge pathways towards improved stewardship and sustainable development in future drylands in a world that will also be buffeted by climate change. Many issues related to transforming and governing drylands have been developed theoretically at the global scale [e.g., sustainable development goals and land degradation neutrality (Orr et al. 2017; Cowie et al. 2018)]. Some plans have been implemented at a national scale (INEGI 2019; UNCCD 2019), but scaling down sustainable development to dryland local communities is still lacking. Furthermore, suitable SES research methods that fully respond to such theoretical developments are required and need to be better defined and promoted.

Therefore, we present the International Network for Dryland Sustainability ("Red Internacional para la Sostenibilidad de Zonas Áridas, RISZA") that tackles the current dryland challenges at the local and regional scale, and supports several activities and goals. These include: (1) Creation of multisectoral partnerships associated with local SESs; (2) facilitation of intercultural exchange and dialogue; (3) weaving of different knowledge systems (Johnson et al. 2016; Tengö et al. 2017); (4) encouragement of transdisciplinary and participatory research (Schuttenberg and Guth 2015; Hickey 2018, Hickey et al. 2018; Willyard et al. 2018) for the co-production of relevant knowledge for action research (Clark et al. 2016; Durose et al. 2018); (5) generation of place-based learning communities (Davidson-Hunt and O'Flaherty 2007); (6) stimulation of the co-design of novel management, assessment, and governance schemes (Whitfield and Reed 2012; Schoon et al. 2015; Bautista et al. 2017; Bodin 2017; de Vente et al. 2017, (7) providing information for sustainable policy and socio-economic development standards in accordance with the United Nations Sustainable Development Goals (Agenda 2030). This network is the first national/international node of a recently founded international platform (see Chap. 13) to coordinate novel research, management, and assessment models in the drylands of Latin America, North Africa, and Europe in response to global environmental change in the Anthropocene.

The RISZA initiative also contributes to the wide range of activities related to the so-called *Global South* to foster the global scientific and research-development agenda on drylands. As a matter of context, the concept "Global South" refers broadly to the regions of Latin America, Asia, Africa, and Oceania. It is a term that has emerged as an alternative to the misconceived and former colonial ideas of "The Third World" and "Periphery" adopted in Europe and North America pointing to low-income and often politically or culturally marginalized countries of the planet (Dados and Conell 2012). The use of the "Global South" idea marks a shift from a central focus on underdevelopment or cultural differences in world countries, towards an emphasis on geopolitical relations of power among more equal nations. This is possible through the economic, political, cultural, and environmental changes that many developing nations in different continents have undergone over the past three decades. The Global South is rather an international political and economic concept that focuses on how world cultures, particularly those from Latin America Africa and Asia, respond to globalization and global processes linked to the environment, poverty, immigration, gender, etc., together with transformation, colonialism and post-colonialism, and modernity.

In the specific case of this book, we address the vision of drylands stewardship through the lens of a group of countries in Latin America (mostly Mexico) and Africa, through the nexus with the Agadir Platform, a transdisciplinary initiative, where countries from the two regions and Southern Europe collaborate on a common scientific agenda on sustainable development in drylands in the light of climate change.

Drylands Vulnerability in the Twenty-First Century

Over millennia the drylands have undergone innumerable transformations in climate, biotic interactions, and human conditions. Pressing current challenges in global drylands include a broad spectrum of issues as shown in Table 1.1.

Hence, these challenges explain why drylands currently cover over 35% of the global biodiversity hotspot area (Davies et al. 2012) and 28% of the total area of World Heritage Sites (Gudka et al. 2014). Past climate warming has been most

Table 1.1 Pressing current challenges in global drylands

Challenges	Some references
Human population growth	Wang et al. (2012), Reid et al. 2014), Cherlet et al. 2018)
Conversion of key rangeland resources to agricultural uses and groundwater exploitation	Chapter 3; Peters et al. (2015)
Sedentarization of pastoralists and other changes in traditional livelihoods	Chapter 2; Marlowe (2005), Reid et al. (2014)
Migration	Coppock et al. (2017)
Privatization of communal land	Reid et al. (2014)
Expanding urbanization	Reid et al. (2014), Peters et al. (2015)
Expansion of infrastructure for renewable energy generation and intensive agriculture	Chapter 5; Matson (2012), Reid et al. (2014), Cherlet et al. (2018)
Extraction of fossil fuels	Reid et al. (2014)
Expansion of mining	Reid et al. (2014)
Overgrazing by domestic livestock	Peters et al. (2015), Cherlet et al. (2018), Middleton (2018)
Invasive species	Reid et al. (2014)
Proliferation of water development	Chapter 3; Wilcox et al. (2011)
Aquifer overexploitation	Chapter 3; Aeschbach and Gleeson (2012)
Imposed or inadequate conservation management plans	Dudley (2008), Dressler et al. (2010) but see Gudka et al. (2014)
Inappropriate restoration and/or afforestation projects to enhance carbon capture	Wilcox et al. (2011), Veldman et al. (2015), Nolan et al. (2018)
Loss of local and indigenous knowledge	Figueroa (2011), Johnson et al. (2016) but see Gómez-Baggethun and Reyes-García (2013) for interpretation
Increased frequency of droughts	Chapter 15; Huang et al. (2017b)

pronounced in drylands, with an average increase of 1.7 °C between the years 1948 and 2008 (Huang et al. 2012); this warming trend is about 2.1 and 1.5 times greater than any increase observed in humid regions and globally, respectively (Huang et al. 2015, 2017a, b). Over a sixty-year period (1948–2008), drylands have expanded to their current extension (Feng and Fu 2013). Drylands are one of the most vulnerable biomes to climate warming, likely unable to tolerate the 2 °C warming threshold of the 2015 Paris agreement (Huang et al. 2017a). When considering high CO_2 emission scenarios (RCP 8.5), global drylands are predicted to expand at an even faster rate in that they will cover up to 56% of the terrestrial surface by 2071–2100 (Huang et al. 2015, 2017b). When considering only the CO_2 fertilization effect, drylands are predicted to increase their productivity. It has been shown that within 28 years (1982–2010) leaf cover has increased by 11% likely attributable to a 14% increase in atmospheric CO_2 concentration (Donohue et al. 2013). Finally, recent simulation models suggest that temperate drylands will shrink by a third and convert to subtropical drylands, and that drought may reduce water availability primarily at deep soil layers during the growing season with obvious implications on vegetation shifts, declines in ecosystem services supply and livelihood options (Schlaepfer et al. 2017).

Such accelerated changes in dryland use can introduce new dynamics in SES and in the transitions between stable and unstable SES states (Huber-Sannwald et al. 2012; Bestelmeyer et al. 2015). A state is characterized by certain vegetation and soil types and ecosystem processes (Bestelmeyer et al. 2015), which supplies a set of ecosystem goods and services in accordance to human demand (Yahdjian et al. 2015). Inherent and new sources of disturbances may cause changes of SES states; these changes can be abrupt, gradual, reversible, or persistent. Hence, unpredictable trends of change will be accompanied by new challenges related to understanding the combined and interacting effects of historic land use change, climate variability, alterations in the functioning of dryland SES, and their resilience and ability to deliver future ecosystem services (Folke et al. 2009, 2010). While extended droughts and increased variability in precipitation directly exacerbate socio-environmental degradation in drylands (Puigdefábregas 1998; Stott 2016), indirect policy-induced desertification also occurs (Geist and Lambin 2004; Adams 2009; Davis 2016b; Huaico Malhue et al. 2018).

Scholars have long debated on how to better manage the inherent variability of drylands to improve human living conditions. Such engineering approaches are grounded on the premise that one can reduce the inherent variability of drylands by adopting agricultural practices that have been successful where water availability is more predictable. A prominent example is crop irrigation, for instance, in the Yaqui valley in Mexico; this desert area has been the cradle of the Green revolution and the worldwide leader in wheat producer (Matson 2012). Environmental uniformity and stability, and the removal of redundancy may guarantee short-term high crop yields and temporarily increase food security, yet at the cost of irreversible loss of biotic and cultural diversity (Holling and Meffe 1996; Safriel et al. 2005; Walker and Salt 2006) along with trade-offs on sustaining ecosystem services (Papanastasis et al. 2017).

Human interventions intended to achieve sustainable development, as defined in the UN Sustainable Development Goals (https://www.un.org/sustainabledevelopment/sustainable-development-goals/), no longer require investment in maximizing commodity production, but rather in diversifying protections afforded to the biota, cultures, and knowledge systems in order to increase the response and adaptation spectra to regional or global socio-environmental change (Chapin III et al. 2009a). This increases the system buffering capacity against unpredictable change (Huber-Sannwald et al. 2012). The role of traditional ecological knowledge in understanding SES is crucial to understand how some local communities have sustained resilient landscapes, but also for the successful stewardship of diverse SES where the division between nature and society is bridged and true ethical multisectoral collaborations are accomplished (Johnson et al. 2016).

Desertification and Land Degradation Versus Drylands Resilience

According to the United Nations Convention to Combat Desertification (UNCCD 1994) desertification refers to land degradation in drylands due to various factors, including climatic variations and/or human activities (Article 1 of the UNCCD). The

Millennium Ecosystem Assessment (2005) defines land degradation as a process that leads to a long-term failure to balance the demand for and the supply of ecosystem goods and services. While there are estimates that about 10–20% of global drylands suffer from desertification (Reynolds et al. 2007; D'Odorico et al. 2013), due to the complexity of the causes of desertification and the impacts of land degradation, we have little understanding of both local expressions and the global extent of this problem (Cherlet et al. 2018). What is the origin of desertification? Where does its legacy originate, in the (false) sense that deserts are the result of deforestation, overgrazing, and excessive burning by indigenous nomadic pastoralist populations (Davis 2016a, b)? How can one explain major investments globally and regionally in strategic projects of "re"forestation and greening that promise to convert deserts into "productive land" (Davis 2016a, b; Stafford Smith 2016; 8000 km of Great Green Wall in the Sahel https://www.greatgreenwall.org/about-great-green-wall)?

The Earth's largest drylands are about 65 million years old, but like other biomes drylands have undergone dramatic changes over time (Goudie 1986). As noted above, in the drylands the scarcity, variability, and unpredictability of water over space and time are unique characteristics that have challenged traditional linear approaches to understanding ecosystem dynamics (Whitfield and Reed 2012). Seminal works by Westoby et al. (1989), Walker (1993), and Holling (1988) have stated that after a disturbance event, ecosystems return to a stable state of "equilibrium" or "climax," with a new "non-equilibrium" paradigm (Westoby et al. 1989). However, in most dryland SES most likely we will find both equilibrium and non-equilibrium features due to an extremely high spatiotemporal heterogeneity in structure, function, and overall system resilience (Coppock and Briske personal comment). What is currently labeled redundant or "noise" may be the source of system stability and resilience in the future under changing and interacting environmental conditions (Folke et al. 2010).

This concept of "non-equilibrium" is not only reflected in multiple stable biophysical states, but necessarily applies also to alternative socio-economic states (Reynolds and Stafford Smith 2002; Huber-Sannwald et al. 2012). While innate natural disturbance regimes have been acknowledged in contributing to the natural dynamics of SES (Pickett and White 1985), these aspects have not been considered in environmental policy formulation, concepts of dryland development, and anti-desertification policies (Behnke and Mortimore 2016; Davis 2016b) with potentially detrimental implications as they do not foresee the unpredictable non-linear nature of SES change (Reynolds et al. 2007; von Wehrden et al. 2012).

Desertification was recognized as one of the first major global change problems (UNCCD 1994; Thomas and Middleton 1992) and since then, it has been on the global UN agenda (Stafford Smith 2016). In 1977, the first United Nations Conference on Desertification (UNCOD) was organized. In parallel, in the second half of the twentieth century global dryland policy was targeted towards dryland restoration to enhance productivity in ways aligned with capitalist development goals (Davis 2016a). Ironically, however, some of the regions, most severely affected by desertification seem to have been related to those inappropriate policies that arose from misperceptions on the origin and (falsely promoted lack of) value of drylands and the supposedly inappropriate traditional uses by local populations (Davis 2016a). While scholars continue to debate how to best distinguish land degradation from desertification and to identify their underlying causes (Reynolds and Stafford Smith 2002; Reed and Stringer 2016; Reed et al. 2011; Behnke and Mortimore 2016; Davis 2016a, b), or measuring how much land loss has occurred (Huaico Malhue et al. 2018), the UN is targeting a land degradation neutral world by 2030 as one of the sustainable development goals (Chasek et al. 2015; Safriel 2017; Cowie et al. 2018), highlighting new challenges and opportunities (Stavi and Lal 2015; Akhtar-Schuster et al. 2017) and ignoring potential pitfalls (Easdale 2016; Okpara et al. 2018). Undoubtedly however, climate change may have contrasting impacts at the regional scale and thereby interact with human effects on land degradation, either by causing mega-droughts reducing vegetation cover and thereby exacerbate land degradation or by enhanced precipitation leading to some re-greening of drylands (for example, in the Sahel, Herrmann and Sop 2016; Behnke and Mortimore 2016).

Despite global awareness of and attention to desertification, success stories about its combating and/or developing the world's drylands are surprisingly scarce (for exception, see Reid et al. 2014). Thus, leading us to question why re- and afforestation projects have failed and have, at times, negatively affected biodiversity, as well as the hydrological and biogeochemical cycles of drylands (Amdan et al. 2013). Both irrigated and rain-fed agricultural schemes in drylands have overall rendered low crop yields and increased soil salinization and land degradation (Southgate 1990; Lambin et al. 2001), while rangeland management programs appear to have had little or no effects on improving land degradation (Dregne and Chou 1992). Conversely, regions formerly claimed to be notoriously and presumably irreversibly degraded by overgrazing have recovered after the end of long drought periods (Donohue et al. 2013; Dardel et al. 2014). In the wake of an accelerated rate of global socio-environmental change (Steffen et al. 2015), it is useful to question whether drylands are doomed to be physically degraded and desertified by humans (Reynolds and Stafford Smith 2002), or whether they instead present an opportunity for sustainable development (Reynolds et al. 2007; Mortimore et al. 2009; Krätli 2015; Behnke and Mortimore 2016).

Dryland Socio-Ecological Systems Are Complex Systems

Socio-ecological systems are complex adaptive systems where the relationships between humans and nature are based on interconnections among system components, whose interlinkages and dynamics create emerging properties with synergistic effects (Berkes et al. 2008; Koontz et al. 2015; Biesbroek et al. 2017; Tàbara et al. 2018). All complex systems have their inherent quantitative measures such as structure, dynamics, evolution, development, and complexity (Bar-Yam 1997; García 2006). Physical, biological, social, cultural, economic, and political components interact and provide feedback at different rates and intensities across different spatial and temporal scales, thus, they undergo non-linear, unpredictable changes and self-organize after disturbance events (Liu et al. 2007).

Understanding the connectedness between humans and nature necessarily requires inter- and transdisciplinary efforts and frameworks, including scholarly expertise from the natural and social sciences, and platforms of communication, negotiation, and decision-making that facilitate the formation of learning communities (similarly to Davidson-Hunt and O'Flaherty 2007; Bautista et al. 2017). The purpose of such learning communities is the sharing of scientific, local, indigenous, and technical knowledge, and ethics, wisdom, and worldviews, to ensure equal dialogue among all involved stakeholders. Utilizing and simultaneously protecting the wealth of natural resources and ecosystem goods and services upon which humanity depends calls for novel, holistic, transboundary designs, analysis, and knowledge co-production (Daily 1997; MEA 2005; Chapin III et al. 2009b). This integrated approach is fundamental for co-management, collaborative governance, and adaptive policy development (Bautista et al. 2017).

With this perspective, in 2002, a global multidisciplinary think tank of dryland specialists developed the Drylands Development Paradigm (DDP) (Stafford Smith and Reynolds 2002; Reynolds et al. 2007). The DDP is an integrative framework for the analysis, restoration, mitigation, and/or prevention of dryland SES as affected by degradation and/or desertification (Reynolds et al. 2007) and for policy development. The DDP is based on complex adaptive systems theory (CAS) (Ashby 1962; von Bertalanffy 1968), in that systems consist of interconnected elements conferring a particular structure following the underlying rules of the specific purpose or function of a system (Meadows 2008). Elements and processes may change at different rates, either "slow" or "fast," and at and across different spatial and/or temporal scales. The CAS has three key properties: (1) order is emergent not pre-determined, where the system adjusts and self-organizes after disturbance events; (2) historic impacts are irreversible in that current dynamics are linked to and influenced by past events (legacy effects); (3) based on (1) and (2) the future of CAS is unpredictable and the past lays the foundation for future changes (Chapin III et al. 2009a; Curtin 2015).

When studying SES, we need to ask the questions what causes overall system dynamics, internal connectedness, SES contexts and feedbacks of system components in response to internal and external long-term drivers (e.g., climate change, human population growth), stressors (e.g., mega-drought, emigration, fluctuations in markets or commodity prices, policy change), or pulsed trigger events (e.g., extreme weather or natural hazards, sudden access to electricity, communication technology), and how do system elements (resources, species, social actors) disperse, migrate, or interact across socio-ecological systems (Biggs et al. 2012). Do the variables and processes change slowly or rapidly and are they connected to the dynamics of events occurring at other times or places (Folke et al. 2009)?

For relevant research questions, clear understanding is required of the spatial and temporal dimensions of connectivity in SES. This implies understanding the connections between landscape units, habitats, species, social groupings, generations, knowledge types, institutions, and policies, among others (Biggs et al. 2015). Comprehending the functional integrity of SES as CAS is daunting yet crucially important, as the younger generations' tolerance to and perception of environmental degradation is changing such that the threshold of acceptance of environmental condition is declining, a psychological and sociological phenomenon laconically coined shifting baseline syndrome (SBS) (Pauli 1985). Hence, understanding SES dynamics and the degree of land degradation and potential human's preventive, reactive, proactive, mitigating, and/or adaptive responses requires a multi-criteria assessment in different contexts (Ocampo-Melgar et al. 2017).

When SES are managed, restored, or protected, necessarily through the lens of CAS theory, they will not remain or can be maintained at a mature or desirable state, respectively, but rather undergo adaptive cycles (Gunderson and Holling 2002). Complex systems may organize around one of several stable states within a desirable (from a stakeholder perspective) regime of the system. Thus, rather than focusing on detailed characteristics of one stable state, one may want to understand the internal and external drivers that cause the transition to alternative states and how systems elements reorganize without losing the underlying interconnectedness, structure, function, and feedback responses of the system, thus maintaining its resilience (Westoby et al. 1989; Scheffer and Carpenter 2003).

In Holling's adaptive cycle, the systems once fully developed are commonly held in the so-called conservation phase (Walker and Salt 2006). Humans tend to interfere in the adaptive cycle and frequently prolong this phase. For instance, by maximizing the production of a single ecosystem service, for example, forage production, water extraction, and so on, thereby eliminating unnecessary system variability and redundancy. However, this comes at the cost of eradicating high levels of diversity including biotic (species and functional groups, ecosystem service bundles), cultural (flexibility to adapt and adopt new livelihoods, high adaptive capacity related to local knowledge), and social diversity (institutional organizations, social networks, social memory, adaptive local governance systems). This diversity is needed as it confers insurance and buffer against unpredictable future changes (e.g., prolonged drought, fire, diseases, pests, drop in prices of commodities, new legislation). Similarly, we may want to ask do high levels of response diversity also provide systems with a potentially broad adaptive capacity to reorganize once a system has collapsed and shifted from the conservation to the release phase. How do SES reorganize, after they have lost the internal connectedness and release all resources, energy, and/or information itself?

Acknowledging and eventually managing the cyclic behavior of SES requires the incorporation of different sources of knowledge. The trajectory of system development follows both the system's memory and the current social–ecological context and conditions, thus conferring new ecological and/or social opportunities characterizing a certain system state within the desired regime (Huber-Sannwald et al. 2012). Lack of ability to respond to or recover after a system has collapsed may trigger the crossing of a threshold (biophysical or socio-economic) or tipping point (social, ecological, or socio-ecological (Milkoreit et al. 2017), and the shift into a new (albeit less desirable) regime. A system's capacity to build and maintain resilience and to re(self)-organize after a shock or severe disturbance event is critical, whether external, internal, or interacting drivers induce system change. Since the 1950s, in the time of Great Acceleration (Steffen et al. 2007), this may occur more rapidly, unpredictably, or irreversibly (global population growth, local and regional migration, land use change, soil erosion), directionally (loss of vegetation cover, change in species composition, climate warming, exploitation of aquifers, fisheries) (Steffen et al. 2015), or as an emerging phenomenon (loss of system resilience, landscape dysfunction, impoverishment of local traditional knowledge by introduction of information technology, migration, land degradation, desertification). Therefore, due to the complexity of SES, monitoring, tracing, and evaluating socio-ecological system change require multiple disciplines, expertise, knowledge systems, concepts, methodologies, frameworks, novel platforms, newly emerging sciences, thus fundamental approaches to do dryland system science achieve the stewardship of future drylands and the sustainable development goals (see Table 1.1).

RISZA and the Conceptual/Operational Model

In March 2017, the *International Network for Drylands Sustainability/Red Internacional para la Sostenibilidad de las Zonas Áridas (RISZA)* (www.risza.com.mx) was launched by the Instituto Potosino de Investigación Científica y Tecnológica (IPICYT) in San Luis Potosi, Mexico, with financial support from the National Council of Science and Technology (CONACYT by its Spanish acronym). The aim of RISZA is to generate and foster research, development, and innovation in partnership at the national level linked to the tripartite alliance between Latin America, Africa, and Europe (see Chap. 13/Agadir Platform), with a strong regional, inter-sectoral emphasis.

RISZA aims to guide and facilitate transdisciplinary and participatory research including academics, governmental and non-governmental organizations, civil societies, local stakeholders, and policy-makers, to foster collective knowledge production, iterative system monitoring and (re) evaluation, and capacity building in dryland stewardship at all levels. The principal research goal is to contribute to accomplish the SDGs in Mexican drylands in synergistic ways, drawing on expertise from other countries of the Platform of Agadir. As a ratifying country, Mexico needs to comply with each of the SDGs and their associated targets. Pre-established global indicators and country-specific indicators serve to monitor each SDG. The role of Mexico's and global drylands in meeting the SDGs both as a national and a global biome is poorly understood (for an exception, see FAO 2018). RISZA as part of the Agadir Platform will contribute with knowledge, technology, and innovation to meet these goals, in particular to the SDGs 13, 14, 15, and 17.

As a product of the inaugural RISZA participatory planning meeting in May 2017, a transdisciplinary group consisting of 80 people co-designed a comprehensive framework consisting of four dimensions: philosophies, study objects, actions, and long-term goals (Fig. 1.1). The framework is also a network, with the four dimensions representing nodes and the participatory or collective nature of the network representing the links, the blue line meaning water (a crucial determinant of drylands) as a transversal main focal point. This framework is flexible and open for feedback and adjustment; its emphasis lies on establishing the basis for inter- and transdisciplinary collaborations in national and international drylands with a current emphasis on Latin America, North Africa, and Southern Europe. It is also necessary to motivate the dialogue and co-production of knowledge by different actors (i.e., academia, government, private sector, civil societies, local communities, indigenous groups).

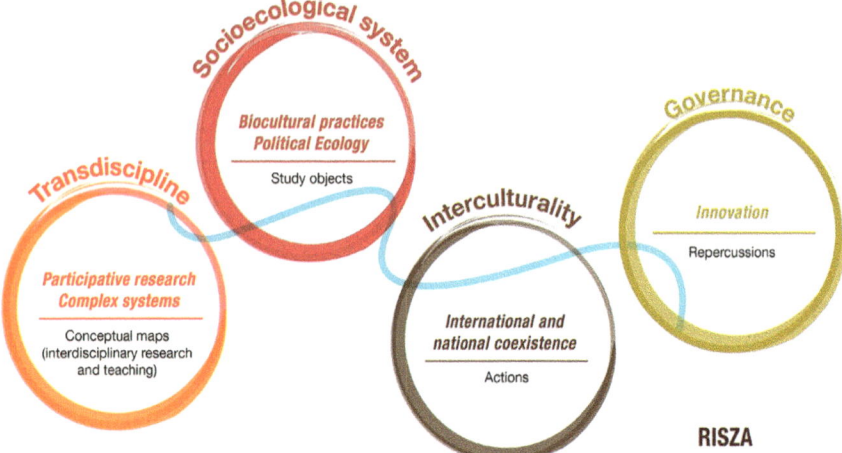

Fig. 1.1 Conceptual and operational framework of the International Network for the Sustainability of Drylands (RISZA—the Spanish acronym)

Part I: Drylands and Socio-Ecological Systems

Places, territories, landscapes, and/or ecosystems where people live and depend on natural resources and regulating forces have been described as social–ecological systems (SES). For this reason, the central study units for RISZA are SES, as people associate important values to these services, many represented by the 17 sustainable development goals (SDGs). When land conditions change caused by external or internal drivers, the quality and quantity of these services change as well.

For example, dryland degradation is advancing rapidly (Safriel et al. 2005; Safriel 2017) affecting the productivity and the functioning of inland and coastal ecosystems. Loss of biotic and cultural diversity, and climate change and socio-economic changes related to globalization enhance the effect of land degradation, thereby eliminating the inherent buffer characteristics that allow SES to resist, mitigate, or adapt to these adverse effects. Biodiversity and cultural diversity are directly related; cultural diversity includes genetic, linguistic, and cognitive diversity and thus has tangible and intangible assets. Cultural diversity often originates from the biotic diversity regularly in places inhabited by indigenous or rural communities (Toledo and Barrera-Bassols 2008). Furthermore, beyond a direct relationship the concept of biocultural eliminates the dichotomy and describes objects like animals, plants, rivers, or mountains as having a corresponding linguistic expression, charged with identity and individual or collective memory, sacred meaning, or ritual importance (Boege 2008).

Therefore, the interactions between humans and the environment are highly complex resulting from various interrelated factors. Research has demonstrated a broad transcultural variability in the environmental consequences of human behavior (Oviedo et al. 2000). In addition to the employment of SES as a study object and framework, political ecology research has analyzed the different actions and perceptions that social actors have and their relations of dependence and influence with their environment, together with the understanding of what causes such interrelationships (Robbins 2012). Thus, the emphasis is placed on the relationship of environmental, ideological, social, economic, and political aspects on each particular spatiotemporal context while articulating their local, regional, and global components (Greenberg and Park 1994). In both approaches, for research to have real repercussions in the objectives of sustainable development, an emphasis is placed on the importance of the role of institutions and social organizations in environmental contexts, where the individual and social diversity must be considered and the power relationships understood through governance studies.

Dryland socio-ecological, biocultural systems, and sustainable land management practices are affected by global socio-environmental change (i.e., land use change, climate change, diversity loss, livestock grazing, mining, migration, among others). Understanding specific historical, socio-cultural, socio-economic, and socio-political contexts will allow interregional, intercultural, and inter-policy comparisons among similar dryland SES and their divergent responses to global environmental change drivers. For RISZA we adopt the drylands development paradigm (Reynolds et al. 2007), as it considers the complexity, diversity, and uncertainty as inherent properties of dryland SES, as well as different system components, slow variables, cross-scale linkages, and diverse knowledge systems both as scientific and operational framework (Reynolds et al. 2007; Stringer et al. 2017).

Part II: Transdisciplinarity in Drylands

Participatory research originated in the 1970s in Latin America as part of social movements and processes of policy transformations related to social and education planning (Freire 1970). New research approaches with active democratic participation of the population via participatory mechanisms were proposed to plan and execute new education and development projects (Durston and Miranda 2002), as well as research efforts for conservation aims (Newing et al. 2011). Participatory research requires the fulfillment of a series of operational procedures to acquire useful knowledge and to eventually induce change in a situation or system. Many different methods have been efficiently used, the most common being action research (Whyte 1989), participatory rural appraisal (Chambers 1983), participatory mapping (Chapin et al. 2005), and participatory workshops and monitoring (Knapp et al. 2011). Due to its participatory nature, dialogue development among participating actors leads to the co-production of knowledge, allows the systematization of experiences, collective wisdom, and local knowledge (Ander-Egg 1990), and generates confidence among participants. All of these components are essential in positive governance (Schuttenberg and Guth 2015).

Problem solving inherently is multi, inter, or transdisciplinarity, requiring necessarily participatory research, because it involves collaboration among different stakeholders and is based on the generation of knowledge that emerges and extends beyond the limits of single scientific disciplines. It starts from the premise that knowledge is constantly developing by weaving scientific knowledge with empirical observations, technical know-how, contemporary scientific facts, local and traditional knowledge, and ancestral wisdom. Bridging and connecting knowledge systems helps understand socio-ecological system dynamics, opens dialogue between different cultures, mental models, institutions, actors, and their practices and influences, and/or informs governance schemes and policies (Tengö et al. 2017; Challenger et al. 2018). Co-production of knowledge opens new learning and teaching opportunities (Gutiérrez Serrano 2016) and contributes to an ever-growing horizontal learning community. The concept of transdisciplinarity originally stems from the idea that thinking is a complex process including biological, cerebral, spiritual, logical, linguistic, cultural, social, and historical processes, with emphasis on the connections and communication between knowledge systems (Morin 1977; García 2006; Castañares Maddox 2009; Díaz et al. 2015, 2018). Furthermore, it emerges with the intention to not only systematically solve a problem, but to make research and thus education more relevant to society (Kockelmans 1979).

Part III: Interculturality in Drylands

The development of transdisciplinary and participatory research is based on continuous knowledge development and dialogue as the basis to develop sustainable development projects with direct participation of local actors. Multiple stakeholder collaborations are fundamental for the co-production of useful knowledge production as different wisdoms, disciplines, foci, and positions can be shared (Gutiérrez Serrano 2016). This transdisciplinary academic context has strongly helped remove unequal power relations, which is the basis for interculturality (Alsina 2003), and thus generates friendly intimate relations built on confidence and trust to their legitimacy (Coppock 2016). Interculturality refers to the process of establishing equitable communication and interaction forms between people and groups with specific cultural identities, which stimulates dialogue and integration between different cultures, knowledge, and worldviews (Alsina 2003; Clark et al. 2016). The process of action through interculturality requires not only a different type of knowledge, but also a novel process of knowledge creation. It requires the creation of collective and participatory wisdom based on equal cognitive and emotional exchanges, providing an emancipatory knowledge that goes beyond colonialism to accomplish solidarity (Santos 2002).

While intercultural dialogues are not always exempt of conflict, strong emphasis is given on mutual respect, horizontality in communication channels, equitable access to information, and joint search for synergies (Alsina 2003). Thus, when looking for engagement by a diverse group of stakeholders, emphasis needs to be given on a diverse spectrum of values with potentially conflicting worldviews, and on different knowledge systems, rules, and norms (Gorddard et al. 2016). This then will guide management practices, power relations, skills, and preferences, which in turn may dramatically influence the perception, definition, and stewardship of socio-ecological systems (Davies et al. 2015). Therefore the diverse values, knowledge, and rules of multiple social actors influence the decision-making context and together with the inherent complexity and interrelatedness of dryland SES components, their feedbacks and non-linear responses to management and climate change across different spatial and temporal scales may trigger uncertainty related to both the vulnerability to increasing risks associated with climate change and the resilience to the long-term provisioning of ecosystem services and human well-being.

These so-called wicked problems (a type III problem according to Rittel and Webber 1973 cited in DeFries and Nagendra 2017) that RISZA will be tackling have no clearly defined solutions, yet require continuous collective learning, integrated collaborative institutional designs, and adaptive, iterative pathways, towards finding solutions both at the management, policy, and governance level (Curtin 2015; DeFries and Nagendra 2017). Ultimately, to guarantee respectful exchange among knowledge types and diverse cultures, RISZA commits to follow the established ethical codes to protect human rights (see United Nations Rights of Indigenous Peoples and the regulations of the WIPO "World Intellectual Property Organization") and to adopt the scheme of basic agreements on human studies established by the Institutional Review Boards (IRBs) in the USA. It is important to note that transdisciplinary and intercultural endeavors must require a commitment to true ethical academic procedures where the individual becomes part of a research team and considers the participation of all actors involved throughout all the project's stages: from the establishment of objectives to the final co-authorship of research results.

Part IV: The Governance of Drylands

Drylands are the home for a large amalgamation of geological, biological, and cultural diversity (Stafford Smith et al. 2009). For local populations, they consist of meaningful places representing history, memory, and identity, where accumulated experiences and knowledge are weaved into objects, songs, rituals, stories, and other social practices (Martínez-Tagüeña and Torres Cubillas 2018). The creation of stable institutions with high institutional capacity implying clearly defined equitable rules, and openness for shared learning as a basis for collective action related to sustainable resource use is essential for the conservation of the biotic and cultural diversity (Ostrom 2000). Hence, the study of governance is fundamental in order to understand the processes of interaction between social actors involved in public affairs that require decision-making and the formulation of public policies. Governance is considered an emerging pattern of interaction among social actors, their objectives, and the instruments used to direct socio-ecological processes within a particular policy area (Kofinas 2009). The governance of complex systems is produced at various scales from local to global and by different sectors including civil, public, and private (Rhodes 1997).

For fisheries, governance proposals have been suggested through the implementation of SES in order to evaluate their sustainability (Leslie et al. 2015). In defined SES, like national protected areas or in specific environmental themes, governance systems have been analyzed to assess their performance and relevance (Martínez and Espejel 2015; Martínez et al. 2015, 2016). In the context of SES in order to maintain certain favorable stable states, it is important to form participatory inter- and transdisciplinary partnerships, who do not only integrate the natural and social sciences, but incorporate knowledge, needs, and interests of all stakeholders, which then allows the co-production of knowledge, diagnostics, monitoring and evaluation systems, and the development of public policies (Ostrom 2009). In Mexico, inventories on biodiversity operated by CONABIO have generated novel forms of knowledge generation through participatory monitoring (www.biodiversidad.gob.mx/sistema_monitoreo), while at the Latin American level citizen observatories have contributed to novel forms of knowledge production (www.desertificación.gob.ar/). In both examples, citizen involvement in knowledge production has transformed attitudes and conferred good governance schemes related to nature, life quality, and the implementation of good practices.

Governance schemes should be flexible and support innovation and risk-taking in research. For instance, the creation of a cross-sectional innovation unit can bridge the gap between public research and private sector initiatives directly engaged in sustainable management and development of the drylands. Thus, RISZA considers that alongside good governance, social innovation is crucial to achieve the sustainable development goals in drylands. Social innovation refers to novel solutions created to solve social problems in a more efficient, effective, sustainable, and fair manner than previous solutions, where the resulting aggregated value corresponds to society at large rather than to a few individuals (Philis et al. 2008). Beyond ample creativity, successful ideas for social innovation have originated from people's needs and dislocations, dissatisfactions, and blockages. And furthermore, from the generation of new knowledge that opens the door to problem solving in innovative ways (Mulgan 2006).

While certain governments have been reticent to invest in social innovation because it entails risk of failure, innovation has a better chance of success when users have choices and contracts for services reward outcomes achieved rather than outputs or activities, or when there is some competition rather than a state monopoly (Díaz Foncea et al. 2012). Other challenges for social innovation come from the typical insights obtained from business innovation. Contrary to following typical market structures these endeavors align with social organizations that have different motives going further than material incentives to include political recognition and support, voluntary labor, compassion, identity, autonomy, and care. According to the authors' experiences, social organizations tend to grow slower than private businesses, but they also tend to be more resilient. However, clearer metrics are needed to understand social innovation since it is complicated to judge their success. Scale or market share may matter little for a social innovation concerned with a very intense but contained need. In some cases, participants' lives are dramatically improved by the act of collaboration (Mulgan 2006, also for an example, see http://www.in-control.org.uk).

In drylands, it is important to document multiple innovation examples that have been implemented to better manage natural resources like water, minerals, new technology for pastoralism and ranching and agricultural practices, among others.

Social technologies are looking to develop novel tools for diagnostic, monitoring, evaluation, and transfer needed in order to have successful social innovation projects that provide potentially better management schemes and in particular deal with water scarcity. It is important to mention the efforts made in the area of technology for social businesses that promote rural development and in social entrepreneurship that join a socio-ecological scope with an economic benefit based on local communities' self-management (Navarro and Climent 2010).

In this book, a series of chapters retake the key concepts and pillars of the RISZA conceptual framework explained throughout this introduction chapter. Examples of innovative dryland policies and case studies in response to climate change and other global change drivers, and collective thinking gave place to this compendium of highly diverse experiences representing transdisciplinary efforts coauthored by a total of over 80 co-authors from different sectors and organizations. It presents novel ideas to solve the continuously evolving and challenging dryland problems of the Global South.

Acknowledgements The authors greatly appreciate the stimulating discussions and dialogues with members of the RISZA network during many workshops and group meetings; a special thank you to Mark Stafford Smith, Chair of the Science Committee of Future Earth. A special thanks to Ana Delia del Pilar Moran Mendoza for logistic support in compiling all chapters of this book. EHS gratefully acknowledges financial support by CONACYT (projects CB 2015-251388B, 293793, PDCPN-2017/5036).

References

Adams WA (2009) Green development: environment and sustainability in a developing world, 3rd edn. Routledge, London, pp 225–228

Aeschbach W, Gleeson T (2012) Regional strategies for the accelerating global problem of ground-water depletion. Nat Geosci 5:853–861

Akhtar-Schuster M, Stringer LC, Erlewein A, Metternicht G, Minelli S, Safriel U, Sommer S (2017) Unpacking the concept of land degradation neutrality and addressing its operation through the Rio conventions. J Environ Manage 195:4–15

Alsina M (2003) La comunicación intercultural Barcelona. Ed. Antropos, Paris

Amdan L, Aragón RM, Jobbágy EG, Volante J, Paruelo JM (2013) Onset of deep drainage and salt mobilization following forest clearing and cultivation in the Chaco plains (Argentina). Water Resour Res 49:6601–6612

Ander-Egg E (1990) Repensando la Investigación-Acción-Participativa, Colección Política, Servicios y Trabajo Social. Grupo Editorial Lumen Humanitas, Buenos Aires

Archibold OW (1995) Ecology of world vegetation. Chapman and Hall, London

Ashby WR (1962) Principles of the self-organizing system. In: Von Foerster H, Zopf GW Jr (eds) Principles of self-organization: transactions of the University of Illinois Symposium. Pergamon Press, London, pp 255–278

Bar-Yam Y (1997) Dynamics of complex systems. Westview Press, New England

Bastin JF, Berrahmouni N, Grainger A, Maniatis D, Mollicone D, Moore R, Patriarca C, Picard N, Sparrow B, Abraham EM, Aloui K, Atesoglu A, Attore F, Bassüllü C, Bey A, Garzuglia M, García Montero LG, Groot N, Guerin G, Lastadius L, Lowe AJ, Mamane B, Marchi G, Patterson P, Rezende M, Ricci S, Salcedo I, Sanchez Paus Diaz A, Stolle F, Surappaeva V, Castro R (2017) The extent of forest in dryland biomes. Science 356:635–638

Bautista S, Llovet J, Ocampo-Melgar A, Vilagrosa A, Mayor ÁG, Murias C, Vallejo VR, Orr BJ (2017) Integrating knowledge exchange and the assessment of dryland management alternatives – a learning-centered participatory approach. J Environ Manag 195:35–45

Behnke RH, Mortimore M (2016) The end of desertification? Springer, New York, p 560

Berkes F, Colding J, Folke C (2008) Navigating social-ecological systems: building resilience for complexity and change. Cambridge University Press, Cambridge, 394 pp

von Bertalanffy L (1968) General systems theory. George Braziller, New York

Bestelmeyer BT, Okin GS, Duniway MC, Archer SR, Sayre NF, Williamson JC, Herrick JE (2015) Desertification, land use, and the transformation of global drylands. Front Ecol Environ 13:28–36

Biesbroek R, Berrang-Ford L, Ford JD, Tanabe A, Austin SE, Lesnikowski A (2017) Data, concepts and methods for large-n comparative climate change adaptation policy research: a systematic literature review. Clim Chang 9:e548

Biggs R, Schlüter M, Biggs D et al (2012) Towards principles for enhancing the resilience of ecosystem services. Annu Rev Environ Resour 37:421–448

Biggs R, Schlüter M, Schoon ML (2015) Principles for building resilience. Cambridge University Press, Cambridge

Bodin Ö (2017) Collaborative environmental governance: achieving collective action in social-ecological systems. Science 357(6352):eaan1114. https://doi.org/10.1126/science.aan1114

Boege E (2008) El Patrimonio biocultural de los pueblos indígenas de México. Hacia la conservación in situ de la biodiversidad y agrodiversidad en los territorios indígenas. INAH, CDI, Ciudad de México

Castañares Maddox EJ (2009) Sistemas complejos y gestión ambiental: el caso del Corredor Biológico Mesoamericano México. Comisión Nacional para el Conocimiento y Uso de la Biodiversidad, Corredor Biológico Mesoamericano México, Ciudad de México

Challenger A, Cordova A, Lazos Chavero E, Equihua M, Maass M (2018) Opportunities and obstacles to socioecosystem based environmental policy in Mexico: expert opinion at the science-policy interface. Ecol Soc 23(2):31

Chambers R (1983) Rural development: putting the last first. Harlow, Prentice Hall

Chapin FS III, Carpenter SR, Kofinas GP, Folke C, Abel N, Clark WC, Olsson P, Stafford Smith DM, Walker B, Younq OR, Berkes F, Biggs R, Grove JM, Naylor RL, Pinkerton E, Steffen W, Swanson FJ (2009c) Ecosystem stewardship: sustainability strategies for a rapidly changing planet. Trends Ecol Evol 25(4):241–249

Chapin FS III, Folke C, Kofinas GP (2009a) A framework for understanding change. In: Chapin FS III, Kofinas GP, Folke C (eds) Principles of ecosystem stewardship. Springer, New York, pp 3–28

Chapin FS III, Kofinas GP, Folke C, Carpenter SR, Olsson P, Abel N, Biggs R, Naylor RL, Pinkerton E, Stafford Smith DM, Steffen W, Walker B, Young OR (2009b) Resilience based stewardship: strategies for navigating sustainable pathways in a changing world. In: Chapin FS III, Kofinas GP, Folke C (eds) Principles of ecosystem stewardship. Springer, New York, pp 319–337

Chapin M, Lamb Z, Threlkeld B (2005) Mapping indigenous lands. Annu Rev Anthropol 34:619–638

Chasek P, Safriel U, Shikongo S, Futran Fuhrman V (2015) Operationalizing zero net land degradation: the next stage in international efforts to combat desertification? J Arid Environ 112:5–13

Cherlet M, Hutchinson C, Reynolds JF, Hill J, Sommer S, von Maltitz G (2018) World atlas of desertification. Publication Office of the European Union, Luxembourg

Clark W, van Kerkhoff L, Lebel L, Gallopin GC (2016) Crafting usable knowledge for sustainable development. Proc Natl Acad Sci 113(17):4570–4578

Coppock DL, Fernández-Giménez F, Hiernaux P, Huber-Sannwald E, Schloeder C, Valdivia C, Arredondo JT, Jacobs M, Turin C, Turner M (2017) Rangeland systems in developing nations: conceptual advances and societal implications. In: Briske DD (ed) Rangeland Systems. Springer, New York, pp 569–641

Coppock LD (2016) Cast off the shackles of academia! Use participatory approaches to tackle real-world problems with underserved populations. Rangelands 38(1):5–13

Cowie AL, Orr BJ, Castillo Sanchez VM, Chasek P, Crossman ND, Erlewein A, Louwagie G, Maron M, Metternicht GI, Minelli S, Tengberg AE, Walter S, Welton S (2018) Land in balance: the scientific conceptual framework for land degradation neutrality. Environ Sci Policy 79:25–35

Curtin CG (2015) The science of open spaces. Island Press, Washington, DC

D'Odorico P, Bhattachan A, Davis K, Ravi S, Runyan C (2013) Global desertification: drivers and feedbacks. Adv Water Resour 51:326–344

Dados N, Conell R (2012) The global south. Contexts 11(1):12–13. https://doi.org/10.1177/1536504212436479

Daily G (1997) Nature's services. Island Press, Washington, DC

Dardel C, Kergoat L, Hiernaux P et al (2014) Re-greening Sahel: 30 years of remote sensing data and field observations (Mali, Niger). Remote Sens Environ 140(1):350–364

Davidson-Hunt IJ, O'Flaherty RM (2007) Researchers, indigenous peoples, and place-based learning communities. Soc Nat Resour 20(4):291–305

Davies J, Poulsen L, Schulte-Herbrüggen B, Mackinnon K, Crawhall N, Henwood WD, Dudley N, Smith J, Gudka M (2012) Conserving dryland biodiversity. IUCN, Nairobi, xii+84 p

Davies J, Robinson LW, Ericksen PJ (2015) Development process resilience and sustainable development: insights from the drylands of eastern Africa. Soc Nat Resour 28(3):328–343

Davis DK (2016a) The arid lands: history, power, knowledge. MIT Press, Cambridge

Davis DK (2016b) Deserts and drylands before the age of desertification. In: Behnke RH, Mortimore M (eds) The end of desertification? Springer, New York, pp 203–223

De Vente J, Bautista S, Orr B (2017) Preface: optimizing science impact for effective implementation of sustainable land management. J Environ Manage 195:1–3

DeFries R, Nagendra H (2017) Ecosystem management as a wicked problem. Science 356:265–270

Díaz Foncea M, Marcuello C, Marcuello C (2012) Empresas sociales y evaluación del impacto social. CIRIEC-España Rev Econ Pública Soc Coop 75:179–198

Díaz S, Demissew S, Carabias J, Joly C, Lonsdale M, Ash N, Larigauderie A, Adhikari JR, Arico S, Báldi A, Bartuska A, Baste I, Bilgin A, Brondizio E, MaChan K, Figuero VE, Duraiappah A, Fischer M, Zlatanova D (2015) The IPBES conceptual framework — connecting nature and people. Curr Opin Environ Sustain 14:1–16

Díaz S, Pascual U, Stenseke M, Martín-López B, Watson RT, Molnár Z, Hill R, Chan KMA, Baste IA, Brauman KA, Polasky S, Church A, Lonsdale M, Larigauderie A, Leadley PW, van Oudenhoven APE, van der Plaat F, Schröter M, Lavorel S, Aumeeruddy-Thomas Y, Bukvareva E, Davies K, Demissew S, Erpul G, Failler P, Guerra CA, Hewitt CL, Keune H, Lindley S, Shirayama Y (2018) Assessing nature's contributions to people. Science 359(6373):270–272

Donohue RJ, Roderick ML, McVicar TR, Farquhar GD (2013) Impact of CO_2 fertilization on maximum foliage cover across the globe's warm, arid environments. Geophys Res Lett 40: 3031–3035

Dregne HE, Chou NT (1992) Global desertification dimensions and costs. In: Degradation and restoration of arid lands. Texas Tech. University, Lubbock, pp 73–92

Dressler W, Büscher B, Schoon M, Brockington D, Hayes T, Kull CA, McCarthy J, Shrestha K (2010) From hope to crisis and back again? A critical history of the global CBNRM narrative. Environ Conserv 37:5–15

Dudley N (2008) Guidelines for applying protected area management categories. IUCN, Gland

Durose C, Richardson L, Perry B (2018) Craft metrics to value co-production. Nature 562:32–33

Durston J, Miranda F (2002) Experiencias y metodología de la investigación participativa. Serie Políticas sociales. División de Desarrollo Social, CEPAL, Publicación Naciones Unidas, Santiago de Chile

Easdale MH (2016) Zero net livelihood degradation – the quest for a multidimensional protocol to combat desertification. Soil 2:129–134

FAO (Food and Agriculture Organization of United Nations) (2018) World livestock: transforming the livestock sector through the sustainable development goals. Rome. 222 pp. Licence: CC BY-NC-SA 3.0 IGO

Feng S, Fu Q (2013) Expansion of global drylands under warming climate. Atmos Chem Phys 13(9):10081–10094

Figueroa M (2011) Indigenous peoples and cultural losses. In: The Oxford handbook of climate change and society. Oxford University Press, New York, pp 232–249

Folke C, Carpenter SR, Walker BH, Scheffer M, Chapin FS III, Rockström J (2010) Resilience thinking: integrating resilience, adaptability and transformability. Ecol Soc 15:20

Folke C, Chapin FS III, Olsson P (2009) Transformations in ecosystem stewardship. In: Chapin FS III, Kofinas PG, Folke C (eds) Principles of ecosystem stewardship: resilience-based natural resource management in a changing world. Springer, New York, pp 103–125

Freire P (1970) Pedagogy of the oppressed. Continuum, New York

García R (2006) Sistemas Complejos. Conceptos, métodos y fundamentación epistemológica de la investigación interdisciplinaria. GEDISA, Barcelona, 200 pp

Geist JJ, Lambin EF (2004) Dynamical causal patterns of desertification. Bioscience 54:817–829

Global Land Project (GLP) (2005) Science plan and implementation strategy, IGBP (international geosphere biosphere program) report no. 53. IHDP report no. 19. IGBP Secretariat, Stockholm

Gómez-Baggethun E, Reyes-García V (2013) Reinterpreting change in traditional ecological knowledge. Hum Ecol 41:643–647

Gorddard R, Colloff MJ, Wise RM, Ware D, Dunlop M (2016) Values, rules and knowledge. Adaptation as change in the decision context. Environ Sci Policy 57:60–69

Goudie AS (1986) The search for Timbuktu: a view of deserts. Clarendon Press, Oxford

Grace J, San José J, Meir P, Miranda HS, Montes RA (2006) Productivity and carbon fluxes of tropical savannas. J Biogeogr 33:387–400

Greenberg JB, Park TK (1994) Political ecology. J Political Ecol 1:1–12

Griggs D, Stafford-Smith DM, Gaffney O, Rockström J, Öhman MC, Shyamsundar P, Steffen W, Glaser G, Kanie N, Noble I (2013) Sustainable development goals for people and planet. Nature 495:305–307

Gudka M, Davies J, Poulsen L, Schulte-Herbrügger B, MacKinnon K, Crawhall N, Henwood WD, Dudley N, Smith J (2014) Conserving dryland biodiversity: a future vision of sustainable dryland development. Biodiversity 15:143–147

Gunderson LH, Holling CS (2002) Panarchy: understanding transformations in human and natural systems. Island Press, Washington, DC

Gutiérrez Serrano NG (2016) Senderos académicos para el encuentro. Conocimiento transdisciplinario y configuraciones en red. Universidad Nacional Autónoma de México, México, 213 pp

Herrmann SM, Sop TK (2016) The map is not the territory: how satellite remote sensing and ground evidence have re-shaped the image of Sahelian desertification. In: Behnke RH, Mortimore M (eds) The end of desertification? Springer, New York, pp 117–145

Hickey G (2018) Co-production from proposal to paper: share power in five ways. Nature 562: 29–30

Hickey G, Brearley S, Coldham T, Denegri S, Green G, Staniszewska S et al (2018) Guidance on co-producing a research project. INVOLVE, Southampton http://www.invo.org.uk/wp-content/uploads/2018/03/Copro_Guidance_Mar18.pdf

Holling CS (1988) Temperate forest insect outbreaks, tropical deforestation and migratory birds. Mem Entomol Soc Can 146:21–32

Holling CS, Meffe GK (1996) Command and control and the pathology of natural resource management. Conserv Biol 10:328–337

Huaico Malhue AI, Romero Díaz A, Espejel Carbajal MI (2018) Evolución de los enfoques en desertificación. Cuad Geogr 57(2):53–71

Huang J, Guan X, Ji F (2012) Enhanced cold-season warming in semi-arid regions. Atmos Chem Phys 12(12):5391–5398

Huang J, Li Y, Fu C, Chen F, Fu Q, Dai A, Shinoda M, Ma Z, Guo W, Li Z, Zhang L, Liu Y, Yu H, He Y, Xie Y, Guan X, Ji M, Lin L, Wang S, Yan H, Wanget G (2017b) Dryland climate change: recent progress and challenges. Rev Geophys 55:719–778

Huang J, Yu H, Dai A, Wei Y, Kang L (2017a) Potential threats over drylands behind 2°C global warming target. Nat Clim Change 7:417–422

Huang J, Yu H, Guan X, Wang G, Guo R (2015) Accelerated dryland expansion under climate change. Nat Clim Chang 6(2):166–171

Huber-Sannwald E, Ribeiro Palacios M, Arredondo JT, Braasch M, Martínez Peña M, García de Alba J, Monzalvo K (2012) Navigating challenges and opportunities of land degradation and sustainable livelihood development. Philos Trans R Soc Lond B Biol Sci 367:3158–3177

Hulme M (1996) Recent climate change in the world's drylands. Geophys Res Lett 23(1):61–64

INEGI (2019) México: Objetivos de Desarrollo Sostenible. http://www.agenda2030.mx/#/home

Jia B, Zhang Z, Ci L, Ren Y, Pan B, Zhang Z (2004) Oasis land-use dynamics and its influence on the oasis environment in Xinjiang, China. J Arid Environ 56(1):11–26

Johnson JT, Howitt R, Cajete G, Berkes F, Pualani Louis R, Kliskey A (2016) Weaving indigenous and sustainability sciences to diversify our methods. Sustain Sci 11:1–11

Jonathan Davies, Lance W. Robinson, Polly J. Ericksen, (2015) Development Process Resilience and Sustainable Development: Insights from the Drylands of Eastern Africa. Society & Natural Resources 28 (3):328–343

Knapp CN, Fernandez-Gimenez M, Kachergis E, Rudeen A (2011) Using participatory workshops to integrate state-and-transition models created with local knowledge and ecological data. Rangel Ecol Manag 64(2):158–170

Kockelmans JJ (1979) Interdisciplinarity and higher education. Penn State University, State College

Kofinas GP (2009) Adaptive co-management in socio-ecological governance. principle of ecosystem stewardship. Springer, New York, pp 77–101

Koontz TM, Gupta D, Mudliar P, Ranjan P (2015) Adaptive institutions in social-ecological systems governance: a synthesis framework. Environ Sci Policy 35(B):139–151

Krätli S (2015) In: de Jode H (ed) Valuing variability: new perspectives on climate-resilient drylands development. International Institute for Environment and Development, London

Lambin EF, Turner BL, Geist HJ, Agbola SB, Angelsen A, Bruce JW, Coomes OT, Dirzo R, Fischer G, Folke C, George PS, Homewood K, Imbernon J, Leemans R, Lin X, Moran EF, Mortimore M, Ramakrishnan PS, Richards JF, Skane H, Steffen W, Stone GD, Svedin U, Veldkamp TA, Vogel C, Xu J (2001) The causes of land-use and land-cover change: moving beyond the myths. Glob Environ Chang 11:261–269

Leslie HM, Basurto X, Nenadovic M, Sievanen L, Cavanaugh KC, Cota-Nieto JJ, Erisman BE, Finkbeiner E, Hinojosa-Arango G, Moreno-Báez M, Nagavarapu S, Reddy SMW, Sánchez-Rodríguez S, Siegel K, Ulibarria-Valenzuela JJ, Hudson Weaver A, Aburto-Oropez O (2015) Operationalizing the social-ecological systems framework to assess sustainability. Proc Natl Acad Sci 112(19):5979–5984

Liu J, Dietz T, Carpenter SR, Alberti M, Folke C, Alberti M, Redman CL, Schneider SH, Ostrom E, Pell AN, Lubchenco J, Taylor WW, Ouyang Z, Deadman P, Kratz T, Provencher W (2007) Coupled human and natural systems. Ambio 36(8):639–649

Lindsay C. Stringer, Mark S. Reed, Luuk Fleskens, Richard J. Thomas, Quang Bao Le, Tana Lala-Pritchard, (2017) A New Dryland Development Paradigm Grounded in Empirical Analysis of Dryland Systems Science. Land Degradation & Development 28 (7):1952–1961

Marlowe FW (2005) Hunter-gatherers and human evolution. Evol Anthropol 14:54–67

Martínez N, Brenner L, Espejel I (2015) Red de participación institucional en las áreas naturales protegidas de la península de Baja California. Reg Soc 27(62):27–62

Martínez N, Espejel I (2015) La investigación de la gobernanza en México y su aplicabilidad ambiental. Econ Soc Territorio 15(47):153–183

Martínez N, Espejel I, Martínez Valdes C (2016) Evaluation of governance in the administration of protected areas on the peninsula of Baja California. Front Norte 28(55):103–129

Martínez-Tagüeña N, Torres Cubillas LA (2018) Walking the desert, paddling the sea: Comcaac mobility in time. J Archaeol Anthropol 49:146–160

Matson P (2012) Seeds of sustainability. Lessons from the birthplace of the Green revolution in agriculture. Island Press, Washington, DC

MEA (2005) Millennium Ecosystem Assessment Board. Ecosystem and human wellbeing. Island Press, Washington, DC

Meadows DH (2008) Thinking in systems. EarthScan, London

Middleton N (2018) Rangeland management and climate hazards in drylands: dust storms, desertification and the overgrazing debate. Nat Hazards 92(1):57–70

Middleton N, Stringer L, Goudie A, Thomas D (2011) The forgotten billion: MDG achievement in the drylands. UNDP-UNCCD, New York

Middleton N, Thomas D (1997) World atlas of desertification, 2nd edn. Arnold, London

Milkoreit M, Hodbod J, Baggio J, Benessaiah K, Calderon Contreras R, Donges JF, Mathias JD, Rocha JC, Schoon M, Werners SE (2017) Defining tipping points for social-ecological systems scholarship – an interdisciplinary literature review. Environ Res Lett 13:033005

Morin E (1977) El Método. La Naturaleza de la Naturaleza. Ediciones Cátedra, Madrid

Mortimore M, Anderson S, Cotula L, Davies J, Faccer K, Hesse C et al (2009) Dryland opportunities: a new paradigm for people, ecosystems and development. IUCN, Gland; IIED, London

Mulgan G (2006) The process of social innovation. Innov Technol Gov Glob 1(2):145–162

Navarro AM, Climent VC (2010) Emprendedurismo y economía social como mecanismos de inserción sociolaboral en tiempos de crisis. REVESCO 100:43–67

Newing H, Eagle CM, Puri RK, Watson CW (2011) Conducting research in conservation: social science methods and practice. Routledge, New York

Nolan RH, Sinclair J, Eldridge DJ, Ramp D (2018) Biophysical risks to carbon sequestration and storage in Australian drylands. J Environ Manag 208:102–111

Noy-Meir I (1973) Desert ecosystems: environment and producers. Annu Rev Ecol Syst 4(1973):25–51

Ocampo-Melgar BS, de Steiguer JG, Orr BJ (2017) Potential of an outranking multi-criteria approach to support the participatory assessment of land management actions. J Environ Manag 195:70–77

Okpara UT, Stringer LC, Akhtar-Schuster M, Metternicht GI, Dallimer M, Requier-Desjardins M (2018) A social-ecological systems approach is necessary to achieve land degradation neutrality. Environ Sci Policy 89:59–66

Orr BJ, Cowie AL, Castill Sanchez VM, Chasek P, Crossman ND, Erlewein A, Louwagie G, Maron M, Metternicht GI, Minelli S, Tengberg AE, Walter S, Welton S (2017) Scientific conceptual framework for land degradation neutrality. A report of the science-policy interface. United Nations Convention to Combat Desertification (UNCCD), Bonn

Ostrom E (2000) El gobierno de los bienes comunes. La evolución de las instituciones de acción colectiva. Fondo de Cultura Económica, Ciudad de México

Ostrom E (2009) A general framework for analyzing sustainability of socio-ecological systems. Science 325(5939):419–422

Oviedo G, Maffi L, Larsen PB (2000) Indigenous and traditional peoples of the world and Ecoregion conservation. An integrated approach to conserving the World's biological and cultural diversity. WWF-World Wide Fund for Nature, Gland

Papanastasis VP, Bautista S, Chouvardas D, Mantzanas K, Papadimitriou M, Mayor AG, Koukioumi P, Papaioannou A, Vallejo RV (2017) Comparative assessment of goods and services provided by grazing regulation and reforestation in degraded Mediterranean rangelands. Land Degrad Dev 28:1178–1187

Pauli D (1985) Anecdotes and the shifting baseline syndrome of fisheries. Trends Ecol Evol 10(10):430

Peters DPC, Havstad KM, Archer SR, Sala OE (2015) Beyond desertification: a new paradigm for dryland landscapes. Front Ecol Environ 13(1):4–12

Philis JA, Deiglmeier K, Miller DT (Fall 2008) Rediscovering social innovation. Stanf Soc Innov Rev. http://ssir.org/articles/entry/rediscovering_social_innovation. Accessed November 2015

Pickett STA, White PS (1985) The ecology of natural disturbance as patch dynamics. Academic, New York

Puigdefábregas J (1998) Ecological impacts of global change on drylands and their implications for desertification. Land Degrad Dev 9:393–406

Reed MS, Buenemann M, Atlhopheng J, Akhtar-Schuster M, Bachmann F, Bastin G, Bigas H, Chanda R, Dougill AJ, Essahli W, Evely AC, Fleskens L, Geeson N, Glass JH, Hessel R,

Holden J, Ioris A, Kruger B, Liniger HP, Mphinyane W, Nainggolan D, Perkins J, Raymond CM, Ritsema CJ, Schwilch G, Sebego R, Seely M, Stringer LC, Thomas R, Twomlow S, Verzandvoort S (2011) Cross-scale monitoring and assessment of land degradation and sustainable land management: a methodological framework for knowledge management. Land Degrad Dev 22:261–271

Reed MS, Stringer LC (2016) Land degradation, desertification and climate change. Anticipating, assessing and adapting to future change. Routledge, London/New York

Reid RS, Fernández-Giménez ME, Galvin KA (2014) Dynamics and resilience of rangelands and pastoral peoples around the globe. Annu Rev Environ Resour 39:217–242

Reynolds JF, Stafford Smith DM (2002) Global desertification: do humans cause deserts. Dahlem University Press, Berlin

Reynolds JF, Stafford Smith DM, Lambin EF, Turner BL II, Mortimore M, Batterbury SPJ, Downing TE, Dowlatabadi H, Fernandez RJ, Herrick JE, Huber-Sannwald E, Leemans R, Lynam T, Maestre FT, Ayarza M, Walker B (2007) Global desertification: building a science for global dryland development. Science 316:847–851

Rhodes RAW (1997) Understanding governance: policy networks, governance, reflexivity and accountability (public policy & management). Open University Press, Philadelphia, 252 pp

Rittel H, Webber MM (1973) Dilemmas in a general theory of planning. Policy Sci 4:155–169

Robbins P (2012) Political ecology: a critical introduction. Wiley-Blackwell, West Sussex

Safriel U (2017) Land degradation neutrality (LDN) in drylands and beyond – where has it come from and where does it go. Silva Fenn 51:1650. https://doi.org/10.14214/sf.1650

Safriel U, Adeel Z, Niemeijer D, Puigdefabregas J, White R, Lal R, Winslow M, Ziedler J, Prince S, Archer E, King C (2005) Dryland systems. In: Hassan R, Scholes R, Ash N (eds) Ecosystems and human Well-being: current state and trends. Island Press, Washington, DC, pp 623–662

Santos B (2002) Hacia una concepción multicultural de los derechos humanos. El Otro Derecho 28:59–83

Scheffer M, Carpenter SR (2003) Catastrophic regime shifts in ecosystems: linking theory to observation. Trends Ecol Evol 18:648–656

Schlaepfer DR, Bradford JB, Lauenroth WK, Munson SM, Tietjen B, Hall SA, Wilson SD, Duniway MC, Jia G, Pyke DA, Lkhagva A, Jamiyansharav K (2017) Climate change reduces extent of temperate drylands and intensifies drought in deep soils. Nat Commun 8:14196

Schoon ML, Robards MD, Brown K, Engle N, Meek CL, Biggs R (2015) Politics and the resilience of ecosystem services. In: Biggs R, Schlüter M, Schoon ML (eds) Principles for building resilience. Cambridge University Press, Cambridge

Schuttenberg HZ, Guth HK (2015) Seeking our shared wisdom: a framework for understanding knowledge coproduction and coproductive capacities. Ecol Soc 20(1):15

Southgate D (1990) The causes of land degradation along "spontaneously" expanding agricultural frontiers in the third world. Land Econ 66(1):93–101

Squires VR (2010) Desert transformation or desertification control? J Rangel Sci 1(1):17–21

Stafford Smith DM (2016) Desertification: reflections on the mirage. In: Behnke RH, Mortimore M (eds) The end of desertification? Springer, New York, pp 539–560

Stafford Smith DM, Abel N, Walker B, Chapin FS III (2009) Drylands: coping with uncertainty, thresholds, and changes in state. In: Chapin FS III, Kofinas GP, Folke C (eds) Principles of ecosystem stewardship. Springer, New York, pp 171–195

Stafford Smith DM, Cribb J (2009) Dry times. CSIRO Publishing, Clayton

Stafford Smith DM, Reynolds JF (2002) Desertification: a new paradigm for an old problem. In: Reynolds JF, Stafford Smith DM (eds) Global desertification: do humans cause deserts? Dahlem workshop reports, vol 88. Dahlem Univ. Press, Berlin, pp 403–424

Stavi I, Lal R (2015) Achieving zero net land degradation: challenges and opportunities. J Arid Environ 112:44–51

Steffen W, Broadgate W, Deutsch L, Gaffney O, Ludwig C (2015) The trajectory of the Anthropocene: the great acceleration. Anthropocene Rev 2(1):81–98

Steffen W, Crutzen PJ, McNeill JR (2007) The Anthropocene: are humans now overwhelming the great forces of nature? Ambio 36:614–621

Stott P (2016) How climate change affects extreme weather events. Science 352(6293):1517–1518

Stringer LC, Reed MS, Fleskens L, Thomas RJ, Le QB, Lala-Pritchard T (2017) A new dryland development paradigm grounded in empirical analysis of dryland systems science. Land Degrad Dev 28(7):1952–1961

Tàbara D, Frantzeskaki N, Hölscher K, Pedde S, Kok K, Lamperti F, Christensen JH, Jäger J, Berry P (2018) Positive tipping points in a rapidly warming world. Curr Opin Environ Sustain 2018(31):120–129

Tengö M, Hill R, Malmer P, Raymond CM, Spierenburg M, Danielsen F, Elmqvist T, Folke C (2017) Weaving knowledge systems in IPBES, CBD and beyond—lessons learned for sustainability. Curr Opin Environ Sustain 26–27:17–25

Thomas D, Middleton N (1992) World atlas of desertification. Edward Arnold for UNEP, London

Toledo V, Barrera-Bassols N (2008) La Memoria Biocultural. La importancia ecológica de las sabidurías tradicionales. Icaria Editorial, Barcelona

UNCCD (1994) United Nations convention to combat desertification. United Nations, Geneva, p 58

UNCCD (2019) United Nations convention to combat desertification. Land degradation neutrality: land profiles. https://www.unccd.int/actionsldn-target-setting-programme/ldn-country-profiles

United Nations Development Programme (UNDP) (2019) Global policy centre on resilient ecosystems and desertification. Nairobi, Kenya. Consulted April 2019

Veldman JW, Buisson E, Durigan G, Fernandes GW, Le Stradie S, Mahy G, Negreiros D, Overbeck GE, Veldman RG, Zaloumis NP, Putz FE, Bond WJ (2015) Toward an old-growth concept for grasslands, savannas, and woodlands. Front Ecol Environ 13(3):154–162

Von Wehrden H, Hanspach J, Kazzensky P, Fischer J, Wesche K (2012) Global assessment of the non-equilibrium concept in rangelands. Ecol Appl 22(2):393–399

Walker B (1993) Rangeland ecology: understanding and managing change. Ambio 22:2–3

Walker B, Salt D (2006) Resilience thinking: sustaining ecosystems and people in a changing world. Island Press, Washington, DC

Wang L, D'Odorico PD, Evans JP, Eldridge DJ, McCabe MF, Caylor KK, King EG (2012) Dryland ecohydrology and climate change: critical issues and technical advances. Hydrol Earth Syst Sci 16:2585–2603

Westoby M, Walker B, Noy-Meir I (1989) Opportunistic management of rangelands not at equilibrium. J Range Manag 42(4):266–274

Whitfield S, Reed MS (2012) Participatory environmental assessment in drylands: introducing a new approach. J Arid Environ 77:1–10

Whitford WG (2002) Ecology of desert systems. Academic, San Diego

Whyte WF (1989) Introduction to action research for the twenty-first century: participation, reflection, and practice. Am Behav Sci 32:502–512

Wilcox BP, Sorice MG, Young MH (2011) Dryland ecohydrology in the Anthropocene: taking stock of human-ecological interactions. Geogr Compass 5(3):112–127

Willyard C, Scudellari M, Nordling L (2018) Partners in science. Nature 562:24–28

Yahdjian L, Sala OE, Havstad KM (2015) Rangeland ecosystem services: shifting focus from supply to reconciling supply and demand. Front Ecol Evol 13:44–51

Part I
Drylands and Socio-Ecological Systems

Chapter 2
Sustainable Development Goals and Drylands: Addressing the Interconnection

S. Lucatello and E. Huber-Sannwald

Abstract Sustainable dryland practices and drought risk management are the centerpieces for sustainable management systems as well as measures to sustainable land use, particularly resilience and ecosystem services. On the other hand, sustainable development goals (SDGs) and target 15.3 provide a general guidance for environmental and socio-ecological dynamics for improving a better-coordinated approach to land management, particularly in the case of drylands. During the period 1998–2013, about one-fifth of the Earth's land surface covered by vegetation showed persistent and declining trends in productivity, particularly in the case of Latin American countries. It is therefore key to reverse advanced stages of land degradation by sustainable land management and improving lives and livelihoods of millions of people currently under threat in the region. Paragraph 33 of the 2030 Agenda for Sustainable Development focuses on the linkage between sustainable management of the planet's natural resources and social and economic development as well as on strengthening "cooperation on desertification, dust storms, land degradation and drought" and promote resilience and disaster risk reduction." In this article, we explore the relation between SDGs and their interconnections with drylands management. SDGs are considered a major instrument to combat desertification, drought, and land degradation together with climate change and the loss of biodiversity by combining and scaling up established socioeconomic principles and practices to reach SDG target 15.3 and the objectives of UNCCD for land degradation neutrality (LDN).

Keywords Sustainable development goals · Land degradation neutrality · Agenda for sustainable development · Interconnection

S. Lucatello (✉)
Estudios Ambientales y Territoriales, Instituto Mora, Mexico City, Mexico
e-mail: slucatello@institutomora.edu.mx

E. Huber-Sannwald
División de Ciencias Ambientales, Instituto Potosino de Investigación Científica y Tecnológica, San Luis Potosi, San Luis Potosí, Mexico

© Springer Nature Switzerland AG 2020
S. Lucatello et al. (eds.), *Stewardship of Future Drylands and Climate Change in the Global South*, Springer Climate,
https://doi.org/10.1007/978-3-030-22464-6_2

Introduction

International figures and literature on drylands converge on the fact that this diverse biome complex covers approximately 40% of the world's land area, and supports two billion people, 90% of whom live in developing countries (UNCCD 2017; Cherlet et al. 2018). Though drylands are mostly prevalent in Africa and Asia, Latin America dryland ecosystems also play an important role in providing essential elements for livelihood development; however, maintaining land quality is currently the single biggest challenge for long-term provision of dryland systems' goods and services for human well-being both in rural and urban dryland environments. Over the past three decades, urbanization in drylands has been increasing at a similar rate as in other ecoregions of the world; it is predicted that by 2025, 58% of the world population will be living in urbanized drylands (Balk 2008). Of the globally 1692 large cities, 35% (586) are situated in drylands (Cherlet et al. 2018).

As previously mentioned in this book, increasing land degradation in drylands poses one of the greatest challenges for land owners, producers, land managers, land use planers, and policy makers, as land quality is at the interface of ecosystem functioning and human security including access to clean water, air, food, energy as the basic elements for livelihood development (UNCCD 2017). Hence, land degradation may present serious threats and risks to communities and ecosystems.

Unsustainable practices of land and water use, together with climate change impacts and climate variability, shape the long-term legacy of continuing drylands degradation potentially triggering socio-ecological collapse expressed as a desertification process (UNCCD 1994). This is mostly accompanied by great damages to the main regional biogeochemical cycles, altering biological diversity and productivity as well as curbing the ability of ecosystems to produce natural goods and provide livelihoods to rural and urban communities.

Climate change is an amplifier of such dramatic changes and it is directly affecting crop yields and agricultural systems among others. According to IPCC (2015) estimates for the region, this is likely to worsen in the next three decades. It is likely that climate change will affect grassland productivity by reducing it between 50 and 80% in semi-arid and arid regions (IPCC 2015); climate scenarios forecast that high levels of desertification and soil salinization, and increasing water stress, will occur in many parts of Latin America. Due to these likely changes, increasing food prices together with rapidly increasing production and transportation costs will have direct impacts on highly vulnerable populations, especially those who are exposed to extreme weather events, such as mega-droughts or short-term floodings. Therefore, as drylands ecosystems are highly diverse and provide all categories of ecosystem services, which benefit local communities, they require increasingly adaptive management approaches to deal with unpredictable system change. Besides they must be integrated in the larger biophysical, socioeconomic, and institutional contexts considering multifunctional landscapes, resilient local socio-ecological systems, land tenure, and the long-term productive potential of land resources including soils, water, plants, animals, and microorganisms for the production of goods and services to meet changing human needs without neglecting the long-term land potential (WOCAT 2010).

Recent efforts have called for the need to scale up local sustainable land use practices and local initiatives. These may eventually transform into larger scale integrated multi-stakeholder assessments to reorient land degradation towards sustainable land management (Reed et al. 2015), with a focus on knowledge exchange and a learning-centered participatory approach (Bautista et al. 2017) to build resilience to sustain ecosystem services in dryland socio-ecological systems (Biggs et al. 2015). This approach prepares for adjusting to new rapidly changing environmental, socioeconomic, and sociopolitical conditions.

However, as we already know, understanding the legacy of land degradation has proven difficult and challenging as it results from complex interactions and feedbacks of multiple exogenous and endogenous drivers undergoing change (Liu et al. 2007; Huber-Sannwald et al. 2012) affecting a large number of ecological and social properties in an unpredictable fashion (Reynolds et al. 2007). Hence, identifying the appropriate restoration mechanisms to induce short-, mid-, and long-term recovery of the biophysical and socioeconomic subsystem is often hampered by contrasting value systems, world views, knowledge types, and institutional ruling systems (Berkes et al. 2006; Gorddard et al. 2016). As the empirical work led by the International Network for Sustainable Drylands (RISZA) showed through several case studies and during international fora, it becomes crucial to promote a transformative framework for action. This framework should include from land degradation to sustainable land management considering multiple SDGs simultaneously including synergies and trade-offs among them (Mainali et al. 2018) and the inclusion of multiple sectors, actors, and regions---countries (Stafford Smith et al. 2017).

The sustainable management of drylands can also increase the potential of local, regional, and global benefits of drylands that have not been fully utilized due to weak public policies, market failures, gender inequalities, and mostly the lack of integrated local policies that involve drylands communities (Stafford Smith et al. 2009). As a matter of fact, there is great concern in the Latin American region that dryland ecosystems and populations face several risks and costs including land tenure reforms, soil conflicts, variable weather and climate, the drain of "community human capital," and the fast speed of erosion of natural areas. Indigenous knowledge and production methods are at risk too, being the regional host of 40% of the international indigenous world heritage.

The IPCC SR15 (2015) urges for effective climate action through the formation of novel partnerships of national and state governments and non-government sectors, including academic institutions, civil societies, the private sector, indigenous groups, and local communities. These types of partnerships have also been advocated for the Sustainable Development Agenda of the United Nations (Sustainable Development Goal 17, UN Agenda 2030). They aim to implement climate actions (SDG 13) like combat desertification by achieving land degradation neutrality

(Safriel 2017, SDG 15) and protect the biosphere's terrestrial and marine biodiversity (SDG 15 and 14, respectively) as fundamental goals to eradicate poverty (SDG 1). But it advocates also to achieve gender equality (SDG 5), food security (SDG 2), health (SDG 3), economic prosperity (SDG 8), safe water, sanitation, and drinking water quality/quantity (SDG 6), among others. Long-term mitigation efforts to combat desertification and restore degraded drylands have rarely been successful (Reynolds et al. 2007; Cherlet et al. 2018). Globally, these efforts have been coordinated by the UNCCD, yet bottom-up governance and management approaches are needed to achieve adaptive livelihoods that built on social and natural capital as the essential pillars of sustainable dryland life-supporting systems (Chapin et al. 2009).

Addressing SDGs and Their Interconnection

The SDGs represent the most ambitious world development agenda in the history of global environmental governance and global development promoted by the international community; it comprises 17 goals and a large set of specific targets that should be reached by 2030.

The agenda is part of a decade-long process of global development, which finds its roots in previous attempts to address inequalities and foster social, economic, and environmental welfare both with the Millennium Development Goals (MDGs) and the United Nations Development Program (UNDP). In the case of the SDGs, however, there is certainly a strong component towards environmental protection, social and economic sustainability which is also part of international efforts dating back to the Brundtland Report in the 1980s, the Rio Earth Summit and Rio+20 conferences in 1992 and 2012, respectively. Broadly speaking, the SDGs are intended to be a guideline for all governments (see Table 2.1). The goals reflect the three pillars of sustainable development that are the environment, economy, and social aspects. Some goals are in fact mainly socio-economic in character (e.g., goals 1, 4, 5, 8–11, 16, 17), while others focus clearly on the environment and the biophysical system, where land quality and natural resources play a seminal role (e.g., goals 2, 3, 6, 7, 12–15). All goals include targets with clear interlinkages among SDGs; however, they are designed to be both mutually inclusive and universal. Therefore, the goals oriented towards socioeconomic and environmental aspects mutually depend on each other. For example, environmental sustainability will depend on the actions of climate change through both mitigation and adaptation at different levels and including local governments, indigenous communities, cities but also other actors who are involved in the SDGs fulfillment.

Please refer to the following table for better understanding of SDGs and their function.

Table 2.1 The UN Sustainable Development Goals for the period 2015–2030 (http://sustainabledevelopment.un.org/focussdgs.html) related to ecosystem services and soil function

SDG topic		Ecosystem services											Relates to soil
	Provision of food, wood and fibre	Provision of raw materials	Provision of support for human infrastructures and animals	Flood mitigation	Filtering of nutrients and contaminants	Carbon storage and greenhouse gas regulation	Detoxification and the recycling of wastes	Regulation of pests and disease populations	Recreation	Aesthetics	Heritage values	Cultural identity	function (Table 2.2)
	1	2	3	4	5	6	7	8	9	10	11	12	
1 End poverty in all its forms everywhere	X	X	X	X									1, 5
2 End hunger, achieve food security and improved nutrition and promote sustainable agriculture	X		X										1, 2, 4
3 Ensure healthy lives and promote well-being for all at all ages	X							X	X	X	X	X	1, 2, 3, 4, 5, 7
4 Ensure inclusive and equitable quality education and promote lifelong learning opportunities for all												X	7
5 Achieve gender equality and empower all women and girls													
6 Ensure availability and sustainable management of water and sanitation for all				X	X		X		X				2
7 Ensure access to affordable, reliable, sustainable and modern energy for all	X	X											1, 5, 6
8 Promote sustained, inclusive and sustainable economic growth, full and productive employment and decent work for all	X	X	X										1, 2, 5, 6
9 Build resilient infrastructure, promote inclusive and sustainable industrialization and foster innovation		X	X										2, 4, 5
10 Reduce inequality within and among countries													
11 Make cities and human settlements inclusive, safe, resilient and sustainable		X	X										2, 4, 5,
12 Ensure sustainable consumption and production patterns	X	X			X	X	X						1, 2
13 Take urgent action to combat climate change and its impacts				X		X							2, 6
14 Conserve and sustainably use the oceans, seas and marine resources for sustainable development													
15 Protect, restore and promote sustainable use of terrestrial ecosystems, sustainably manage forests, combat desertification, and halt and reverse land degradation and halt biodiversity loss	X	X	X	X	X	X	X	X	X		X	X	1, 2, 3, 4, 5, 6
16 Promote peaceful and inclusive societies for sustainable development, provide access to justice for all and build effective, accountable and inclusive institutions at all levels			X						X		X	X	4, 7
17 Strengthen the means of implementation and revitalize the global partnership for sustainable development													

Inclusion of the Sustainability Pillars into the SDGs and Their Interconnection

Since international development agencies, international financial institutions, or specialized agencies—such as the OECD—and the UN system launched global initiatives to foster SDGs, they face the dilemma of how to ensure the coherence of development and development policies within the 2030 agenda.

The issue of political cohesion for development is a persistent problem that affects the governance of development and the environment as well as its performance (OECD 2019). The answer to this dilemma goes through different actions that international organizations try to carry out together with the creation of institutional mechanisms, accompanied by a battery of political declarations, local commitments, and systems of monitoring, analysis, and reporting of results. All this focuses on both national and international political framework, coupled with the necessary changes to local framework with the intention of doing "no harm" to the country development and communities. In this sense, the experience of the member countries of the OECD reminds us that there are no specific and miraculous practices for a country, but everything depends on the ways in which governments function and how they adopt public policy measures, institutional, administrative, among others, to meet their development goals.

In general, as the OECD recalls, the main achievements when establishing these cohesion mechanisms are limited to modest results such as the increase of sensitivity and interest to the problems of the global development agenda or to the creation of institutional commitments by the government in turn (OECD 2019). According to the global index of development commitments which is meant to track and monitor progresses of the Development Assistance Committee (DAC) member countries, they are very efficient when establishing solid cohesion policies on issues of security, technology, environment, trade, and migration, among others (OECD 2019). However, such results are not enough to ensure that a complex agenda such as the SDGs may trigger a robust process of cohesion and comprehensive political cohesion. Systemic and persistent difficulties in many countries to implement SDGs actions continue to affect the understanding of the concept of sustainability and its cohesion in different areas and thus its implementation. In the specific case of the concept of sustainability, the following table shows how the three pillars of the concept (economic, social, and environmental) are distributed in the current SDG agenda.

Although the SDGs consider a balance of the pillars of sustainability, it is crucial to undertake an analysis of how sustainability and its interconnections and implications are reflected in an agenda where diverse stakeholders from different levels (local, national, international) and inside and outside the government understand and apply sustainability itself (ICFC 2017). One of the key challenges facing policymakers is to ensure an integrated approach by implementing the concept of sustainable development. In this sense, a crucial error in this agenda is to think and recognize that all countries accept and understand the SD as dogmatic and monolithic. Reports such as Brundtland's (1987) and others that were published later and promoted by many multilateral institutions constitute the basis of the current international architecture under which the narrative of sustainability has been built.

As demonstrated in several case studies at the global and local level, there are countless visions about sustainable development and its understanding (Lucatello and Vera 2016). Despite having more than three decades of narrative effervescence about sustainability many public policy makers in the field of sustainable development need information and analysis to know what kind of development they are adopting, their inconsistencies, and how these could result from decisions in different sectors to achieve the objectives of national political plans.

Therefore, it is essential to analyze how sustainability turns out to be the element of cohesion for the SDG agenda. In this sense, some elements to consider for a more detailed analysis of the "mainstreaming" of sustainability in the agenda include:

(a) Roles and functions of the different actors that participate in the SDG agenda at different levels (governments, organizations, private sector, and non-governmental organizations), as well as the diverse sources of financing, public and private, national and international, to achieve sustainable development results.
(b) The policy of interconnections between the economic, social sectors, and environmental areas, including identification of synergies, contradictions, and exchanges, as well as the interactions between national and international policies (OECD 2019).
(c) Other local and contextual factors that help to understand the enablers (who can contribute) and the deactivators (who hinder) of sustainable development at its different levels.

To analyze the interconnections of SD and its relationship with the SDGs, we recall that the 2030 Agenda and the SDGs represent an extensive and complex "global development strategy." SDGs are deeply intertwined with each other, and in order to carry out such an ambitious agenda, different "multilevel" and multisectoral efforts are needed for a comprehensive application of the 17 SDGs. It also requires guidance on strategic public policies and the identification of practical tools at the level of governments and at the local level.

The new architecture of development, based on the SDGs, can have profound implications for governments. The compliance of the 2030 Agenda at the local level will also require avoiding the fragmentation of purposes and development objectives. So far, the main problem we face is that we are just beginning to understand how to interrelate the objectives within the SDG agenda. The mapping of interconnections between objectives and goals and the analysis of social networks is a strategy to investigate social structures with theories about networks among others (Collste et al. 2017). The Department of Economic and Social Affairs (DESA) at the UN uses this kind of analysis to map the interconnectedness between the 17 SDGs and their 169 targets as it provides valuable information on the integration and coherence of policies when applied in the national context.

However, there are no comprehensive studies on the same interrelationships. Some materials produced by the International Council for Science (ICSU) are an example of interactions within SDGs, but it is a work in progress and is not yet available. The quantification of SDG interconnections is limited in the existing literature, although there are works on how to categorize the different types of interrelationships. Most of the existing works are limited to the study of the general structure of the SDGs, on how they are interwoven through the identification of interconnections in general, but the identification and quantification of these interconnections at the national level are still nil (OECD 2019).

The work of the OECD and the issue of coherence (and cohesion) of the SDGs in terms of public policies is an important step forward in understanding the relationship of the SDGs and their interrelation. Based on the work of the Institute for Global Environmental Strategies (IGES, for its acronym in English), a project entitled "Sustainable Development Goals, Targets and Indicators" was launched to present an integrated analytical framework on network analysis of SDG interconnections between goals that are then applied to the analysis and visualization of inter-links between objectives and goals. The following map shows an example of the degree of interconnection between the different SDGs, based on the IGES study. As noted above, although the SDGs and targets address specific issues related to sustainable development, many of them are directly or indirectly related to each other.

In Fig. 2.1 each node indicates one sustainable development goal and each direct link indicates the causal relationships between a pair of goals. The size and color of a node indicate the number of links that a node has with others, that is, one goal with another. The thickness of the connection lines indicates the intensity of the relationship between goals. For example, measures that guarantee inclusive and quality education (SDG 4, in red) can reinforce progress in many—if not all—SDGs (IGES 2017).

The causality and interrelationship can generate employment and promote economic growth (SDG 8), help reduce poverty and increase access to food (SDG 1 and SDG 2, respectively), contribute to improving health and well-being (SDG 3), and reduce inequalities (SDG 10), to mention a few cases. On the other hand, measures to promote access to food (SDG 2), water (SDG 6), and energy (SDG 7), for example, if applied in an unsustainable manner, could be counterproductive in carrying out actions of sustainable consumption and production (SDG 12), aggravating climate change (SDG 13), and endangering marine life (SDG 14), as well as life on land (SDG 15).

In this way, the logic of the interconnection between goals allows maintaining a sustainability mainstreaming exercise and always keeping in mind the degree of interrelation and cohesion of the SDGs when applying them. The focus of interconnection

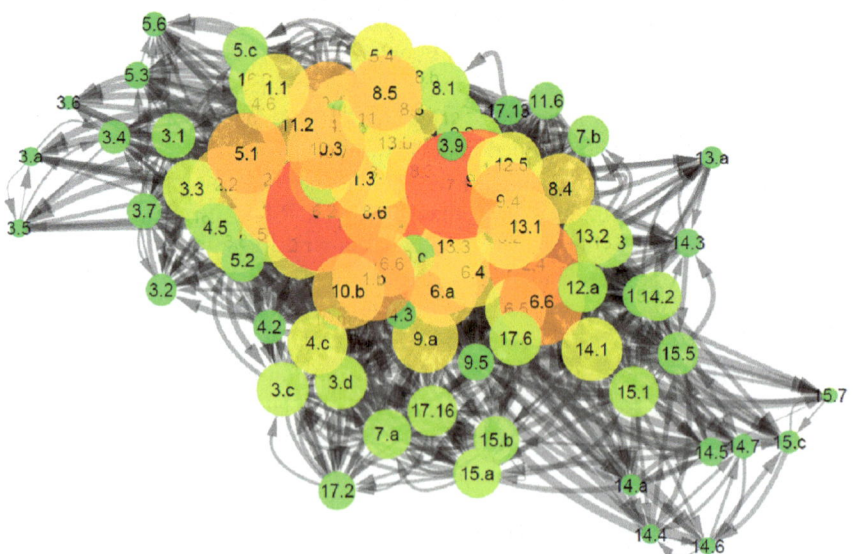

Fig. 2.1 Grade of interconnection between SDGs (IGES 2017)

analysis of the SDGs therefore emphasizes the interrelationships between the SDGs and their goals and is a relevant exercise for understanding the complexity of the development model.

In the case of the interlinkage between areas of importance to drylands development, there is a special focus on SDG target 15.3 which states that: "By 2030, combat desertification, restore degraded land and soil, including land affected by desertification, drought and floods, and strive to achieve a land degradation neutral world." As widely acknowledged, SDG target 15.3 sets out a new global ambition never posed before: to achieve land degradation neutrality (LDN) by the year 2030. LDN has been defined by the Parties to the Convention as: "A state whereby the amount and quality of land resources, necessary to support ecosystem functions and services and enhance food security, remains stable or increases within specified temporal and spatial scales and ecosystems" (Decision 3/COP.12, UNCCD 2015). The goal is maintaining or enhancing the conservation of natural capital stocks associated with land resources and ecosystem services that come from them.

It is worth reminding that the concept of "zero net land degradation" was proposed at the 2012 UN Conference on Sustainable Development (Rio+20). This was reformulated as "strive to achieve a land degradation neutral world" in the final outcome document, The Future We Want, and subsequently adopted by the United Nations General Assembly as part of the sustainable development goals (SDGs), specifically SDG target 15.3 (IGES 2017).

The objectives of LDN are defined as:

- Improve the sustainable delivery of ecosystem services;
- Maintain productivity to foster food security;
- Increase resilience of the land and populations;
- Establish synergies with other social, economic, and environmental SDGs objectives;
- Promote responsible and inclusive land governance.

Thus far, more than 100 countries worldwide have begun to pursue LDN target programs in different ways. LDN represents therefore a paradigm that may shift land management policies and practices in favor of reducing loss of productive land with recovery of degraded areas. However, the implementation of LDN requires multi-stakeholder engagement and planning across scales and sectors, supported by national-scale coordination that should work with and incorporate existing local and regional governance structures (UNDP 2017).

Through LDN, SDG target 15.3 could better link sustainable livelihoods in the context of SDG 1 (End poverty in all its forms everywhere); and Sustainable land management policies and practices in the context of SDG 15, on Terrestrial Ecosystems. Promoting SD in the drylands may also contribute to other SDGs, such as:

SDG 2: Achieving food security
SDG 5: Gender equality and women empowerment
SDG 6: Sustainable water management

SDG 10: Reducing inequity among countries
SDG 13: Combating climate change.

Considering that the relationship between sustainable land management, climate change, water, food security, and other goals is a comprehensive effort to simultaneously meet different goals and targets, it remains to be seen what could be the degree of linkages between drylands and other SDGs' targets in order to promote a coherent development policy for drylands.

SDGs and Drylands: An Overview

Drylands constitute a significant resource, both cross-cutting and critical for achieving the 2030 Agenda. The interlinkage between dryland development and the SDGs is the already mentioned target 15.3 which states that: "By 2030, combat desertification, restore degraded land and soil, including land affected by desertification, drought and floods, and strive to achieve a land degradation neutral world" (Tóth et al. 2018). As widely acknowledged, SDG target 15.3 sets out a new global ambition never posed before: to achieve land degradation neutrality (LDN) by the year 2030. The goal is maintaining or enhancing the conservation of natural capital stocks associated with land resources and ecosystem services. It is worth reminding that the concept of "zero net land degradation" was proposed at the 2012 UN Conference on Sustainable Development (Rio+20). This was reformulated as "strive to achieve a land degradation neutral world" in the final outcome document entitled, The Future We Want, and subsequently adopted by the UN General Assembly as part of the sustainable development goals (SDGs). At the basis of the LDN concept is the difference between two alternative phases of land degradation. One is the ongoing process of degradation through an intense land use that reduces productivity within certain time, and the second one is at a state that can be already abandoned degraded land with a recovery potential (Safriel 2017).

If we take into account that the land degradation process induces a persistent decline in productivity among other factors, it is crucial to promote and scale up a sustained land management and monitoring system to reduce degradation and increase restoration. In the case of drylands, it is therefore necessary to set a frame with realistic milestones for LDN and SDGs in order to measure the improved conditions for a better land management within different local contexts (Bouma 2019). When wanting to adapt the sustainable development goals to drylands, this should be done by considering the dryland syndrome (Reynolds et al. 2007; Stafford Smith et al. 2009). These are causally linked climatic, edaphic, demographic, geographic drivers that ultimately lead to a state of marginalization of dryland dwellers. Highly variable, unpredictable climate and scarcity of soil resources permit only low vegetation cover and patchy distribution of natural resources, which in turn inhibits otherwise rapid human population growth and leads to a mobile lifestyle (e.g., pastoralism) in remote areas. The remoteness also considers markets, centers of political and economic decision-making, and centers of education and mental model development (Stafford Smith and Cribb 2009). While the dryland social-ecological context has contributed to the formation of distinct local/traditional knowledge systems and world views by distinct social and cultural groups, their geographic isolation often triggers an extremely high level of social uncertainty, and little possibility for a prosperous life, economic growth and/or development (Stafford Smith et al. 2009). Furthermore, local particularities and differences often generate unstable and disconnected governance between the local, regional, national, and international levels. Based on this syndrome Reynolds et al. (2007) proposed the drylands development paradigm (DDP) a useful integrative analytical framework for SESs to assess, mitigate, and inhibit land degradation and desertification consisting of five principles:

1. Social-ecological systems (SESs) are highly coupled, co-adapted, and interconnected and their dynamics can only be understood in a spatiotemporal context.
2. The underlying dynamics of SESs are controlled by a limited number of key slow variables.
3. Crossing thresholds of these slow variables may move SES into different states with similar structures and functions or into new regimes with changed structures and functions.
4. SESs are hierarchical, nested, and networked across multiple spatial and temporal scales.
5. Different knowledge systems including local environmental knowledge need to be considered to allow functional coadaptation of SES.

To operationalize dryland development science in the context of the SDGS, Stringer et al. (2017) propose to "upgrade" the DDP; they add three principles that supposedly switch focus from research for development to research in development:

Unpacking relationships and interactions in dryland systems and livelihood portfolios can help to identify opportunities and risks for socio-technical innovation and investment to adapt to multiple interacting drivers of change at different spatial and temporal scales.

Traversing scales and sectors can improve co-creation, availability of and access to options, shaped and owned by land users and other value chain actors. This enables more contextual, people-centered focus in assessing risks, trade-offs, and vulnerabilities, supporting sustainable, resilient, and efficient pro-poor value chains. A networked approach to value chains can enable context-specific analysis and facilitate more inclusive, participatory governance reform.

Sharing knowledge, learning, and experience to empower dryland communities, researchers, policymakers, and other stakeholders is important to reduce trade-offs and externalities, leverage no-regrets options, and avoid unintended consequences. This is especially important in drylands where feedbacks, uncertainties, and non-linearities characterize the system. Current knowledge is weakest in terms of understanding social processes such as social learning, decision-making behavior, and power balances within coupled social-ecological systems. As the DDP of Reynolds et al. (2007) advocates a robust empirical analysis of cutting edge dryland science and development, the added principles by Stringer et al. (2017) can be used to advance sustainable drylands livelihoods in the context of the 2030 Sustainable Development Goals.

Conclusions

With good policies and adequate support, the drylands can be productive, and the livelihoods of dryland communities can be greatly improved with gains to the global development agenda in terms of poverty alleviation, moving towards land degradation neutrality and mitigating and adapting to the effects of climate change. SDGs and their interconnections is a trending topic and capitalizing on synergies between them offers a number of potential advantages to achieve short-and long-term goals within the broader drylands agenda. Such an interconnecting strategy may allow the recognition of a variety of possibilities to reach trade-offs among goal 15 and its targets but also it may allow to better integrate LDN into existing frameworks, such as the ecosystem ones, the UNCCD, and others. However, there is a huge challenge in terms of designing appropriate ways to address the interconnection of SDGs with drylands at local level. A key challenge is to identify a strategy that may work locally and find a balance that triggers governance mechanisms that can achieve the SDGs targets as well as fostering the diverse environmental and social outcomes from different drylands contexts.

Acknowledgments The authors appreciate the invaluable intellectual input by Gerardo Arroyo-O'Grady, Director of the Sustainable Development Project, UNDP, Mexico. EHS gratefully acknowledges the financial support from CONACYT (projects CB 2015-251388B, 293793, PDCPN-2017/5036).

References

Balk D (2008) Urban population distribution and the rising risks of climate change. Presentation at the United Nations Population Division. (CUNY Institute for Demographic Research & School of Public Affairs, Baruch College), New York

Bautista S, Llovet J, Ocampo-Melgar A, Vilagrosa A, Mayor AG, Murias C, Vallejo VR, Orr BJ (2017) Integrating knowledge exchange and the assessment of dryland management alternatives - a learning-centered participatory approach. J Environ Manag 195:35–45

Berkes F, Reid WV, Wilbanks TJ, Capistrano D (2006) Conclusions. Bridging scales and knowledge systems. In: Reid WV, Berkes F, Wilbanks TJ, Capistrano D (eds) Bridging scales and knowledge systems: concepts and applications in ecosystem assessment, 1st edn. Island Press, Washington, D.C., pp 315–331

Biggs R, Schlüter M, Schoon ML (2015) Principles for building resilience. Cambridge University Press, Cambridge

Bouma T (2019) How to communicate soil expertise more effectively in the information age when aiming at the UN sustainable development goals. Soil Use Manag 35:32–38

Brundtland (1987) Our common future. United Nations report of the world commission on environment and development. http://netzwerk-n.org/wp-content/uploads/2017/04/0_Brundtland_Report-1987-Our_Common_Future.pdf. Consulted July 2019

Chapin FS III, Folke C, Kofinas GP (2009) A framework for understanding change. In: Chapin FS III, Kofinas GP, Folke C (eds) Principles of ecosystem stewardship. Springer Verlag, New York, NY, pp 3–28

Cherlet M, Hutchinson C, Reynolds J, Hill J, Sommer S, von Maltitz G (2018) World atlas of desertification. Publication Office of the European Union, Luxembourg

Collste D, Pedercini M, Cornell SE (2017) Policy coherence to achieve the SDGs: using integrated simulation models to assess effective policies. Sustain Sci 12(6):921–931

Gorddard R, Colloff MJ, Wise RM, Ware D, Dunlop M (2016) Values, rules, knowledge: adaptation as change in the decision context. Environ Sci Policy 57:60–69

Huber-Sannwald E, Ribeiro Palacios M, Arredondo JT, Braasch M, Martínez Peña M, García de Alba J, Monzalvo K (2012) Navigating challenges and opportunities of land degradation and sustainable livelihood development. Philos Trans R Soc B 367:3158–3177

Institute for Global Environmental Strategies (IGES). (2017). Sustainable development goals interlinkages and network analysis: a practical tool for SDG integration and policy coherence. IGES Research Report. https://pub.iges.or.jp/pub/sustainable-development-goals-interlinkages. Accessed 20 May 2018

International Council for Science (2017) A guide to SDGs interactions: from science to implementation. https://council.science/cms/2017/05/SDGs-Guide-to-Interactions.pdf

IPCC SR15 (2015) Special report, global warming of 1.5°C. https://www.ipcc.ch/sr15/. Consulted July 2019

Liu J, Dietz T, Carpenter SR, Alberti M, Folke C, Moran E, Deadman AN, Kratz T, Lubchenco J, Ostrom E, Ouyang Z, Provencher W, Redman CL, Schneider SH, Taylor WW (2007) Complexity of coupled human and natural systems. Science 317:1513–1516

Lucatello S, Vera L (2016) La implementación de la Agenda 21 en México: Aportes críticos a la sustentabilidad local. Instituto Mora, México, D.F.

Mainali B, Luukkanen J, Silveira S, Kaivo-ojo J (2018) Evaluating synergies and trade-offs among sustainable development goals (SDGs): explorative analysis of development paths in South Asia and Sub-Saharan Africa. Sustainability 10:815–840

OECD (2019) Policy coherence in the SDGs. http://www.oecd.org/governance/pcsd/Note%20on%20Shaping%20Targets.pdf

Reed MS, Stringer LC, Dougill AJ, Perkins JS, Atlhopheng JR, Mulale K, Favretto N (2015) Reorienting land degradation towards sustainable land management: linking sustainable livelihoods with ecosystem services in rangeland systems. J Environ Manag 151:472–485

Reynolds JF, Stafford Smith DM, Lambin EF, Turner BL II, Mortimore M, Batterbury SPJ, Downing TE, Dowlatabadi H, Fernandez RJ, Herrick JE, Huber-Sannwald E, Leemans R, Lynam T, Maestre FT, Ayarza M, Walker B (2007) Global desertification: building a science for global dryland development. Science 316:847–851

Safriel U (2017) Land Degradation Neutrality (LDN) in drylands and beyond – where has it come from and where does it go. Silva Fennica 51(xx):1650. https://doi.org/10.14214/sf.1650

Stafford Smith, M. and J. Cribb (2009). Dry Times – Blueprint for a Red Land. CSIRO Publishing, Australia. p. 176

Stafford Smith DM, Abel N, Walker B, Chapin FS III (2009) Drylands: coping with uncertainty, thresholds, and changes in state. In: Chapin IIIFS, Kofinas GP, Folke C (eds) Principles of ecosystem stewardship. Springer, New York, pp 171–195

Stafford Smith DM, Griggs D, Gaffney O, Ullah F, Reyers B, Kanie N, Stigson B, Shrivastava P, Leach M, O'Connell D (2017) Integration: the key to implementing the sustainable development goals. Sustain Sci 12:911–919

Stringer LC, Reed MS, Fleskens L, Thomas RJ, Bao Le Q, Lala-Pritchard T (2017) A new dryland development paradigm grounded in empirical analysis of dryland systems science. Land Degrad Dev 28:1952–1961

Tóth G, Hermann T, Ravina da Silva M, Luca Montanarella L (2018) Monitoring soil for sustainable development and land degradation neutrality. Environ Monit Assess 190(2):57

UN General Assembly (12 May 2016) Resolution adopted by the General Assembly on 27 April 2016. A/RES/70/262. Review of the United Nations peacebuilding architecture. United Nations. http://www.un.org/en/development/desa/population/migration/generalassembly/docs/globalcompact/A_RES_70_262.pdf. Accessed 20 June 2018

UNCCD (1994) United Nations convention to combat desertification. United Nations, Geneva, Switzerland, p 58

UNCCD (2015) Climate change and land degradation: bridging knowledge and stakeholders. https://www.unccd.int/publications/climate-change-and-land-degradation-bridging-knowledge-and-stakeholders. Consulted July 2019

UNCCD (2017) Land degradation neutrality: transformative action, tapping opportunities. https://wwwunccdint/publications/land-degradation-neutrality-transformative-action-tapping-opportunities. Consulted July 2019

United Nations Development Programme (UNDP) (2017) Global policy Centre on resilient ecosystems and desertification. Nairobi, Kenya. Consulted April 2019

WOCAT (2010) World overview of conservation approaches and technologies

Chapter 3
Pastoralism and Achievement of the 2030 Agenda for Sustainable Development: A Missing Piece of the Global Puzzle

M. Niamir-Fuller and E. Huber-Sannwald

Abstract Integrity and productivity of ecosystem goods and services from rangelands are critical to the livelihoods of over a billion people worldwide. Pastoralists play a key role in the conservation and sustainable management of these important life-support systems. Yet, an increasing number of environmental, socio-economic, and political threats have jeopardized the integrity of rangelands and the security of this livelihood type, and hence stewardship of global natural rangeland systems. Sustainable pastoralism can synergistically link several sustainable development goals (SDG) as they are critical to achieving food security, resilient local and regional economies, cultural diversity, conservation and sustainable use of biological diversity, carbon sequestration, land and water rehabilitation. Pastoralism has a cross-cutting role for stewardship of drylands and natural rangelands while also achieving the 2030 Agenda for Sustainable Development. We show how single-sector economic development approaches of the past have led to loss of livelihoods, increased inequality and land degradation for many pastoral communities in developing countries. How can the SDGs help induce more integrated approaches, and what are the challenges and barriers to doing so? We also highlight recent tools and approaches that can help, including the benefits of developing integrated indicators for monitoring and measuring progress towards sustainable pastoralism.

Keywords Pastoralism · Rangelands · Sustainability · SDG · Indicators

M. Niamir-Fuller (✉)
Vice-Chair International Support Group, International Year of Rangelands and Pastoralists, Purcellville, VA, USA

E. Huber-Sannwald
División de Ciencias Ambientales, Instituto Potosino de Investigación Científica y Tecnológica, San Luis Potosi, San Luis Potosí, Mexico

© Springer Nature Switzerland AG 2020   41
S. Lucatello et al. (eds.), *Stewardship of Future Drylands and Climate Change in the Global South*, Springer Climate,
https://doi.org/10.1007/978-3-030-22464-6_3

Introduction: Pastoralism, Natural Rangelands, and Benign Neglect

Pastoralism has been practiced for millennia in the drylands, tundra, and alpine regions of the world. As there are many different forms of pastoralism, so are there many different definitions of it. The people who practice it rarely call themselves pastoralist—they are herders, graziers, cowboys, shepherds, criançeros, livestock farmers, and more. Up to now the term "pastoralist" has been used primarily in francophone countries, but is gaining global recognition as a term to describe the many varied forms of raising domesticated or semi-domesticated ruminant animals on natural rangelands (Reid et al. 2008).

The definition of "rangelands" is also variable depending on the language and culture of different countries. The term is more popular in North America and Australia, whereas other regions use terms such as natural pasture, grassland, savanna, veld, pampa, llanos, cerrado, campos, etc. The term is gaining global recognition to describe those lands on which the indigenous vegetation consists predominantly of grasses, grass-like plants, forbs, shrubs, or trees that are grazed or have the potential to be grazed or browsed, and which are used as a natural ecosystem for raising grazing livestock and wildlife (Allen et al. 2011). Rangelands provide a broad spectrum of ecosystem goods and services (Sala et al. 2017). They are intricate social-ecological systems with a long history of co-adaptation with ruminant animals. Pastoralist communities and their resource management strategies have co-evolved with the inherently dynamic rangelands and their high spatial heterogeneity, biodiversity, and with climate-related spatial and temporal variability (Little 1996).

Most researchers agree that the current extent of pastoralism and rangelands has been globally underestimated. A recent report concluded that there is low confidence in global statistics of rangelands and pastoralists because of the incomparability of different definitions and terminologies, whereas there is usually higher confidence of such statistics at the local level where only one definition is used (Johnsen et al. 2019). The population of pastoralists has been reported varying from 200 million to over 1 billion. The geographic extent of rangelands has been reported varying from 18% to over 80% of the world's land surface (Allen et al. 2011).

The World Atlas of Desertification (WAD), third edition, is the most recent effort to harmonize terminologies and statistics on rangelands and drylands. It estimates that global rangeland area is 29 million km^2, of which 63% are in drylands. It also estimates that 18.5% of global rangelands, and 14% of rangelands in drylands are experiencing a decrease in land productivity (Cherlet et al. 2018). Over the past 50 years, rangelands have been reduced in size and fragmented. This observation comes from individual research sites rather than any global assessment. Reduction and fragmentation has been noted especially as a result of expansion of rainfed agriculture into local wetlands and other fertile spots of rangelands. Other causes are the widespread use of irrigation and greenhouse technologies expanding croplands into drylands less suited to crops, mining exploration, fracking, infrastructure for generation of renewable energy (wind and photovoltaic farms, large-scale dams), urbanization, and expansion of state-protected areas. As a result, what usable rangelands remain are considerably more arid and less productive than in the past, but are expected to support the same or sometimes higher densities of livestock with likely implications on the functional integrity and resilience of rangelands as life-support systems. But the lack of consensus on the definition of "land degradation" and "desertification" continues to challenge estimates of this important parameter (Cherlet et al. 2018).

There are many myths and misunderstandings of pastoralism dating back to colonial times, leading to deeply entrenched prejudices against them (Davis 2016). Most developing country government policies in the past decades have promoted the industrialization of agriculture. In the livestock sector, this usually meant giving incentives, subsidies, and other support to the development of large-scale commercial feedlots (Niamir-Fuller 2016). For instance, it is estimated that in 2012 Argentina, where grass-feed beef production has dominated for decades, 50% of cattle production terminated in feedlot-type systems (Deblitz 2012). Smallholder pastoralism was seen as outmoded and inefficient. Their lands were converted by fiat to large-scale irrigation systems, national parks and reserves, or to rainfed agriculture as population of farmers increased. For example, the establishment of the Park du W in Niger led to displacement, poverty, conflict, and radicalization among pastoralists (Ben Hounet et al. 2016). Poverty among pastoralists also led them to sedentarize. For example, in Kenya, pastoral areas have a poverty index twice the national average and in Asia infant mortality rates in drylands are 50% above the mean (McGahey et al. 2014). Between 1900 and 1990 the average livestock holding per household in Karamoja (Uganda) decreased from 100 to 28 (Niamir-Fuller 1999). Poverty is driving millions of pastoralists out of their land and out of the livestock sector, leaving many destitute, dependent on food aid, or joining the growing tide of international migrants. An estimated 30% of pastoralists have migrated to cities and towns in Tanzania, and transhumance has almost completely disappeared in Tunisia (Niamir-Fuller and Turner 1999).

Privatization of rangelands, the blocking of traditional livestock migration routes, and breakdown of common property systems of natural resources management have been linked to increased land degradation (Swallow and Bromley 1991). Many countries do not adequately protect pastoral land tenure, and chronically underinvest in these areas. What has been described as "benign neglect" (Swift 1993) led to an unbalanced development scenario, where farmers were favored over pastoralists. Recently however, there is growing recognition of the value of healthy rangelands and sustainable pastoralism and their interdependency. Pastoralism has been shown to produce from two to ten times more output per unit area than some of the alternative land uses that have been proposed as a replacement (McGahey et al. 2014). Inherent variability of productivity of rangelands means that these areas have a cost–benefit ratio that does not favor exclusionary, privatized investments (Dyson-Hudson and Smith 1978). Variability is expected to increase in an unprecedented fashion due to the high vulnerability of drylands to predicted climate change (Huang et al. 2017).

Pastoral mobility (the movement or rotation of livestock over large areas) has been shown to promote healthier ecosystems (McGahey et al. 2014), and greater wildlife compatibility (Niamir-Fuller et al. 2012), and is better adjusted to the high levels of climatic unpredictability in the drylands. Pastoralism is a natural adaptation to climatically uncertain and variable environments, and generally more resilient and adaptive than sedentary farming in drylands. For example, a study in Morocco predicts that mobile pastoralists will be barely affected by reduction in precipitation by 2050, while the income of sedentary pastoralists will fall by nearly 19% (Freier et al. 2014). Another study in the Limpopo Basin in South Africa concludes that pastoralists will be more resilient to climate change than households that do not raise livestock (Shewmake 2008). Hence these traditional systems may be the best adaptation pathways towards future climate change.

From Production Maximization to Integrated Development

Development approaches typically followed a command and control approach maximizing economic production and efficiency centered around a single ecosystem good or service. Progress was measured in terms of production output (number of livestock; kilograms of meat or milk). In theory, production maximization would lead to higher and more reliable incomes and therefore greater development. However, decades later we are seeing the adverse effects and socio-ecological costs of such single-sector approaches.

Research on sedentarization (where pastoralists were forced or encouraged to settle in villages to receive health and other services) shows livelihood vulnerability and enhanced land degradation from sedentary grazing (Galaty et al. 1981). For example, sedentarization and concentration of animals in the Ferlo of Senegal resulted in overgrazing around settlements and waterholes, as well as degradation of distance pastures due to lower grazing (Niamir-Fuller 1997). The economic benefits of conversion of riparian areas of rangelands into irrigation schemes were often short-lived, and resulted in degradation of adjacent nonriparian areas given that pastoralists no longer had access to high-quality natural fodder (Hogg 1987).

Encouragement of confined livestock systems (smallholder or large holder) through subsidies and technical support has also had negative impacts on the environment. Today, 40% of all arable land is being used to produce animal feed for intensive systems and the figure is rising. The impact of deforestation on the Amazon and savannas of the world from conversion to the production of feed has been well documented (Fearnside 2005). UNEP estimates that the cereals used to produce feed for animals could instead feed 3.5 billion people. Fifty percent of fertilizer applied to agricultural land, and 70% of herbicide used in agriculture are attributed to animal feed production. In China, as much as 40% of the chemicals are in excess and released to the environment, resulting in dead zones and marine hypoxia (UNEP 2016).

Tse-tse eradication programs and other veterinarian measures in Sub-Saharan Africa were primarily aimed at allowing the adoption of non-native "improved" livestock breeds and increasing the international trade of livestock. Such efforts have had mixed success in reducing the disease burden or increasing productivity and offtake (Roeder 1996). The construction of deep wells (known as "boreholes") in the Sahel in the 1970s and 1980s led to land degradation because it attracted far more livestock to the water than the available rangeland fodder could handle, and also to destruction of common property systems. At the root cause was the fact that these water developments did not have local institutions in place to maintain and manage them (Thébaud 1988). Many pastoralist communities are facing social and cultural breakdown or have already disbanded. For example, there has been increasing concentration of wealth in the hands of a few in Senegal (Sutter 1987) or in Kenya (Little 1985). Such trends are also seen in the USA and Europe. In Romania a fall in the number of small-scale family farms (and decrease in farms owned by younger farmers) as well as a fall in national dairy cow numbers since 2009 was a result of government policies that provided subsidies only for larger farms (Page and Popa 2013). Rangeland degradation can occur not just from overgrazing but also from insufficient grazing because of the grazing dependence of many rangeland ecosystems (Seligman and Perevolotsky 1994).

A single-sector approach to development cannot work in pastoral systems today because as socio-ecological systems they require a multi-sectoral integrated approach. It may work with farmers who have been subsidized over decades, and whose villages and communities are still socially viable and robust, but not pastoralists who have been neglected and whose governance and social systems have broken down, and who are almost entirely dependent on naturally healthy rangelands that are degrading (Digard et al. 1993). In the late 1970s and into the 1980s, there was a movement by donors to finance "integrated rural development" approaches. This was due to the increasing recognition that development cannot work in isolation—that projects must involve more than one ministry, and must coordinate between each other. There was a call to go away from sectoral approaches to spatially more integrated approaches (Niamir-Fuller and Turner 1999). However, these early trials with integrated approaches largely failed, for at least two reasons. Firstly, it proved extremely challenging for sectoral ministries to coordinate with each other (Kamuzora and Franks 2001). There were no mechanisms to address trade-offs and competition for financing and resources between the sectors. Information was rarely shared across sectors, perhaps due to the competition, but also probably due to the lack of effective communication tools. Secondly, such projects and programs continued to maximize production and economic gain, and a lack of performance in such indicators automatically meant failure.

In the early years of the twenty-first century, with the renewed attention to sustainable development, different approaches have been proposed, including "community based natural resource management," "sustainable livelihood approaches" (Kamuzora and Franks 2001), "the landscape approach" (Scherr et al. 2013), and "ecosystem stewardship" (Chapin et al. 2009). These approaches draw from a few common lessons: to move from sectoral to multi-sectoral and spatial approaches; to promote not just economic development but also social and cultural progress and ecological improvement; to build on local initiatives and solutions by adopting real participation and to manage social and ecological properties in a system context, while fostering variability and diversity.

Potential of SDGs to Achieve Sustainable Pastoralism

The most quoted definition of "sustainable development" first appeared in the Brundtland Report (Brundtland Commission 1987). But it was not until 1992 when the global community took practical steps towards its achievement. During the UN Conference on Environment and Development (the Rio Summit) governments committed to negotiating global environmental goals on climate change, biodiversity conservation, and combatting desertification. It took another 28 years for the sustainable development goals to be adopted by 198 countries in 2015 as the operational part of the 2030 Agenda for Sustainable Development (UN 2015). It covers 17 major goals, 169 targets, and 232 indicators. The process of adoption of these goals was an excellent example of how multi-sectoral dialogue can be promoted. The package also shows how the complexity of issues today cannot be distilled into three or four primary goals (as some politicians and pundits had wished for). Although most of the 17 goals and 169 targets are focusing on one sector, many are cross-cutting (e.g., inequality, capacity building), and the challenge is to target synergies among several SDGs (Mainali et al. 2018).

SDG Strategy, Goals, and Targets

The preambular clause of the 2030 Agenda (paragraph 16) calls for integrated solutions, and commits nations to a new approach to sustainable development, recognizing that eradicating poverty and inequality, preserving the planet, and creating sustained and inclusive economic growth are linked and interdependent. But the negotiations that led to the creation of the SDGs also highlighted some of the challenges, namely the difficulties in reconciling regional priorities and national sovereignty with a global consensus; the political sensitivity or skepticism of certain concepts and issues such as inequality, climate change, ecosystem services, or governance; and the aggregation and generalization of issues through diplomatic language. The word "pastoralist" is only mentioned once in the SDGs, in connection with promoting sustainable agriculture (Goal 2, Target 2.3). It took considerable lobbying by many countries with large populations of pastoralists, and by civil society, to succeed in gaining this recognition. The 2030 Agenda preambular clause (paragraphs 18, 22, 23) especially singles out indigenous peoples, and for some countries that includes pastoralists. The term "rangeland" is not mentioned explicitly but is expected to be included in the term "ecosystems."

Sustainable pastoralism can potentially help achieve all SDG goals. But also implementing the SDGs can help transform pastoralism towards sustainability, especially if pastoralists are not forgotten or ignored. Below we give examples of this codependent process, which are elaborated further in Fig. 3.1. An examination of the innovative solutions shows that with a few exceptions, each can achieve more than one SDG target.

Sustainable pastoralism can promote healthier rangelands and help achieve Targets 15.2 (degraded forests), Target 15.3 (degraded lands), Target 15.4 (mountains), Target 15.5 (natural habitats and biodiversity), and Target 15.8 (invasive species). After decades of promoting sedentarization and privatization of common lands, some governments are recognizing the importance of animal mobility for maintaining ecosystem functions, higher biological diversity, lower incidence of invasive alien species, and stronger compatibility with wildlife. An FAO study reports that livestock grazing is frequently used to promote ecosystems services in protected areas (FAO 2014). Socio-cultural rules and regulations for such mobility (e.g., transhumance between summer and winter pastures) are being recognized legally in some countries, for example, the Spanish Cañada (Jefatura del Estado 1995), and

Fig. 3.1 Selected initiatives and links to achieving targets of the sustainable development goals in pastoralist social-ecological systems (links can be to specific targets, or to the entire goal and thereby all of its listed targets)

transhumance routes held in trust for pastoralists in Senegal's Férlo (Ly and Niamir-Fuller 2005).

Goal 2 calls for promoting sustainable agriculture. Increasing the sustainability of livestock production through pastoral mobility, improvement of degraded natural rangelands, rebalancing of incentives and subsidies to support sustainable livestock production, and reducing the environmental impacts of confined livestock systems can substantially reduce the negative footprint of the livestock sector. It can also make pastoral organic products more competitive in national, regional, and global markets, thus helping many countries to achieve Target 17.11 (to double the least developing countries' share of global exports by 2020).

There are many countries where pastoral production is a major share of the national GDP. For example, the share of livestock in agricultural GDP is above 50% in Mauritania, Kenya, and Kazakhstan, and well above 80% in Mongolia.[1] Livestock constituted 40% of global agricultural GDP between 2005 and 2014 (ILRI 2018). Goal 8, Target 8.1 calls for countries to sustain GDP per annum growth at 7% or more. Such countries cannot afford to ignore their pastoral populations.

Goal 10, Target 10.7 while focusing on migration issues does include the term "mobility of people" which some have interpreted to include mobility of pastoralists within country and internationally. For example, FAO calls for enacting migration policies that "takes into account people moving with their animals" (FAO 2018). Economic Community of West African States (ECOWAS) has established International Transhumance Certificates to improve the security of transboundary movement of pastoralists (ECOWAS 2017).

Other SDG targets offer indirect guidance. For example, Target 8.2 calls for achieving higher levels of economic productivity through diversification, "particularly in high-value added and labor-intensive sectors." Pastoralism is thought to benefit around 1.3 billion people along the value chain worldwide (Ouedraogo and Davies 2016). Countries that seriously implement the SDGs will have to give equal attention to farmers as to pastoralists, fisherfolks, and other land users (Target 2.3). Furthermore, countries are committed to better disaggregation—whether in their statistics (specifically showing pastoralists as a group) or in their policies (specifically monitoring the impacts on pastoralists).

SDG Indicators

Monitoring is a very important tool, not only for providing feedback on the impact of actions and for understanding temporal dynamics of system change, but more significantly, for focusing attention on an issue. Ensuring that indicators are appropriately disaggregated can encourage (even require) governments to pay attention to

---

[1] Data extracted and analyzed from FAO Stats (2016). www.fao.org/faostat.

pastoral needs. The majority of SDG indicators are single-sector or single-variable and therefore incapable of substantially measuring an integrated implementation of the SDGs. Furthermore, among the 232 SDG indicators officially adopted in 2016 only 7% can be said to be relevant to pastoralists and rangelands (Table 2.1). However, of note is that indicator 17.18.1 encourages countries to develop new and disaggregated indicators.

Most of the indicators do not explicitly ask for disaggregation of livelihood systems. As examples, the indicator for the target on sustainable agriculture (2.4.1. land area under sustainable agriculture), the indicator for the target on reducing land degradation (15.3.1 proportion of degraded land), or the indicator for universal access to safe drinking water (6.1.1) does not require data specifically on pastoralists unless governments make a concerted effort to separate out the statistics between farmers and pastoralists. Other indicators do not fully address the targets they are aimed at. For example, Target 2.a calls for investment in extension services and specifically mentions livestock gene banks, whereas its two indicators only measure financial flows to the agricultural sector as a whole.

Some indicators will likely bypass pastoralists and other rural populations. For example, the indicator for the target on increasing knowledge of sustainability (Target 4.7) focuses only on formal education, whereas pastoralists and farmers would benefit the most from other types of knowledge, including valuation of local and indigenous knowledge and technology (LITK).

Some indicators fail to leverage the value of pastoralism. For example, the indicator for the target on sustainable consumption and production (8.4.1 material footprint of GDP) would only be relevant if the GDP index itself were to adequately value the contributions of pastoralism to GDP. The indicator for the target on transborder sustainable infrastructure (9.1.1 population living within 2 km of an all weather road) is inadequate to capture the benefits and values of pastoral mobility and need for sustainable water, veterinary services, markets, and other infrastructure along transhumance routes. The indicators for Goal 13 (climate change) refer only to mainstreaming of climate change into government planning, which is inadequate for valuation of the benefits of sustainable rangelands for sequestration of carbon and as carbon sinks, nor does it value the benefits of pastoralism for adaptation to climate variability or for mitigation of climate change.

Finally, some indicators may perpetuate ongoing discrimination, marginalization, and victimization of pastoralists. For example, the indicator for the target on promoting equal access to the rule of law (16.3.1 reports of victimization) does not necessarily assist pastoralists in situations where the existing justice system is heavily weighted against them.

Table 3.1 suggests that the subset of directly relevant SDG indicators (those rated as high to medium relevance) could be measured in a way that addresses sustainable pastoralism and rangelands, but in many countries this would require policy changes (e.g., statutory recognition of common property tenure), statistical disaggregation of existing data, and investment in collective data collection among pastoralists and specifically in rangelands.

Table 3.1 Monitoring challenges of pastoralist socio-ecological systems for selected SDG indicators (high to medium relevance)

SDG target	SDG indicator	Monitoring challenges
1.1 Eliminate extreme poverty	1.1.1 Population (by employment status and urban or rural)	Disaggregate "rural" population into livelihood type (pastoralists, farmer, fisher, etc.)
1.4 Equal rights to economic resources	1.4.1 Access to basic services 1.4.2 Adult population with secure land rights … by type of tenure	Disaggregate data by livelihood type Full recognition of different land tenure systems, including common property
2.3 Small scale producers	2.3.1 Production per labor unit by pastoral enterprise size 2.3.2 Income by indigenous status	Definition of "pastoral enterprise size" (by number of livestock? Area of rangeland?) Some pastoralists are not considered indigenous
2.4 Sustainable agriculture	2.4.1 Area under sustainable agriculture	Data collection among pastoralists and on rangelands
2.5 Genetic diversity of domesticated animals and intellectual property rights	2.5.1 Risk level of local breeds	Participation of pastoral herders in assessing risk levels of local breeds
4.5 Gender disparity	4.5.1 Parity indices highly disaggregated	Data collection and assessment among women pastoralists
5.2 Violence against women	5.2.2 Proportion of women and girls subjected to sexual violence, by place of occurrence	Disaggregation of data; collection, analysis, and transparent information dissemination of data on women pastoralists in conflict zones
5.a Equal rights to economic resources	5.a.1 Share of women among right-bearers of agricultural land, by type of tenure 5.a.2 Proportion of countries where law guarantees women's equal rights to land	Cultural challenges for inheritance by women Full recognition of common property tenure Respecting role of women pastoralists in women-headed households
6.3 Eliminate waste dumping	6.3.2 Proportion of bodies of water with good ambient water quality (not disaggregated)	Disaggregation of data between rangelands and landscapes of other ecosystem types
6.6 Protect water ecosystems	6.6.1 Change in the extent of water-related ecosystems over time (not disaggregated)	Disaggregation of data between rangelands and landscapes of other ecosystem types
8.10 Financial services to all	8.10.2 Population with account at a bank or mobile money service provider	Data collection and disaggregation among all especially including remote pastoral populations
13.1 Resilience and adaptive capacity	13.1.1 Number of affected people attributed to a disaster	Disaggregation and collection of data among all including remote and mobile pastoralists
15.5 Natural habitats and biodiversity	15.5.1 Red list index	Identification of red list species in particular in dryland rangelands

(continued)

Table 3.1 (continued)

SDG target	SDG indicator	Monitoring challenges
17.18 Disaggregate data by ethnicity, migratory status, geographic location	17.18.1 Proportion of sustainable development indicators produced at national level with full disaggregation	Comparability of indicators across nations and aggregation of information at global level

Challenges and Way Forward

The SDGs offer an ambitious and innovative template for development particularly for drylands. However, simply mentioning "pastoralist" in the SDGs does not mean that development will flow to them and that the rangelands will continue delivering ecosystem goods and services. The Millennium Development Goals (MDG) did little to achieve sustainable development among pastoralists (WISP 2006), and the SDGs run the same risk. Rangelands and pastoralists are a very significant section of both terrestrial ecosystems and global population. By not addressing their needs, the SDGs are missing a very large piece of the global puzzle.

A major problem is how policies have largely failed to disaggregate "the livestock sector." In so doing, policies have focused on intensification of the sector (confined large-scale systems, feed production), rather than supporting the diverse nature-based smallholder systems. The way most countries subsidize their livestock sector helps the intensive sector to the detriment of the nature-based systems. The popular opinion against meat consumption fails to distinguish the positive benefits of nature-based rangeland ecosystems and smallholder pastoralism from the negative effects of intensive, confined large-scale livestock systems.

Researchers have also failed to focus on rangelands and pastoralism as socio-ecological systems, with only 0.1% of all research publications in SCOPUS since the year 2000 being on rangelands and/or pastoralists—although the trend is increasing (Johnsen et al. 2019). Another problem is that the SDGs largely evade the question of the informal economy—and pastoralists make up a large part of that economy in many countries (McGahey et al. 2014). Despite many arguments to recognize the informal economy, the final compromise on the SDGs focused on language that refers primarily to formal employment (for example, see Goal 8). The general lack of relevance of many of the targets and indicators of the SDGs, and the lack of disaggregation of data for monitoring the SDG indicators may mean that pastoralists, their culturally rich livelihoods, and their fundamental role as rangeland stewards and thus managers of ecosystem goods and services to millions of people will once again be left behind. However, the SDGs demand that new and better national indicators be developed (indicator 17.18.1 specifically calls for this), and therefore, efforts must be strengthened to develop indicators that are relevant to pastoralists and rangelands in the appropriate socio-cultural and political contexts.

Past experience with implementing integrated approaches needs to be studied carefully. For example, not enough attention has been given to addressing the root causes of an issue, and most projects focus only on the symptoms. This is largely because donors want quick fixes and visible gains in a short period of time. The Global Environment Facility (GEF) was among the early efforts to focus on root causes, and according to its most recent independent evaluation, it has been fairly successful (IEO 2018). Most integrated efforts continue to have one dominant sector and several subsectors; they lack both designers and implementers who can think integrally and laterally and transdisciplinary partnerships, which consider a broad spectrum of knowledge systems and world-views, and co-generate useful knowledge in participatory settings. The recent advances in "nexus" approaches provide a way to encourage collaboration and integration among a few sectors on an equal footing, such as combining the goals of Zero Hunger (SDG 1), Clean Water and Sanitation (SDG 6), and Affordable and Clean Energy (SDG 7) (NRDP 2019).

Such efforts would benefit from new integrated tools. Multi-criteria and trade-off analysis are receiving growing attention for their ability to reconcile differences and contradictions between the aims and actions of different sectors. Integrated participatory assessments of projects (before during and after) can also make an impact at the project level. The FAO Voluntary Guidelines on Governance of Tenure: Improving Governance of Pastoral Lands (Davies et al. 2016) is an important policy document providing a checklist of do's and don'ts in relation to land use and security of tenure. These tools should be very helpful in encouraging ministers and different sectors and actors to work together and form continuing learning communities. Furthermore, there is insufficient attention given to the underlying politics and power relations among local stakeholders. Evidence shows that the introduction of "group ranches" in Kenya resulted in greater inequality among pastoralists because the better off and more powerful elements of the community were able to capture most of the benefits. Power relations between governments and pastoralists are also often ignored. The SDGs recognize the importance of the "systemic" issues at the national and global levels in Goal 17, buried under diplomatic negotiated language, such as "respect each country's policy space" or encourage "policy coherence." However, this guidance is insufficient to tackle the prejudice against mobile pastoralists at the local levels.

Conclusion

The SDGs offer an innovative and transformational framework or toolkit through which sustainable pastoralism and healthy natural rangelands can be attained. However, their implementation will need innovative approaches, including:

(a) using recent integrated and participatory tools and approaches so as to develop cross-cutting multi-sectoral projects, programs, and policies that target interculturality and reconcile contradictions and competition between sectors; and

(b) developing additional indicators (national and global) to complement the current SDG indicators that specifically measure the success of integrated approaches and progress towards sustainable pastoralism and healthy natural rangelands—one cannot manage what one does not measure.

Acknowledgements The authors greatly appreciate the stimulating discussions and support from the International Rangeland Congress Steering Committee and the IYRP support group. EHS gratefully acknowledges the financial support from CONACYT (projects CB 2015-251388B, 293793, PDCPN-2017/5036).

References

Allen VG, Batello C, Berretta E et al (2011) An international terminology for grazing lands and grazing animals. Grass Forage Sci 66(1):2–28

Ben Hounet Y, Brisebarre A-M, Guinand S (2016) Le patrimoine culturel du pastoralisme: perspective globale, identité étatique et saviors locaux au prisme des races locales au Maroc. Rev Sci Tech Off Int Epiz 35(2):357–363

Brundtland Commission (1987) Our Common Future. UN World Commission on Environment and Development, New York

Chapin FS, Kofinas GP, Folke C (2009) Principles of ecosystem stewardship. Resilience-based natural resource management in a changing world. Springer, New York

Cherlet M, Hutchinson C, Reynolds J et al (eds) (2018) World atlas of desertification. Publication Office of the European Union, Luxembourg, p 180

Davis DK (2016) The arid land: history, power, knowledge. MIT Press, Cambridge MA

Davies J, Herrera P, Ruiz-Mirazo J, Mohamed-Katerere J, Hannam I, & Nuesri E (2016) Improving governance of pastoral lands: implementing the voluntary guidelines of responsible governance of tenure of lands, fisheries and forests in the context of national food security. FAO Governance of Tenure Technical Guide 6

Deblitz C (2012) Feedlots a new tendency in beef production. Working Paper 2. Agri Benchmark Sheep and Beef Network, USA

Digard J-P, Landais E, Lhoste P (1993) La crise des societies pastorals: unregard pluridisciplinaire. Rév Élev Méd Vét Pays Trop 46:683–692

Dyson-Hudson R, Smith EA (1978) Human territoriality: an ecological reassessment. Am Anthropol 80:21–41

ECOWAS (2017) ECOWAS stresses the need to obtain international transhumance certificates. Press release 30 march 2017. Economic Community of West African States, Niamey

FAO (2014) Ecosystem services provided by livestock species and breeds, with special consideration to the contributions of small-scale livestock keepers and pastoralists. Commission on Genetic Resources for Food and Agriculture. Background Study Paper No. 66. Rome. www.fao.org/3/aat598e.pdf/

FAO (2016) Database of country statistics on agriculture in 2016. www.fao.org/faostat

FAO (2018) World livestock: transforming the livestock sector through the sustainable development goals. Rome. License: CC BY-NC-SA 3.0 IGO

Fearnside PM (2005) Deforestation in Brazilian Amazonia: history, rates, and consequences. Conserv Biol 19(3):680–688

Freier KP, Finck M, Schneider UA (2014) Adaptation to new climate by an old strategy? Modeling sedentary and mobile pastoralists in semi-arid Morocco. Land 3(3):917–940

Galaty G, Aronson D, Salzman PC (1981) The future of pastoral peoples. In: Proceedings of conference 4-8 August 1980, Kenya. IDRC, Ottawa

Hogg R (1987) Settlement, pastoralism and the commons: the ideology and practice of irrigation development in northern Kenya. In: Anderson D, Grove R (eds) Conservation in Africa: people, policies and practice. Cambridge University Press, Cambridge

Huang J, Yu H, Dai A, Wei Y, Kang L (2017) Drylands face potential threat under 2 °C global warming target. Nat Clim Chang 7:417–422

IEO (2018) Overall performance review 6. The Global Environment Facility (GEF) Independent Evaluation Office, Washington DC

ILRI (2018) Making the case for livestock: economic opportunity. International Livestock Research Institute, Nairobi. www.whylivestockmatter.org

Johnsen KI, Niamir-Fuller M, Bensada A, Waters-Bayer A (2019) A case of benign neglect: knowledge gaps in the sustainability of pastoralism and rangelands. United Nations Environment Programme, Nairobi

Kamuzora F, Franks T (2001) Dynamics of development interventions in Tanzania: from projects to direct budgetary support and livelihoods approaches. Unpublished report Institute of Development Management Mzumbe, Tanzania and Bradford Centre for International Development, London

Jefatura del Estado (1995) Ley 3/95 de 23 de marzo, de Vias Pecurias. Official State Gazette no. 71 of 24 March 1995

Little PD (1985) Absentee herd owners and part-time pastoralists: the political economy of resource use in northern Kenya. Hum Ecol 13:131–151

Little PD (1996) Pastoralism, biodiversity and the shaping of savanna landscapes in East Africa. Africa: J Int Afr Inst 66(1):37–51. https://doi.org/10.2307/1161510

Ly A, Niamir-Fuller M (2005) La propriété collective et la mobilité pastorale en tant qu'alliées de la conservation: éxperiences et politiques innovatrices au Férlo, Senegal. IUCN/CEESP. IUCN Policy Matters No. 13, Gland, Switzerland

Mainali B, Luukkanen J, Silveira S, Kaivo-oja J (2018) Evaluating synergies and trade-offs among sustainable development goals (SDGs): explorative analyses of development paths in South Asia and Sub-Saharan Africa. Sustainability 10:815. https://doi.org/10.3390/su10030815

McGahey D, Davies J, Hagelberg N, Ouedraogo R (2014) Pastoralism and the green economy – a natural nexus? IUCN and UNEP, Nairobi. http://iucn.org/wisp

Nexus Regional Dialogue Programme (2019). Water-energy-food nexus serves the SDGs. GIZ Berlin. https://www.nexus-dialogue-programme.eu/about/nexus-and-the-sdgs/. Accessed 15 Feb 2019

Niamir-Fuller M (1997) The resilience of pastoral herding in Sahelian Africa. In: Berkes F, Folke C (eds) Linking social and ecological systems: institutional learning for resilience. Cambridge University Press, Cambridge, pp 250–284

Niamir-Fuller M (1999) Conflict management and mobility among pastoralists in Karamoja, Uganda. In: Niamir-Fuller M (ed) Managing mobility in African rangelands: the legitimization of transhumance. FAO and Beijer Institute for Ecological Economics. Intermediate Technology Publications, London, pp 149–183

Niamir-Fuller M (2016) Towards sustainability in the extensive and intensive livestock sectors. Rev Sci Tech Off Int Epiz 35(2):371–387. https://doi.org/10.20506/rst.issue.35.2.2521

Niamir-Fuller M, Kerven C, Reid R, Millner-Guillland E (2012) Co-existence of wildlife and pastoralism on extensive rangelands: competition or compatibility? Pastor Res Pol Pract 2(1):8. https://doi.org/10.1186/20141-7136-2-8

Niamir-Fuller M, Turner MD (1999) A review of recent literature on pastoralism and transhumance in Africa. In: Niamir-Fuller M (ed) Managing mobility in African rangelands: the legitimization of transhumance. FAO and Beijer Institute for Ecological Economics. Intermediate Technology Publications, London, pp 18–46

Ouedraogo R, Davies J (2016) Enabling sustainable pastoralism: policies and investments that optimise livestock production and rangeland stewardship. Rév Sci téch 35(2):619–630. https://doi.org/10.20506/rst.35.2.2544

Page N, Popa R (2013) Family farming in Romania. European Commission Consultations on Family Farming, Saschiz www.fundatia-adept.org

Reid R, Galvin K, Kruska R (2008) Global significance of extensive grazing lands and pastoral societies: an introduction. In: Galvin K et al (eds) Fragmentation in arid and semi-arid landscapes: consequences for human and natural systems. Springer, Dordrecht. https://doi.org/10.1007/978-1-4020-4906-4-1

Roeder P (1996) Livestock disease scenarios of mobile vs. sedentary pastoral systems. In: Proceedings of 3rd international technical consultations on pastoral development in Brussels. UNSO, New York

Sala OE, Yahdijan L, Havstad K, Aguilar MR (2017) Rangeland ecosystem services: nature's supply and humans' demand. In: Systems R (ed) Briske DD. Springer, New York, pp 467–489

Scherr SJ, Shames SA, Friedman R (2013) Defining integrated landscape management for policy makers. EcoAgriculture policy focus no.10. EcoAgriculture Partners, Washington, DC

Seligman N, Perevolotsky A (1994) Has intensive grazing by domestic livestock degraded the Old World Mediterranean rangelands? In: Arianoutsou M, Groves RH (eds) Plant-animal interactions in Mediterranean-type ecosystems. Kluwer, Dordrecht

Shewmake S (2008) Vulnerability and impact of climate change in South Africa's Limpopo River Basin. IFPRI discussion paper. IFPRI, Washington, DC

Sutter JW (1987) Cattle and inequality: herd size differences and pastoral production among the Fulani of northeastern Senegal. Africa 57:196–218

Swallow BM, Bromley DW (1991) Co-management or no management: the prospects for internal governance of common property regimes through dynamic contracts. Oxf Agrar Stud 22:3–16

Swift J, (1993) Dynamic ecological systems and pastoral administration. Abstracts of the workshop on new directions in African range management and policy. ODI/IIED/Commonwealth Secretariat, Woburn, London

Thébaud B (1988) Élevage et développement au Niger. International Labor Organization, Geneva

UNEP (2016) Sustainable pastoralism. United Nations environment assembly background paper for UNEA resolution L.24. United Nations Environment Programme, Nairobi

United Nations (2015) The 2030 sustainable development agenda. United Nations, New York. https://sustainabledevelopment.un.org/content/documents/21252030%20 Agenda%20for%20Sustainable%20Development%20web.pdf.

United Nations (2015) The SDG indicators. United Nations, New York. https://undocs.org/A/RES/71/313

World Initiative for Sustainable Pastoralism (2006) Pastoralism and the millennium development goals. WISP policy brief no. 1. WISP/IUCN/UNDP/GEF, Nairobi

Chapter 4
Changes in the Vegetation Cover and Quality of Aquifers in the Drylands of Mexico: Trends in an Urbanized Complex of Three Socio-Ecological Systems Within the Chihuahuan Desert

V. M. Reyes Gómez, D. Núñez López, and M. Gutiérrez

Abstract The increase in the change of land use in many areas has resulted in an excessive extraction of groundwater, often in volumes that exceed the total recharge in the underlying aquifers. These are expected to be accentuated by climate change. Monitoring and regulating aquifers is a challenging task, especially in developing countries where data are scarce. Over an 11-year interval, the water level (WL) and water quality in wells of three contiguous aquifers in Northern Mexico were measured. An actual average WL drop rate of 2.012 m year^{-1} rendered unsustainable considering that the area receives 0.35 m year^{-1} and negligible horizontal flows. Agricultural land use increased fivefold in the last few years at the expense of rangeland, increasing 11.7–76.2% in the three aquifers, and producing a water demand threefold the aquifer recharge. The permanent presence of As and F above guidelines in several wells makes communities vulnerable to ingestion toxicity. The results of this study stress the inability of these aquifers to supply additional water to a large city nearby and the need of immediate corrective actions, e.g., promoting water-efficient irrigation, artificial aquifer recharge, and an efficient and sustained management policy.

Keywords Socio-ecological systems · Water level · Loss and gain of vegetation cover · Contamination · Geogenic and anthropogenic · As–F co-occurrence

V. M. Reyes Gómez (✉)
Red Ambiente y Sustentabilidad, Instituto de Ecología, A.C., Chihuahua, Mexico
e-mail: victor.reyes@inecol.mx

D. Núñez López
Centro de Investigación en Materiales Avanzados, S.C., Durango, Mexico

M. Gutiérrez
Department of Geography, Geology and Planning, Missouri State University, Springfield, MO, USA

© Springer Nature Switzerland AG 2020   57
S. Lucatello et al. (eds.), *Stewardship of Future Drylands and Climate Change in the Global South*, Springer Climate,
https://doi.org/10.1007/978-3-030-22464-6_4

Introduction: Changes in Land Cover Use (CLCU)

The CLCU are the result of complex interactions between physical, biological, social, economic, and environmental factors that develop at different spatial and temporal scales (Lambin et al. 2001; Longmire et al. 2016). Overpopulation, the scarcity of resources that leads to growing pressure for the production of food and other basic products, either through the intensification of existing agricultural land or the incorporation of new production areas, changes in the regional economy, and global markets, are examples of the multiple factors that can lead to transformations in land cover and use (Aide and Grau 2004; Barbier et al. 2010). Hydrological processes in drylands like precipitation, evapotranspiration, runoff, and aquifer recharge are affected by CLCU and climate change (CC). These changes can occur at global and regional scales. Research on CC and CUCV has allowed simulating the effects of changes in precipitation, runoff, and evapotranspiration, but predicting the recharge and depletion of aquifers (DWL) have been more difficult (Kumar 2012; Scanlon et al. 2016). The relatively high interannual variability in climatic conditions in socio-ecological systems (SES) of drylands exerts a crucial influence on their dynamics and stability equilibrium (D'Odorico and Abinash 2012).

The impacts of CLCU processes have been widely discussed at length in tropical and temperate forest ecosystems (Rosete et al. 2008), and despite the importance and extent of arid and semi-arid ecosystems in Mexico, there are few studies aimed at determining the factors that determine CLCU processes. An exhaustive literature review (Rosete et al. 2014) revealed that just over 85% of the studies developed in Mexico on changes in CLCU focused on determining rates of degradation and deforestation exclusively in tropical and temperate forest ecosystems and only in nine of the 92 studies examined the factors that modulate or determine the processes of change. Drylands, due to their geographic location, climatic characteristics, geological and topographical varieties, host a large number of endemic species, biological richness, and mosaics of plant communities ranging from xerophilous shrub, vegetation of sandy deserts, as well as extensive areas covered by grassland (Hyot 2002; CONABIO 2006). Despite being seriously threatened by extreme climatic variations and constant anthropogenic pressures (agricultural expansion, overgrazing, and urbanization), there are few studies aimed at understanding the socio-environmental conditions that determine changes in CLCU.

In the Chihuahuan Desert in Mexico, CLCU processes show a constant dynamic in the annual losses of surface covered by primary vegetation. The change in land use between 1993 and 2013 reflects an increase of 14% in agricultural land, 63% in urban areas, and 16–27% in primary and secondary temperate forests, respectively (Table 4.1). Conversely, the loss of surface area covered by primary vegetation of grassland and shrubland decreased by 178% and 421%, respectively. Secondary grasslands also lost about 69% during the 20 years of monitoring and assessment, while the secondary shrublands had an increase in 175%.

Table 4.1 Land use change during the 1993–2013 period for the surface area of desert Chihuahuan

Land use	1993–2013 (km²)		Exchange rate (km²)	
			Total	Annual
Agriculture	57,523.1	65,832.7	8309.6	415.5
Non-vegetated	1635.3	1823.5	188.2	9.4
Urban	1672.7	2726.4	1053.7	52.7
Water body	1120.5	1089.1	−31.4	−1.6
Primary temperate forest	12,431.6	12,760.0	328.4	16.4
Secondary temperate forest	8092.3	8633.0	540.8	27.0
Primary tropical forest	11.2	8.4	−2.8	−0.1
Secondary tropical forest	983.8	909.8	−73.9	−3.7
Primary desert shrubland	244,592.9	236,169.4	−8423.5	−421.2
Secondary desert shrubland	23,398.3	26,773.9	3375.6	168.8
Primary grassland	62,442.2	58,880.1	−3562.1	−178.1
Secondary grassland	39,239.8	37,901.7	−1338.2	−66.9
Other types of vegetation	20,717.3	20,352.9	−364.5	−18.2
Total area evaluated	473,861.1	473,861.1		

Positive values indicate increase in land use coverage, negative values indicate a loss

Sustainability of Groundwater

The intensive increase of the CLCU to supply the food demand of the human populations, for example, the change of desert scrubland to agricultural drylands and the increasing urbanization and industrialization in the landscape, has led to an excessive extraction of groundwater that exceeds the total recharge to the underlying aquifers, subjecting water stress to SES, their population, and consequently to the economy and sustainability. These dynamics of exploitation of aquifers are related to anthropogenic process, which consist of changing the use of the soil, which is different from what is known to have been its original use. For example, pristine soil supports pastures, forests, or forests that humans have transformed for their benefit into urban, agricultural, and tourist areas, to name a few. During the last two decades, all these processes have led to plans of sustainable development of aquifers (Reyes Gómez et al. 2017).

The concept of groundwater sustainability is of utmost interest for those in charge of coordinating and deciding how to use water in drylands. Most of the groups that decide and manage groundwater in dryland SES consider that in order to achieve water sustainability, good practices in water use must be adopted and promoted, and that they must be followed following strict guidelines to the very long-term (Wang and Wu 2006). Water sustainability at a global and local scale is achieved through the maintenance and protection of groundwater resources in balance with the economic, environmental, and human (social) benefits (Hiscock et al. 2002). The interpretation of groundwater sustainability considers methodologies that contemplate management practices that ensure the preservation of water, the conservation of ecosystems and resulting social benefits of SES; these practices are currently emerging in Europe (the EU Water Framework Directive), in England and Wales (Abstraction Management Strategies), and in some parts of America (Hiscock et al. 2002; Wang and Wu 2006).

In 2018, in Mexico, of the 653 aquifers delimited throughout the country 106 are classified as overexploited (Fig. 4.1). The distribution of aquifers with this condition encompasses the arid and semi-arid zones of México; they are the southern extensions of the ecoregions of the North American Sonora and Chihuahua deserts, Southern semi-arid elevations, and some of Mediterranean California and Temperate Saws in the southern center of Mexico (INEGI 1993). The condition of overexploitation refers to the fact that the total recharges from rainwater infiltration, irrigation return flows, industrial and urban origin in the aquifer are lower than the extraction for agriculture, industry, and urbanized areas (NOM-011-CNA 2000). When these conditions prevail for periods longer than 3 years, abrupt declines in water table levels occur due to the fact that extractions are maintained, which translates into large economic and natural impacts (higher investment for well drilling, subsidence, and contamination events).

Groundwater resources in dryland SES face negative impacts as a result of aquifers becoming increasingly depleted, the latter evidenced by a steadily drop of groundwater levels (Bredehoeft and Alley 2014; Steward and Allen 2016). Groundwater depletion is a common occurrence in drylands since they are, for the

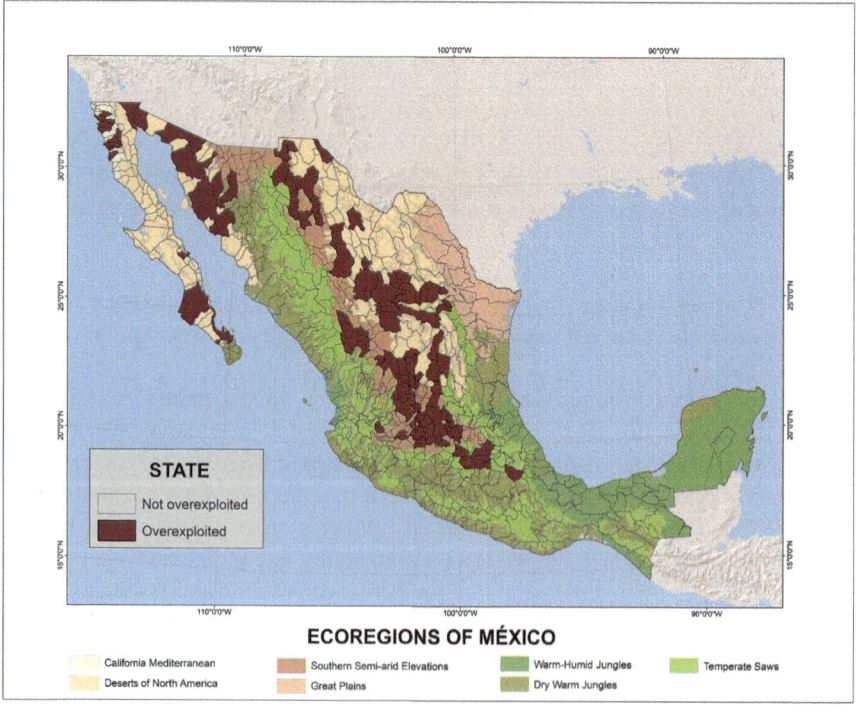

Fig. 4.1 National distribution of overexploited aquifers in Mexico

most part, overly stressed by the water demands posed by a growing population and by the various economic sectors such as agriculture and industry (Scanlon et al. 2012). In addition to a WL drop produced after heavy groundwater withdrawals, the water quality of depleted aquifers deteriorates in many cases posing a threat to the health of communities that use this water (López et al. 2012; Alarcón-Herrera et al. 2013; Reyes Gómez et al. 2015). Since a substantial drop in WL indicates that the sustainability of a vital resource is at risk, the rate of the WL drop has commonly been used as an indicator of water availability and sustainability of aquifers. Efforts towards reducing the negative effects of aquifer depletion require knowing the natural baseline of the hydrogeological behavior for that particular aquifer according to urban, agricultural, and industrial uses and its water quality evolution. Hydrological models can then use these data to characterize the response of an aquifer to impact sources and to identify any potential threats to the quantity and quality of groundwater (El Alfy 2014). Investigators may go a step further and recommend actions that will help achieving sustainability, such as direct artificial recharge, implementation of water-saving irrigation systems, and integral planning of best management practices (Gale 2005). The WL continues to drop due to recurring intensive and long-lasting droughts as well as poor groundwater management practices. Examples include the Comarca Lagunera aquifer, which experienced an average WL drop of over 30 m between 1975 and 1999 (1.25 m year^{-1}), with a 65 m drop in some parts and the Meoqui-Delicias aquifer, where the average WL drop was 30 m. In both cases, the WL dropped as a result of water being used to irrigate crops (Esteller et al. 2012).

In Mexico, the water policy states that the water is to be distributed among users based on its availability, but its implementation has been lacking for the past 30 years or so, as evidenced by water reports showing that the withdrawal volumes are almost always larger than the approved volumes (Scott 2003). One method to correct this deficiency is to take into account WL drop rates operating in the region prior to water allocation, which would become more effective if this information was available for all water uses on a permanent and continuous basis. WL drop rates can also be used to help identify threats and also to identify if these variations are cyclical in nature (Scanlon et al. 2005).

Artificial recharge of aquifers has been considered as an action to counteract the WL drop in systems already undergoing a certain degree of depletion, by means of retention of surface water, injection to aquifer, or by reusing treated or pluvial water to irrigate crops. These measures have successfully improved water sustainability in some other regions where water is scarce (UNESCO 2007).

Although studies about aquifer recharge are common in Mexico, actions towards achieving this goal at a pilot or at large scale are rare and those few studies have operated for only a short interval of time (Esteller et al. 2012). One of these projects aimed at recharging the aquifer of the Comarca Lagunera, a depleted aquifer with high content of arsenic (As), by diverting water from a reservoir into the Nazas and Aguanaval dry riverbeds. The attempt produced mixed results after recharged water flow caused an obstruction in the deeper zones of the aquifer (Rosas et al. 1999). Ongoing pilot tests to recharge aquifers in Mexico with treated water have been reported as promising.

The objectives of this study are to (1) determine the water quantity/quality status of three contiguous aquifers in semi-urbanized areas in Northern Mexico, (2) explore a possible relationship between WL and water quality to changes in land use for scales of the Chihuahuan Desert and the socio-ecological system in study, and (3) draft recommendations towards the implementation of measures that will enhance sustainability.

Materials and Methods

Study Area

The three aquifers comprising the complex semi-urbanized socio-ecological systems are Tabalaopa-Aldama (TA), Aldama-San Diego (SD), and Laguna de Hormigas (LH) (Fig. 4.2), which together occupy an area of 8488 km². All three are open aquifers and consist of basins filled with alluvial material between 300 and 800 m

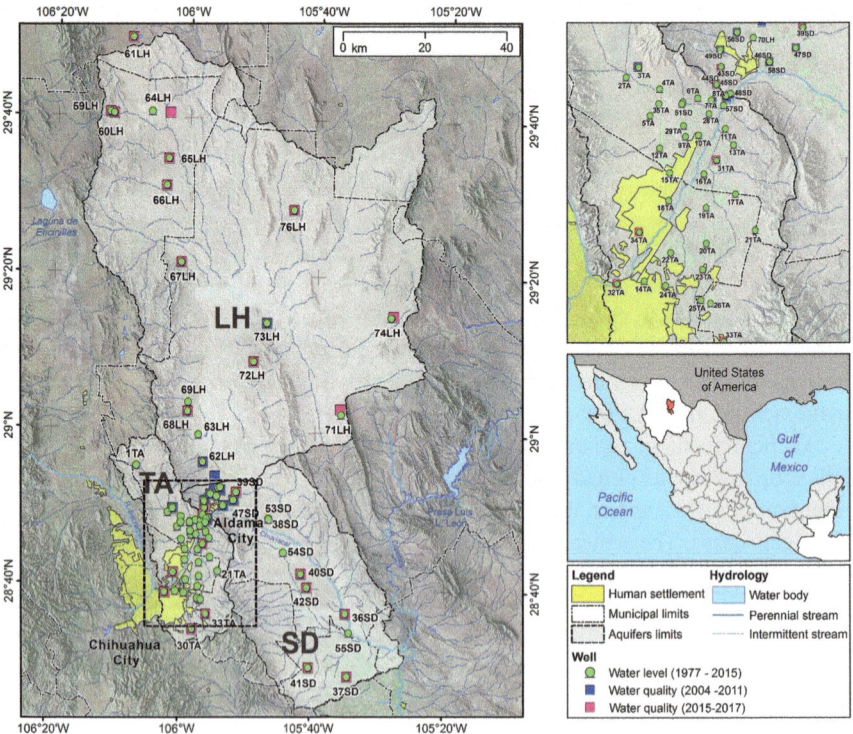

Fig. 4.2 System of monitoring of quality and level water in a network wells of three aquifers semiurbanized in Chihuahua. Aquifers Tabalaopa-Aldama = TA, Laguna de Hormigas = LH, and Aldama-San Diego = SD

thickness. Clay lenses may produce local confinement in a few places, whereas the aquifers may connect with each other under certain conditions (Mahlknecht et al. 2008). The climate in the area is semi-arid, semi-hot with summer monsoon rains (García 2003), with highest temperatures occurring between June and August when the mean temperature reaches 27 °C and the lowest temperatures occurring in January when the mean temperature is 10 °C. The mean annual precipitation is 338 mm and the evaporation rate exceeds 2400 mm. The vegetation cover belongs to scrubland and grassland that support livestock grazing. A small fraction of the land is irrigated to grow crops such as pecan trees, alfalfa, and cotton.

Water Level (WL)

The evolution of the WL was measured in 2004, 2007, 2010, 2011, and 2015. We complemented these data with those reported by CONAGUA from 1978 to 1999. Historical water data, WL measurement, and water sample collection are described in more detail in Alarcón-Herrera et al. (2013) and Reyes Gómez et al. (2015). The WL measurements and water samples were obtained from 31 wells in TA, 14 in SD, and seven in LH (see Fig. 4.2). The WL drop rate was calculated as the difference between the oldest WL measurements minus the most recent value, divided by the number of years between the two measurements. At the same time, WL was measured in seven of the wells (spring, summer, and fall 2015, fall 2016, and winter 2016–2017 in order to appreciate the seasonal variations in WL).

Changes in Land Cover Use (CLCU)

CLCU rates were determined at the scale of the Chihuahuan Desert, a region representative of arid and semi-arid zones of Mexico. Then, we proceeded in the same way but considering surface cover of three semi-urbanized aquifers of the state of Chihuahua, in the center of the Chihuahuan Desert. This way we were able to relate changes in land use with the chemical quality of the water of a well network monitored in that zone of aquifers of Chihuahua. CLCU was determined from a 1:250,000 digital model obtained from Series II (INEGI, 1993) and V (INEGI, 2013). Land use types were classified as: agricultural, urban, water bodies, and non-vegetated land, while vegetation cover grown as a result of primary and secondary succession was classified according to the following three general classes: (1) desert scrubland, (2) grassland, and (3) others. Primary succession indicates vegetation that has suffered little alteration after it has been established and secondary succession an area that has been disturbed by either natural or man-made processes.

CLCU layers were processed using the ArcGIS 10.3 TM software. The loss of vegetation cover was determined after comparing areas that had been previously assigned to a different class, either agricultural, urban, water body, or areas with no apparent vegetation, whereas the opposite change was used to identify areas of recovered vegetation cover. The losses in vegetation cover were identified when the original classification represented different types of vegetation cover: Temperate forest, desert scrub, grasslands, or other types of vegetation were transformed to land uses such as agriculture, urban areas, bodies of water or areas without apparent vegetation. With inverse transitions it was possible to detect the surface recovered by the vegetation cover. Finally, the dynamics of change from a condition of primary to secondary succession in the vegetation cover allowed identifying processes of degradation of vegetation cover.

Water Quality

Water quality was determined from chemical analyses of water samples collected in 2004, 2007, 2010, 2011, and 2015. In 2015, 20 new wells were added to the original ten in order to expand and improve data coverage, as well as to have a better view of temporal variations (Table 4.2). The 2015 samples were collected in spring, summer, and fall and analyzed for hardness, electrical conductivity, and potential contaminants As, F, Na, and nitrate.

In order to determine the chemical quality of the water in wells, a water quality index (WQI) was considered that contemplates the maximum permissible limits (MPL) according to Mexican guidelines (NOM 127 SSA1 1994 mod. 2005). For each of the parameters examined, the WQI was determined by dividing the current concentration over the NML concentration, the result reflects wells with pollution

Table 4.2 Chemical parameters analyzed of water samples

Parameter	Recommended limits	Analytical methods
	Mexican norm NOM 127 SSA1 1994 mod. 2005 (NOM); World Health Organization (WHO)	Mexican norm NOM-230-SSA1–2002
Arsenic, As	10 ppb WHO; 25 ppb NOM	ICP-MS spectrometer
Fluoride, F	1.5 mg L^{-1} NOM, WHO	Ion selective electrode
Calcium, magnesium, potassium, sodium (Ca, Mg, K, Na)	Na 200 mg L^{-1} NOM	ICP-MS spectrometer
Nitrate, N-NO$_3$	10 mg L^{-1} NOM	HACH spectrometer, method 8171
Sulfate, SO$_4$	400 mg L^{-1} NOM	HACH spectrometer, method 8051
Total dissolved solids, TDS	1000 mg L^{-1} NOM	Calculated from EC
Chloride, Cl	250 mg L^{-1} NOM	Titration
pH	6.5–8.5 NOM	pH meter
Total hardness, TH	500 mg CaCO$_3$ L^{-1} NOM	Titration
Salinity	3.32 g L^{-1} NOM	Calculated from EC

levels above the Mexican norm NOM 127 SSA1 1994 (Reyes Gómez et al. 2015). Statistical analyses performed on the WL (and significance test) consisted of linear regression and correlation to test for association among parameters (co-occurrence). For the spatial analyses in land use change, efficacy of the model (EM) was determined according to Mayer and Butler (1993).

Results

Water Level (WL) Drop Rate

All but four of the 52 sampled wells experienced a drop in the WL during the time of this study that varied between 0.02 and 89.44 m (column Dt, Table 4.3) with an annual rate of depletion of 1.49, 2.59, and 1.95 m year^{-1} for TA, SD, and LH aquifers, respectively (column Da). The drop rates of 80–88% of wells in TA, SD, and LH were greater than the mean precipitation (0.350 mm year^{-1}). Between aquifers, the WL drop rate values were found to be not significantly different ($p < 0.05$); however, some wells had drop rates around 4 m year^{-1} (wells 7TA, 35TA, 50SD, 62LH, and 76LH, see Table 4.3), whereas some exceeded 10 m year^{-1} (wells 28TA, 46SD, and 52SD). The latter may have been wells within cones of depression. Interestingly, the wells with higher WL drop rate did not concentrate in a particular part of the basin but seemed to be at random locations, which may be due to heterogeneity of the basin fill and/or particular usage by the well's owner (Fig. 4.3).

Table 4.3 Data used to estimate the annual depletion of water levels in wells of the study aquifers

Well	First record Year	First record WL (m)	Last record Year	Last record WL (m)	n (years)	Dt (m)	Da (m year^{-1})
1TA	1999	9.73	2004	9.55	5	0.18	0.036
2TA	1990	61.88	2004	71.70	14	−9.82	−0.701
3TA[a]	2004	80.00	2011	90.79	7	−10.79	−1.541
4TA	1978	23.00	2004	70.90	26	−47.90	−1.842
5TA	1999	70.07	2004	78.30	5	−8.23	−1.646
6TA	1978	4.50	2007	36.68	29	−32.18	−1.110
7TA	1990	25.00	2004	45.16	14	−20.18	−4.036
8TA	1990	3.78	2007	5.32	29	−1.54	−0.053
9TA	1990	33.67	1999	36.29	8	−2.62	−0.328
10TA	1991	15.21	2000	22.66	8	−7.45	−0.931
11TA	1992	11.35	2001	14.64	8	−3.29	−0.411
12TA	1978	30.52	1999	56.96	21	−26.44	−1.259
13TA	1990	39.15	1999	44.14	8	−4.99	−0.624
14TA	1999	62.80	1999	64.47	8	−1.67	−0.209
15TA	1990	47.70	1999	59.31	8	−11.61	−1.451

(continued)

Table 4.3 (continued)

Well	First record		Last record		n (years)	Dt (m)	Da (m year^{-1})
	Year	WL (m)	Year	WL (m)			
16TA	1978	15.00	1999	28.05	21	−13.05	−0.621
17TA	1990	59.07	1999	66.80	8	−7.73	−0.966
18TA	1978	28.90	1999	57.29	21	−28.39	−1.352
19TA	1999	37.09	2004	39.30	5	−2.21	−0.442
20TA	1978	51.00	1999	65.62	21	−14.62	−0.696
21TA	1990	110.29	1999	112.58	8	−2.29	−0.286
22TA	1978	51.00	1999	66.27	21	−15.27	−0.727
23TA	1999	83.47	2004	81.11	5	2.37	0.474
24TA	1990	82.96	1999	89.80	8	−6.84	−0.855
25TA	1978	90.63	1999	103.20	21	−12.57	−0.599
26TA	1999	114.40	2004	117.75	5	−3.35	−0.670
27TA	2004	19.40	2015	25.64	11	−6.24	−0.567
28TA	2004	15.16	2007	69.82	3	−54.66	−18.220
29TA	2004	31.00	2007	33.00	3	−2.00	−0.667
30TA[a]	2015.4	110.90	2017	113.00	3	−2.10	−1.31
35TA	2004	67.70	2015	96.14	11	−28.44	−2.585
					Ada		−1.49
36SD[a]	2015.6	4.79	2017	4.77	1.4	0.02	0.01
37SD[a]	2015.4	3.00	2015.8	2.35	0.4	0.65	1.63
42SD[a]	2015.4	22.08	2015.8	22.35	0.4	−0.27	−0.68
43SD	2004	86.26	2011	96.00	7	−9.74	−1.145
46SD	2004	40.90	2011	130.34	7	−89.44	−11.18
47SD	2004	29.00	2011	32.54	7	−3.54	−0.44
48SD	2004	44.71	2015	57.31	11	−12.60	−1.133
49SD	2004	86.26	2011	96.00	7	−9.74	−1.39
50SD	2004	90.90	2011	119.30	7	−28.40	−3.55
51SD	2004	51.40	2007	57.47	3	−6.07	−2.02
52SD	2007	100.00	2011	150.00	4	−50.00	−16.67
53SD	2011	19.97	2015	20.10	4	−0.13	−0.03
54SD	2011	13.25	2015	16.00	4	−2.75	−0.69
55SD	2011	6.14	2015	7.00	4	−0.86	−0.22
					Ada		−2.59
62LH	2010	80.75	2011	85.00	1	−4.29	−4.29
63LH	2005	36.65	2015	50.00	10	−13.35	−1.48
64LH	2005	19.19	2015	50.80	10	−31.61	−0.35
69LH	2010	51.34	2015	61.56	5	−9.22	−1.84
71LH[a]	2011	25.60	2015	27.30	4	−3.80	−0.95
74LH	2005	15.10	2015	19.00	10	−3.90	−0.39
76LH[a]	2015.6	27.50	2015.8	28.40	0.2	−0.90	−4.50
					Ada		−1.95

1TA–35TA wells within Tabalaopa-Aldama aquifer, *36SD–55SD* wells within the Aldama-San Diego aquifer, *62LH–73LH* wells within the Laguna Hormigas aquifer, *n* number years between two dates, *WL* depth to water level in wells by two dates (m), *Dt* depth difference between two dates (m), *Da* annual rate of WL drop (m year^{-1}), *Ada* average annual of Da drop rate obtained in this study (m year^{-1})

[a]Wells measured to seasonal scale

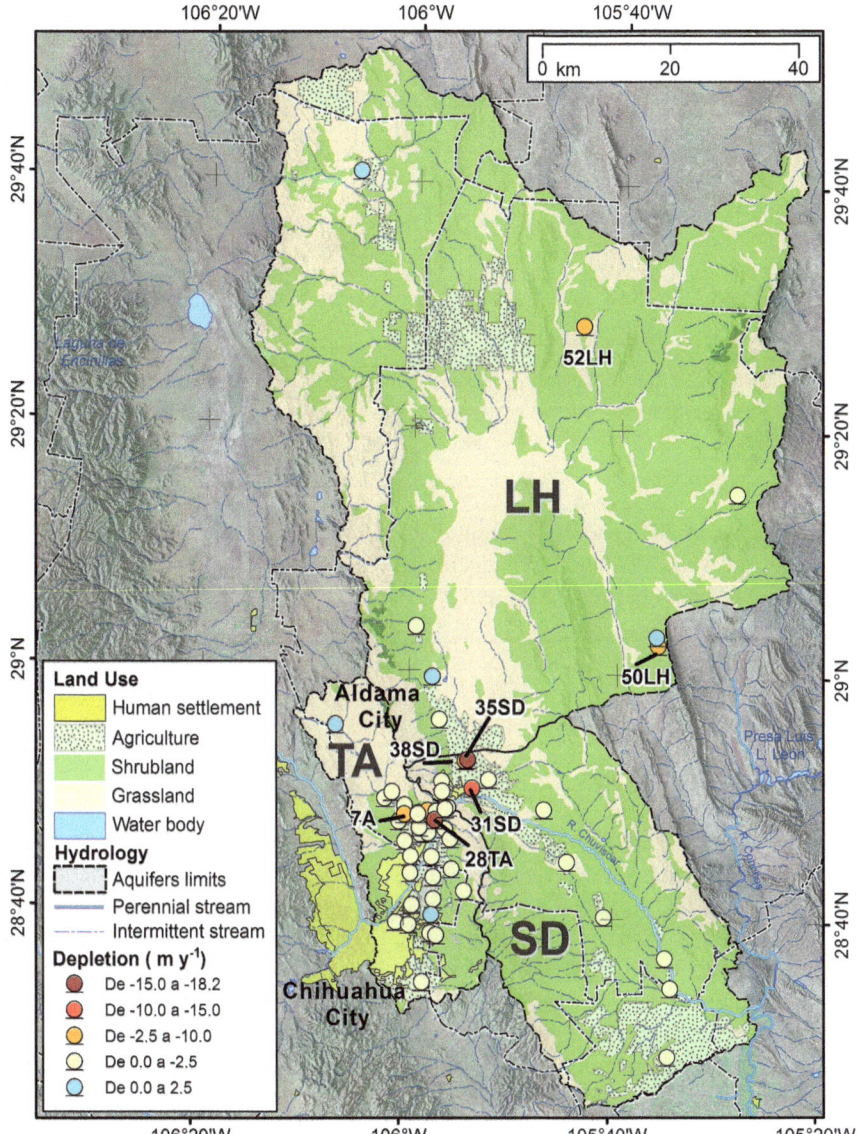

Fig. 4.3 Distribution of depletion of water level (WL) in aquifers semi-urbanized in the surroundings of the city of Chihuahua

The changes of WL per years (1999–2015) and per season for the year 2015 were compared in some of the wells to observe their evolution at annual and seasonal scales (Fig. 4.4a, 4.4b respectively). It can be seen that in four of these wells the level changes are imperceptible, while for the other four the level fluctuates and

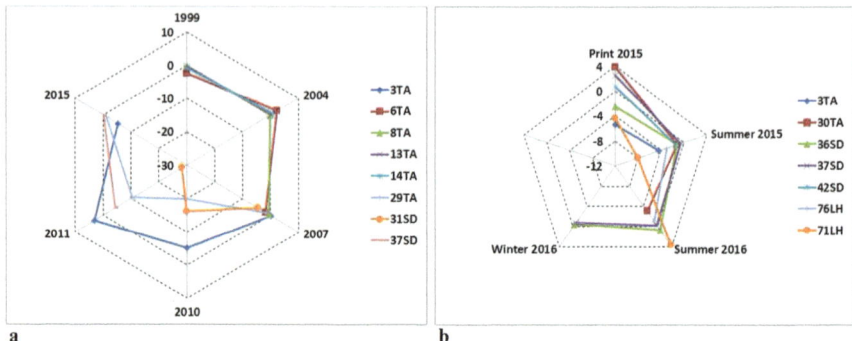

Fig. 4.4 Changes of WL (m year^{-1}) in two scales of time: (**a**) year over year, (**b**) seasonal. *TA* wells aquifer Tabalaopa-Aldama, *SD* wells aquifer Aldama San Diego, *LH* wells aquifer Laguna de Hormigas

there are lows in 2010 and 2011 of up to 15 m. Well 31SD shows the largest drop, while WL of wells 3TA and 29TA decreases but recovers in subsequent years and well 37SD recovers between 2011 and 2015.

CLCU

Scale Chihuahuan Desert

The processes measured in the dynamics of CLCU during these 20 years of evalua-tion, in the Chihuahuan desert, have mainly been loss, gain, and degradation of vegetation cover. Of the total area evaluated throughout the desert (473,861 km²), almost 45,000 km² have been transformed, representing 9.45% of the total Chihuahuan desert (Fig. 4.5). Of that area of change, 41.5% is attributed to the pro-cess of deforestation by transforming forests, shrubs, and grassland to uses such as agriculture and urban planning. Somewhat smaller in surface, but more serious in effects was the degradation of 12.9% of the primary vegetation cover of shrubland and grassland, which could hardly return to its original state in <10 years. The trans-formations of land use mentioned above (Table 4.4), match the increases observed in agricultural areas (18.5%), urbanized areas (2.4%) and the areas with secondary vegetation (23%).

Scale of a Complex Urbanized Socio-Ecological System

During the 19-year interval of CLCU observations, a significant loss of grassland and desert scrubland (containing both primary and secondary vegetation) was recorded, amounting to 454.7 km²; 11.7% in TA, 12.1% in SD, and 76.2% in LH

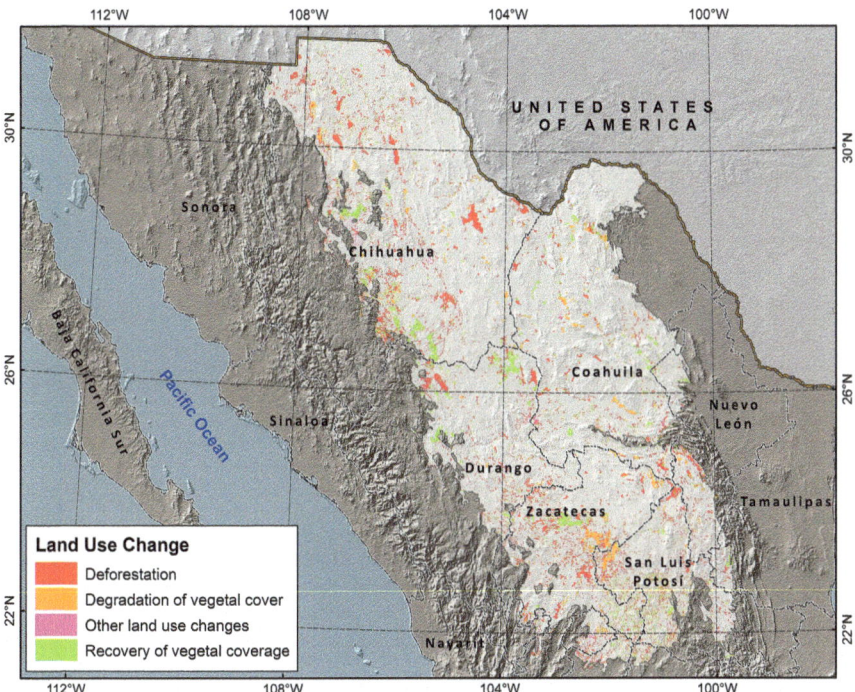

Fig. 4.5 Distribution of the impact of the CLCU processes at the Chihuahuan desert scale

Table 4.4 Changes in land use and vegetation cover in the Chihuahuan desert

Processes evaluated	Surface (km^2)	Surface (%)	Annual change rate (km^2 year^{-1})
Deforestation	18,589.59	41.50	929.48
Degradation	5762.13	12.86	288.11
Recovery (recuperation)	10,307.24	23.01	515.36
Other changes	775.59	1.73	38.78
Agriculture	8309.60	18.55	415.48
Urban (human settlements)	1053.74	2.35	52.69
Total	44,797.89	100.00	

surface areas (Table 4.5). The land use that gained area at the expense of scrubland and grassland was agricultural (365.9 km^2, 0.15% in TA, 14.9% in SD, and 85% in LH) and urban (54.93 km^2, 8% in SD, 92% in TA, and 0% in LH). There was also a slight increase in areas covered with secondary scrub vegetation, 60.6 km^2 in LH (8.25% of the land use change). As shown in Table 4.4, the largest land use changes for agricultural plots replacing rangeland with primary vegetation coverage occurred in LH, followed by SD and TA. Similarly, the loss of primary and secondary vegetation coverage to urban use was detected in the TA area, at the outskirts of the city of Chihuahua, but this change was negligible for SD and LH areas.

Table 4.5 Land use change during the 1993–2013 period for the TA, SD, and LH aquifers; positive values indicate increase in land use coverage, negative values indicate a loss

Land use	SD km²	TA	LH	Total
Agriculture	53.15	0.54	303.16	365.85
Non-vegetated	−0.76	−0.88	0.00	−1.64
Urban	4.41	50.52	0.00	54.93
Water body	−0.16	0.00	0.00	−0.16
Primary desert shrubland	−36.23	−44.05	−208.09	−288.37
Secondary desert shrubland	−8.77	−3.53	24.47	12.17
Primary grassland	−13.20	4.17	−157.34	−166.37
Secondary grassland	0.00	−8.59	35.85	27.23
Other types of vegetation	0.00	0.00	1.91	1.91

Water Quality

Contaminants of higher significance because of their toxicity are As and F. As and F levels above recommended limit in Mexico (0.025 mg L^{-1} As and 1.5 mg L^{-1} F) were found to be common and widespread in all three aquifers, with the highest levels of WQI recorded in LH (Fig. 4.6). The LH aquifer showed high content of SO_4 and Na in some wells, a common occurrence in endorheic basins in an arid environment. The water quality data show concentrations above recommended limit of F in most wells. The values obtained for As also indicated conditions above its recommended limit (29% times in TA, 42% in SD, and 68% in LH). These two contaminants co-occur ($r = 0.66$, $p < 0.001$). Cl on the other hand stayed under the limiting value of 250 mg L^{-1} in all wells, while pH, TH, TDS, SO_4, and Na content remained also below the recommended limits for most samples.

The location of wells with higher concentration of the contaminant of likely anthropogenic origin, nitrate, was superimposed on land use change (Fig. 4.7) to better visualize possible relationships. Nitrate was found within the recommended limit (<10 mg L^{-1}) in all but one well, and while not a health threat, its presence suggests discharge of contaminated water from the surface and lixiviation into the aquifer. Nitrate concentrations >4 mg L^{-1} were detected on nine sampling occasions in three out of the 15 wells sampled for nitrate (wells 57SD, 67H, and 72H in Fig. 4.7). Nitrate high values correspond to agricultural areas or green areas within the city watered with reclaimed water; however, this relationship is inconclusive due to the small number of data. More measurements are needed to test this assumption.

Discussion

The mean annual rates of average depletion obtained for the TA, SD, and LH of 1.49, 2.38, and 1.97 m $year^{-1}$ (Ada, Table 4.3) indicate severe overexploitation, considering that horizontal groundwater flows are almost negligible, precipitation

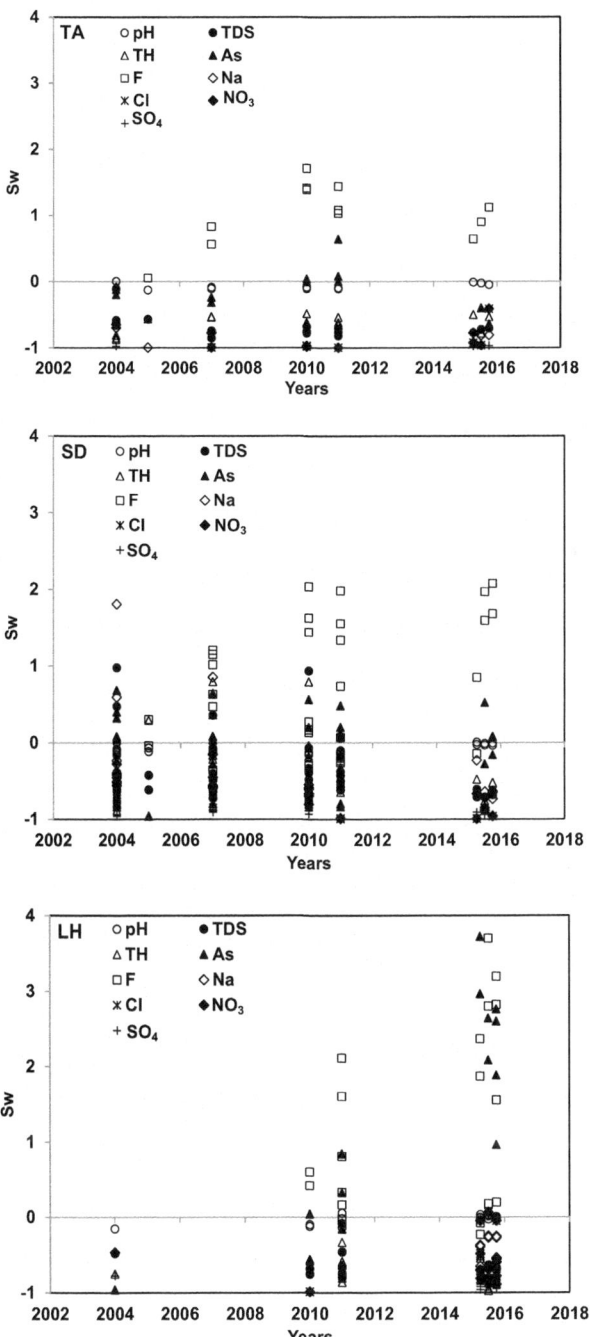

Fig. 4.6 Water quality index (WQI) in wells for nine chemical parameters

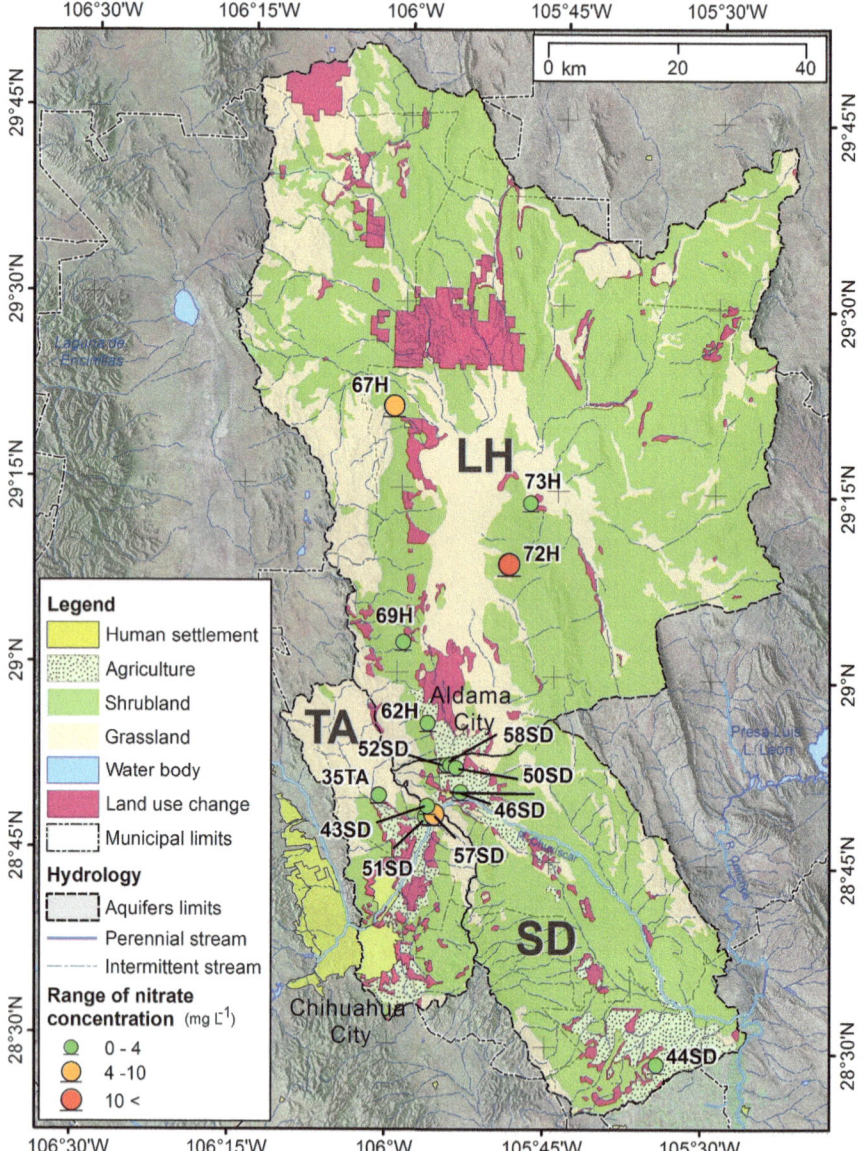

Fig. 4.7 Water quality index (WQI) in wells, for nine of the chemical parameters used in the analysis

averages 0.35 m year^{-1} and long-lasting droughts are a common occurrence in the region (Núñez López et al. 2007; Castle et al. 2014).

In the wells with WL records of several years (interannual), variations between -0.83 and 1.02 m year^{-1} (wells 8TA, 13TA, 14TA, see Fig. 4.4a) were observed, while in other areas larger changes were observed with a range between −4.94 and 3.21 m year⁻¹ WL (wells 3TA and 6TA). Significant values of water level depletion of wells of the order of 0.82 to −28.34 m year⁻¹ were observed (31SD, 37SD, and 29TA wells of the TA and SD aquifers).

Considering the average annual abatement rate estimated in this analysis (2.01 m year⁻¹) as a metric basis to diagnose seasonal trends, it was found that in 43% of the wells for this purpose (3TA, 71LH, and 76LH, Fig. 4.4b), more than 75% of that average was obtained; about 30% of the other wells reached between 20 and 40% (36SD and 42SD), in the rest wells <5%. This can mean a continuing trend of depletion that may lead to a condition of water risk assessment (Brauman et al. 2016). In some wells, seasonally the rate of depletion can be converted into the recharge rate, for example, between summer 2015 and 2016 at well 71LH, the Da value changes from −8.1 to 3.7.

What we can see in these very diverse WL trends is the great need to continue to monitor the depth of WL, but permanently and over a wider network of wells that ensure reliable results in order to make projections and recommendations for sustainable management of aquifers to decision makers. Based on the results presented here and assuming that the drop rates remain constant, the results point to an aggressive depletion of the aquifers that, if occurring, would bring serious economic and social consequences to the region, such as an increase in the extraction costs and conflicts among water users, e.g., agriculture vs. urban (Korus and Burbach 2009). The large number of wells with signs of unsustainable depletion, mainly due to the use of agricultural and domestic areas in urban areas in the aquifer TA (see Fig. 4.3), as well as the opening of new agricultural wells in SD and LH by agricultural and industrial activities, suggests that in a year ahead the level of wells will decrease further, worsening the problem of groundwater availability.

To offset the depletion of aquifers in drylands, several options have been proposed, including reusing reclaimed water, artificial aquifer recharge, and water-efficient irrigation and plumbing fixtures (Gale 2005; Brauman et al. 2016). Even though these options are given serious consideration, it is unlikely that the water management practices in the study area will change rapidly enough to counteract the rapid rate at which the aquifers are being depleted. The implementation of effective policies that would add sustainable practices is therefore a key factor for an integral planning of water resources in the region. Despite these uncertainties, there are numerous attempts at all levels of government and society to develop favorable policy environments, build resilient management institutions, synthesize available science, and increase the consciousness and participation of communities to protect and promote the sustainable use of groundwater in SES of drylands (Morsy et al. 2017). An important component of aquifer management is the volume utilized for irrigation in the agricultural sector, and a deciding factor for farmers to switch to water-efficient irrigation systems water is the offering of incentives towards their implementation. Since such incentives are not in place, farmers grow the more profitable crops, e.g., alfalfa, cotton, and pecans, which are water-intensive WL. Proper water-saving incentives could significantly slow down the present rate of aquifer abatement (Steward and Allen 2016; Scanlon et al. 2012). The land use changes also modify the amount of infiltration and run off (Robertson and Sharp Jr. 2015), and in the case of agricultural land, other parameters that may change are the amount of water gained as irrigation returns, water lost as evapotranspiration, and discharges that contribute salinity, phosphates, and nitrates among other contaminants present in irrigation water return flows (Dávila Pórcel et al. 2014; Opazo et al. 2016).

The distribution of nitrate concentrations superimposed onto a land use map helps visualize a possible relationship between them (see Fig. 4.7). Although higher nitrate concentrations occur near areas of recent land use change, the occurrences of contamination are not high enough as to show a clear relationship. Further monitoring is therefore needed to confirm a connection between the impacts of land use, land cover, and groundwater resource (Scanlon et al. 2005). The contaminants As and F exceeded the Mexican norms (25 μg As L^{-1}, 1.5 mg F L^{-1}) in 50% of the wells for F and 30% of the wells for As (Fig. 4.6). Although As and F are natural geogenic contaminants (Mahlknecht et al. 2008), excessive pumping may reach to deeper parts of the aquifer where water is presumably richer in dissolved salts as well as in As and F (Currell et al. 2011). Regarding pH, TDS, and TH, fewer wells exceeded the norms, but a high WL drop rate could nevertheless pose a problem if the trends for land use change and groundwater extraction continue. Nitrate high values correspond to agricultural areas or green areas within the city watered with reclaimed water; however, this relationship is inconclusive due to the small number of data (Yager and Heywood 2014). More measurements are needed to test this assumption (see Fig. 4.7). The lowering of WL can also negatively affect biodiversity by altering the type and abundance of riparian vegetation, levels of water bodies, and spring flows that sustain wildlife (Granados-Sánchez et al. 2011; Dávila Pórcel et al. 2014; Opazo et al. 2016).

Conclusions

The aquifers were found at risk of losing balance based on (a) lowering of WL, (b) land use change, and (c) water quality. Therefore, continued monitoring of these indicators is recommended, a task that would greatly benefit from automated data loggers to measure and record WL and water quality, especially if these data were made available to all water users. Offering the proper incentives to farmers towards the implementation of water-saving irrigation systems was identified as an action that would greatly attenuate the projected depletion of these aquifers.

Other measures to reduce evaporation and increase artificial aquifer recharge should be considered, as well as the promotion of sustainable actions such as the planting of WL that are drought-resistant and to restrict fertilizer application to only the amount needed by the plant. Injection of treated wastewater into the aquifer has been mentioned as a viable alternative to recharge these aquifers; however, this should be done with care and under high scientific scrutiny that this procedure is safe and that it will not produce clogging that may affect the permeability nor the groundwater quality in the long run, preferably testing at pilot scale first before injection takes place. Other proposed actions leading towards sustainability include rainwater harvesting, water storage in containers that minimize evaporation, contamination prevention, formation of a grassroots and/or technical committee to increase community participation in discussions about sustainability of local resources, and environmental education outreach programs that would emphasize the functions of aquifers and how to manage them sustainably.

References

Aide TM, Grau HR (2004) Globalization, migration, and Latin American ecosystems. Science 305:1915–1916

Alarcón-Herrera MT, Bundschuh J, Nath B et al (2013) Co-occurrence of arsenic and fluoride in groundwater of semi-arid regions in Latin America: genesis, mobility and remediation. J Hazard Mater 262:960–966

Barbier EB, Burgess JC, Grainger A (2010) The forest transition: towards a more comprehensive theoretical framework. Land Use Policy 27:98–107

Brauman KA, Richter BD, Postel S et al (2016) Water depletion: an improved metric for incorporating seasonal and dry-year water scarcity into water risk assessments. Elementa Sci Anthropos 4:000083

Bredehoeft JD, Alley WM (2014) Mining groundwater for sustained yield. Bridge Link Eng Soc 44(1):33–41

Castle SL, Thomas BF, Reager JT et al (2014) Groundwater depletion during drought threatens future water security of the Colorado River Basin. Geophys Res Lett 4:5904–5911

CONABIO (2006) Capital natural y bienestar social. CONABIO, Mexico City

Currell M, Cartwright I, Raveggi M et al (2011) Controls on elevated fluoride and arsenic concentrations in groundwater from the Yuncheng Bas China. Appl Geochem 26(4):540–552

D'Odorico P, Abinash B (2012) Hydrologic variability in dryland regions: impacts on ecosystem dynamics and food security. Philos Trans R Soc Lond B 367:3145–3157

Dávila Pórcel RA, Shuth C, De León-Gómez H et al (2014) Land-use impact and nitrate analysis to validate DRASTIC vulnerability MaWL using a GIS platform of Pablillo River Basin, Linares, N.L., Mexico. Int J Geosci 5:1468–1489

El Alfy M (2014) Numerical groundwater modelling as an effective tool for management of water resources in arid areas. Hydrol Sci J 59:1259–1274

Esteller MV, Rodríguez R, Cardona A, Padilla-Sánchez L (2012) Evaluation of hydrochemical changes due to intensive aquifer exploitation: case studies from Mexico. Environ Monit Assess 184:5725–5741

Gale I (2005) Strategies for managed aquifer recharge (MAR) in semi-arid areas. UNESCO, Paris. http://unesdoc.unesco.org/images/0014/001438/143819e.pdf. Accessed 31 Oct 2016

García E (2003) Distribución de la precipitación en la República Mexicana. Investig Geogr Bol 50:67–76

Granados-Sánchez D, Sánchez González A, Granados-Vitorino RL et al (2011) Vegetation ecology of the Chihuahuan Desert. Rev Chapingo Ser Cienc For Ambiente XVII:111–130

Hiscock KM, Rivet MO, Davison RM (2002) Sustainable groundwater development. Geol Soc Lond Spec Publ 193:1–14

Hyot CA (2002) The Chihuahuan Desert: diversity at risk. Endanger Species Bull XXVII(2):16–17

INEGI (1993) Mapas de Uso del Suelo y Vegetación. Escala 1:250 000. Serie II. México. www3.inegi.org.mx/sistemas/biblioteca/ficha.aspx?upc=702825229344. Accessed 29 Oct 2016

INEGI (2013) Mapas de Uso del Suelo y Vegetación. Escala 1:250 000. Serie V. México. http://catalogo.datos.gob.mx/dataset/mapas-de-uso-del-suelo-y-vegetacion-escala-1-250-000-serie-v-mexico. Accessed 29 Oct 2016

Korus JT, Burbach ME (2009) Analysis of aquifer depletion criteria with implications for groundwater management. Gt Plains Res 19:187–200

Kumar CP (2012) Climate change and its impact on groundwater resources. Int J Eng Sci 1(5): 43–60

Lambin EF, Turner BL, Geist HJ et al (2001) The causes of land-use and land-cover change: moving beyond the myths. Glob Environ Chang 11:261–269

Longmire P, Rearick M, McQuillan D et al (2016) Water quality and hydrogeochemistry of a basin and range watershed in a semi-arid region of northern New Mexico. Environ Earth Sci 75:640

López DL, Bundschuh J, Birkle P et al (2012) Arsenic in volcanic geothermal fluids of Latin America. Sci Total Environ 429:57–75

Mahlknecht J, Horst A, Hernández-Limón G et al (2008) Groundwater geochemistry of the Chihuahua City region in the Rio Conchos Basin (northern Mexico) and implications for water resources management. Hydrol Process 22:4736–4751

Mayer DG, Butler DG (1993) Statistical validation. Ecol Model 68:21–32

Morsy KM, Alenezi A, Alrukaibi DS (2017) Groundwater and dependent ecosystems: revealing the impacts of climate change. Int J Appl Eng Res 12(13):3919–3926

NOM-011-CNA, Norma Oficial Mexicana (2000) Conservación del recurso agua que establece el método para determinar la disponibilidad media anual de las aguas nacionales, México. http://www.conagua.gob.mx/CONAGUA07/Noticias/NOM-011-CNA-2000.pdf. Accessed 19 Oct 2016

NOM-127-SSAI-1994, Modificación a la Norma Oficial Mexicana NOM-127-SSAI-(2005) Salud ambiental. Agua para usos y consumo humano. Límites permisibles de calidad y tratamientos a que debe someterse el agua para su potabilización. http://www.cofepris.gob.mx/MJ/Documents/Normas/mod127ssa1.pdf. Accessed 19 Oct 2016

Núñez López D, Muñoz Robles C, Reyes Gómez VM et al (2007) Characterization of drought at different time scales in Chihuahua, México. Agrociencia 41:253–261

Opazo T, Aravena R, Parker B (2016) Nitrate distribution and potential attenuation mechanisms of a municipal water supply bedrock aquifer. Appl Geochem 73:157–168

Reyes Gómez VM, Alarcón Herrera MT, Gutiérrez M, Núñez López D (2015) Arsenic and fluoride contamination in groundwater of an endorheic basin undergoing land use changes. Arch Environ Contam Toxicol 68:292–304

Reyes Gómez VM, Gutiérrez M, Nájera Haro B et al (2017) Groundwater quality impacted by land use/land cover change in a semiarid region of Mexico. Groundw Sustain Dev 5:160–167

Robertson WM, Sharp JM Jr (2015) Estimates of net infiltration in arid basins and potential impacts on recharge and solute flux due to land use and vegetation change. J Hydrol 522:211–227

Rosas I, Belmont R, Armienta MA (1999) Arsenic concentrations in water, soil, milk and forage in Comarca Lagunera, Mexico. Water Air Soil Pollut 112:133–149

Rosete VF, Pérez-Damián JL, Bocco G (2008) Cambio de uso del suelo y vegetación en la Península de Baja California, México (Land use and vegetation change in Baja California Península, México). Investig Geogr 67:39–58

Rosete VF, Velázquez A, Bocco G, Espejel I (2014) Multi-scale land cover dynamics of semiarid scrubland in Baja California, México. Reg Environ Chang 14:1315–1328

Scanlon BR, Faunt CC, Longuevergne L et al (2012) Groundwater depletion and sustainability of irrigation in the US High Plains and Central Valley. Proc Natl Acad Sci 109:9320–9325

Scanlon BR, Levitt DG, Reedy RC (2005) Ecological controls on water-cycle response to climate variability in deserts. Proc Natl Acad Sci 102(17):6033–6038

Scanlon BR, Zhang Z, Save H et al (2016) Global evaluation of new GRACE mascon products for hydrologic applications. Water Resour Res 52:9412–9429

Scott C (2003) Sustainable groundwater management: have property rights reforms helped in Mexico? In: World Water Forum 3, Kyoto, Japan, International Water Management Institute (IWMI). http://publications.iwmi.org/pdf/H044092.pdf. Accessed 1 Nov 2016

Steward DR, Allen AJ (2016) Peak groundwater depletion in the High Plains Aquifer, projections from 1930 to 2110. Agric Water Manag 170:36–48

UNESCO (2007) Groundwater resources sustainability indicators. IHP-VI, series on groundwater 14. http://unesdoc.unesco.org/images/0014/001497/149754e.pdf. Accessed Oct 2017

Wang Z-Q, Wu Q (2006) Development of groundwater sustainability indicators. In: Sustainability of groundwater resources and its indicators (proceedings of symposium S3 held during the seventh IAHS scientific assembly at Foz do Iguaçu, Brazil, April 2005), vol 302. IAHS, Wallingford, pp 29–34

Yager RM, Heywood CE (2014) Simulation of the effects of seasonally varying pumping on intraborehole flow and the vulnerability of public-supply wells to contamination. Groundwater 52:40–52

Chapter 5
The Socio-Ecological Systems Approach to Research the Integrated Groundwater Management in an Agricultural Dryland in Mexico

M. Villada-Canela, R. Camacho-López, and D. M. Muñoz-Pizza

Abstract Strengthening integrated and sustainable water management is one of the most important objectives of 2030 Sustainable Development Agenda and Mexico's water policy, especially in drylands where the groundwater is the primary source of water. However, 18% of the total aquifers in the country (653) are overexploited, 5% with presence of saline soils and brackish water and 3% with marine intrusion. In this regard, there are still multiple challenges in facing dryland socio-ecological systems (DSES): land degradation, desertification, climatic variations, population growth, water supply and demand, economic activities and land-use changes, cultural perceptions, water policies and governance and management practices that threaten the availability and quality of groundwater. These aspects affect land livelihood systems and the sustainability of human communities. Based on this, we explore a way to research the integrated groundwater management in Maneadero Valley, an agricultural DSES in Baja California, Mexico. We conducted 52 surveys on farmers with a concession for groundwater supply to gain clear understanding of the current water management and farmers' participation. We found a need for real involvement from farmers as stakeholders, better knowledge exchange mechanisms and possibilities to integrate local knowledge in groundwater decision-making. Finally, we make recommendations to improve the groundwater management in the agricultural DSES of Maneadero Valley.

M. Villada-Canela (✉)
Instituto de Investigaciones Oceanológicas, Universidad Autónoma de Baja California
UABC, Ensenada, México
e-mail: mvilladac@uabc.edu.mx

R. Camacho-López
Programa de Maestría en Manejo de Ecosistemas de Zonas Áridas, Universidad Autónoma de Baja California UABC, Ensenada, Mexico

D. M. Muñoz-Pizza
Programa de Doctorado en Medio Ambiente y Desarrollo, Universidad Autónoma de Baja California, Ensenada UABC, México

© Springer Nature Switzerland AG 2020   79
S. Lucatello et al. (eds.), *Stewardship of Future Drylands and Climate Change in the Global South*, Springer Climate,
https://doi.org/10.1007/978-3-030-22464-6_5

Keywords Maneadero Valley · Public participation · Coupled human–water system

Introduction

Soils and freshwater are being rapidly degraded, due to climatic variations that put even more pressure on the resources we depend on, increasing risks associated with disasters, droughts and floods (UN 2015). Therefore, population growth, economic development, water scarcity and pollution, land-use change, land degradation and desertification are some of the real challenges drylands face and in reaching an integrated and sustainable natural resources management. In order to contribute to the drylands science, in this chapter we use the social-ecological systems (SES) approach that considers the link between social (human) and ecological (water) systems (Berkes and Folke 1998), regarding elements such as a resource system (e.g., an agricultural valley), resource units (aquifer, groundwater), users (farmers) and governance systems (Ostrom 2009; Rica et al. 2018), creating a drylands SES (DSES) approach.

DSES can take into account systems-based, intercultural and governance implications to support policy decisions within a framework of stewardship (Huber-Sannwald et al. 2012). Moreover, DSES offers a compelling approach to improve natural resources management through public participation, collaborative processes and transdisciplinarity in a framework that includes public participation (Virapongse et al. 2016; Bautista et al. 2017). The SES approach is a way to understand that the social and ecological dimensions of the groundwater management problem rest in combining analysis of data on the ecological function of the coupled human–water system with the perspectives of land and water users concerning the observed changes and future management options (King and Salem 2012). Social (e.g., gender, education, institutions) and local ecological (e.g., land-use and climate) characteristics play roles in influencing people's perceptions of which water services and coastal aquifers (Herrera-Franco et al. 2018) are important. In this regard, it is relevant to inquire the role played by the local communities' engagement and how the coupled human–water system works in an agricultural dryland.

We selected Baja California, a dryland ecosystem that is located in one of the most arid regions in Mexico (average rainfall of 168 mm a^{-1} and temperature of 20 °C) (CONAGUA 2018), where despite receiving water from the binational basin of the Colorado River, groundwater is the main water supply for agriculture in the coastal zone. Likewise, extreme drought events have occurred in 2016 that have limited social welfare and economic development of agricultural drylands. This is the case of Maneadero Valley, in the municipality of Ensenada, where little is known about the farmers' perception, the social, economic and political context, institutions and policies that govern groundwater use.

Based on the DSES approach, in this chapter we investigate how the human subsystem in the coupled human–water system works to reach integrated groundwater management in Maneadero Valley. We examine the perception of the main water users (farmers), regarding socioeconomic attributes, attitudes towards public participation and agricultural and water management practices. Then, we propose a participative strategy to improve the integrated groundwater management in this agricultural dryland. At the end, we draw some conclusions on challenges faced in DSES research.

The Socio-Ecological Systems Approach on Integrated Groundwater Management in Agricultural Drylands

Dryland Socio-Ecological Systems (DSES) Approach

Close to 50% of Mexico's land area is characterized by arid and semi-arid climate where land degradation and desertification are extensive and growing problems. Hence, a dryland stewardship strategy at local level in Mexico must recognize that drylands are SES where some of the largest and most far-reaching global environmental and social change problems occur, and thus are a daunting challenge for science and society (Huber-Sannwald et al. 2012).

The DSES approach conceptualizes the environment as an open system consisting of ecological and social processes and components. These processes are integrated through interactions (e.g., management practices, adaptation, resource use) that occur on multiple scales and through cycles, and are influenced by broad scale forces (political and economic conditions) and large-scale biogeochemical conditions. System components interact within a dynamic, web-like structure that facilitates interdependencies and feedbacks influenced by direct and indirect drivers at different temporal and spatial scales (Virapongse et al. 2016).

Disruptions to the hydrological cycles in drylands result in less and more erratic rainfall that will exacerbate the already critical state of water scarcity and conflicts over water allocation (Thomas 2008). Reduced land productivity, socioeconomic problems, uncertainty in food security, migration and limited development are some consequences of dryland degradation. When drylands are severely degraded, the inhabitants of dryland regions are forced to make various adaptations, especially water users (Ouessar et al. 2017).

The DSES implies to overcome a multidisciplinary view of the hydrological, ecological, meteorological or socioeconomic aspects of land degradation as isolated issues, to focus on their interactions and creating meaningful collaborations between academic and non-academic people. That is why assessment of how both water and human subsystems simultaneously affect, and are affected by land degradation has been recognized as one of the most important and challenging topics for drylands science (Huber-Sannwald et al. 2006; Holzer et al. 2018).

Integrated Groundwater Management in Dryland Socio-Ecological Systems

The science that links the human subsystem and the water subsystem is called socio-hydrology, which is aimed at understanding the dynamics and co-evolution of coupled human–water systems (Sivapalan et al. 2011). In the case of socio-ecological or biophysical-social conceptualization of groundwater systems (the coupled human–water system) in drylands and their management, the fact that environmental and socioeconomic processes are intrinsically inter-connected should be taken into account, even when hydrological, ecological, meteorological or socioeconomic aspects coexist in a contested area with multiple actors and interests, therefore in a transdisciplinary way.

Drylands science and governance involves complex interactions of multiple drivers, and affects a plurality of SES and functions, often in complex, non-linear ways, accompanied by multiple and often conflicting values, perspectives, attitudes and underlying sources of knowledge, where public participation and social learning are relevant (Bautista et al. 2017) (Whitfield and Reed 2012). This is important if countries are following the global model called integrated water resources management (IWRM) (GWP 2000), through which administrative decentralization, public participation and democracy in decision-making and water utilities privatization promise to mitigate the problems associated with groundwater scarcity and pollution.

Aligned with IWRM, the integrated groundwater management (IGM) seeks to incorporate the human and the water subsystems: issues of concern, management options and governance arrangements, stakeholders, natural and human settings, spatial and temporal scales, disciplines, methods, models, tools, data, sources and types of uncertainty (Foster and Ait-Kadi 2012; Jakeman et al. 2016). Groundwater management and governance, which comprises the enabling framework and guiding principles for collective management of groundwater for sustainability, equity and efficiency, can be framed from a DSES perspective. This approach is useful to acknowledge the complexity of the interactions between groundwater use and society (Rica et al. 2018).

A framework for analysing, comparing and diagnosing SES in agricultural drylands is proposed by McGinnis and Ostrom (2014). The main components of the SES analytical framework can be adapted for groundwater governance taking into account first and second-tier variables applied at local level (Table 5.1).

Groundwater Users' Participation in Dryland Development

Even when international agreements demand more participatory processes or recognize public participation as an indirect driver of change (Millennium Ecosystem Assessment 2005; UNCCD 2009; UN 2015), stakeholder participation in the evaluation of dryland restoration and sustainable dryland management is very scarce (Bautista et al. 2017). However, the existing participatory and "integrated local and scientific knowledge" methodologies have been conceived, tested and published in an attempt to offer solutions to the challenges of conducting dryland research from a socio-ecological perspective (Whitfield and Reed 2012).

First, different stakeholder perspectives must be identified and acknowledged before any management decisions are made. Consequently, building relationships with stakeholders requires an understanding of each other's culture (Virapongse et al. 2016), and it is a way to gain knowledge on how governance systems operate

Table 5.1 Components and variables of a SES in an agricultural dryland (modified from McGinnis and Ostrom (2014))

Component		Variables
Water subsystem	Resource systems and units	– Sector (agriculture, water) – Groundwater and ecosystem services – Clarity of the system boundaries – Size of the resource system – Type of human-constructed facilities – Productivity of aquifer – Predictability of aquifer dynamics – Storage characteristics – Recharge and discharge flow rates – Economic value – Spatial and temporal distribution
Human subsystem	Social, economic and political settings	– Economic development and economic sectors – Demographic dynamics or trends – Political stability – Governance and governmental frameworks – Land-use and agriculture trends and policies – Infrastructure and technological development – Market influence – Media interest on social or environmental issues
	Governance systems	– Governmental organizations – Non-governmental organizations (NGOs) – Actors' network structure – Information sharing – Property rights systems and bundle of rights – Rules: operational-choice, collective-choice, constitutional-choice, monitoring and sanctioning
	Actors	– Socioeconomic attributes – History or past experiences – Location – Leadership/entrepreneurship – Norms (trust-reciprocity)/social capital – Knowledge of SES/mental models – Importance of resource (dependence) – Technologies available

(Lopez-Porras et al. 2018) and how to enhance trust among different stakeholders (De Vente et al. 2016). Second, implementing a participatory process for assessing dryland agricultural sustainability requires inventorying farmers' (land or water users) perceptions. It has the potential to contribute to sustainable development by generating an understanding of agro-ecosystem dynamics, priority constraints and opportunities, and in developing interventions based on farmers' perceptions and knowledge, and formal science linkages (Onduru and Du Preez 2008).

Third, common aspects in researching the integrated resources management and governance in drylands, based on a SES approach and considering users' participation and perception, must include the analysis of stakeholders' involvement, legitimate representation, non-state actor influence, communication to non-state actors, information exchange through face-to-face contact, facilitated knowledge exchange, structured information aggregation, process moderation/facilitation, among others (De Vente et al. 2016).

Thus, the reviewed literature opens an opportunity to research groundwater users' participation and perception of drylands to improve the integrated groundwater management where conflicts among actors or exclusionary decision-making processes appear, and how the components of the coupled human–water system can enable or impede public participation in Maneadero Valley and facilitate a stewardship of drylands social-ecological systems.

The Maneadero Valley as an Agricultural Dryland

Maneadero Valley is located in the northern part of the Baja California Peninsula in Mexico. This valley has an agricultural surface of more than 3700 hectares, focused on the production of vegetables and flowers for exportation where more than 90% of the products are commercialized in the USA, primarily California. From the total sowed area, 67.4% are irrigated crops, while 32.6% are seasonal crops (SEDAGRO 2015).

Various factors have promoted the economic activity in Maneadero Valley. On the one hand, the class situation, the migrant character, the ethnic origin and the gender condition of a segment of the work force have promoted the exploitation of indigenous people. On the other hand, the cost of electric power for water exploitation is subsidized. Thus, the profitability of horticultural production has resulted in low investment by producers in housing and workers' services (Garduño et al. 2011) and low awareness of groundwater over-extraction. Water provides goods and services which are interrelated, determined by the quantity and quality of available water. While the Maneadero-Las Animas aquifer is the main water supply for the city of Ensenada (30% of its water supply), it is one of the eight overexploited coastal aquifers in Baja California and one of the five coastal aquifers affected by saltwater intrusion (CONAGUA 2018), with total dissolved solids (TDS) concentrations between 1080 and 26,950 g L^{-1} (Gilabert-Alarcón et al. 2018a, b), exceeding the maximum permissible limit of 1000 mg/L (Official Mexican Standard NOM-127-SSA).

The Maneadero-Las Animas aquifer belongs to the number one hydrologic region in north-western Baja California. The aquifer is located 15 km south of the city of Ensenada ($31°$ $41'$ N and $116°$ $30'$ W). It is a hydrologic sub-basin covering 1795 km^2, of which 75 km^2 belongs to a coastal plain. In addition to rainfall, the Maneadero coastal aquifer receives water from the runoff of the basin, mainly from the San Carlos and Las Animas creeks (Fig. 5.1).

The Maneadero aquifer has an average annual recharge of 33.7 Mm^3, volume allocated of 39.09 Mm^3 and a deficit of 5.37 Mm^3 (DOF 2018). For 2015, a designated extraction (allowed) of 38.37 Mm^3 and a deficit of 17.57 Mm^3 were registered (DOF 2015). Currently, there is no availability of new wells or volume increases, because of the legal restriction on May 15th, 1965 (DOF 1965).

The consumption and water demand of the Maneadero Valley has increased every year due to the population growth and the economic activities that take place

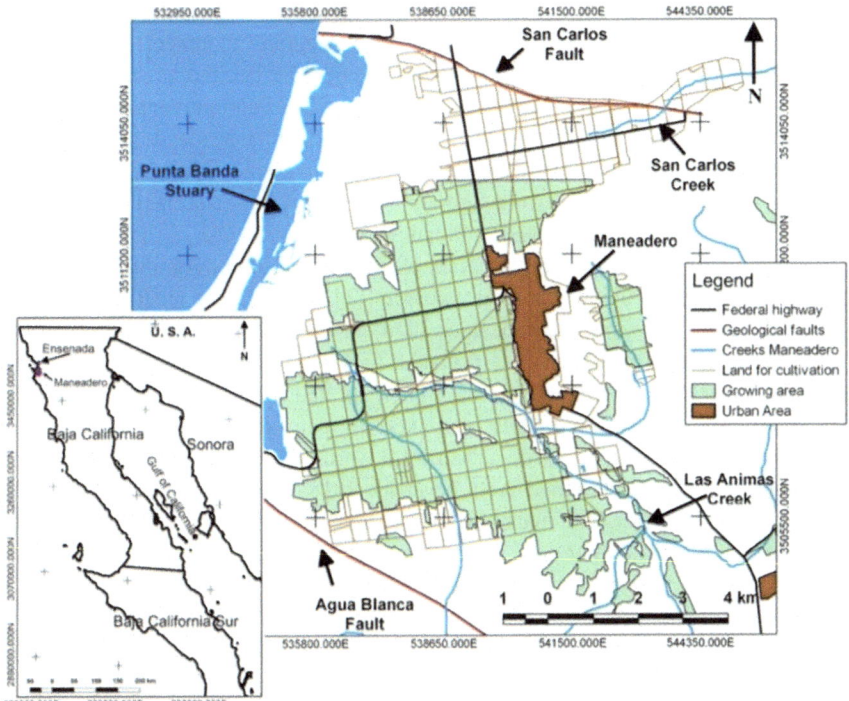

Fig. 5.1 Localization of the Maneadero Valley, Baja California, Mexico

inside and outside the valley. Currently, there are more than 300 water users regis-tered in the Technical Groundwater Committee of Maneadero aquifer (COTAS, in its Spanish acronym) from different water uses: agricultural, multiple, public-urban, services, domestic, livestock and industrial. The largest water consumptive group in Maneadero Valley is agriculture (commercial export agriculture, greenhouse opera-tions and ranchos), which utilizes 71% of the groundwater resources, through a water-well concession. The multiple consumptive uses represent mostly small-scale agriculture, livestock and domestic uses and are a large proportion (33%) of the total concession volume. The water allocation of the water utility company in the municipality of Ensenada (CESPE, in its Spanish acronym) is the largest water con-sumer in the public-urban use, which constitutes 17% of the category's total (Medellin-Azuara et al. 2013).

The Public Registry of Water Deeds database (REPDA) from the National Water Commission (CONAGUA, in its Spanish acronym) exhibits 360 wells located in the geographical limits of the Maneadero aquifer: 180 for agricultural use, 114 for mul-tiple use, 20 for public-urban use, 18 for use in services, 17 for domestic use, 7 for livestock use and 4 for industrial use. Nevertheless, the groundwater assigned by concession or allocation of rights according to the REPDA is not what is really being extracted.

In the Maneadero Valley, a wastewater treatment plant (called Maneadero) was built to avoid more groundwater pollution. Also, programs to cancel some groundwater wells were implemented, and piezometric nets are being operated to monitor the aquifer depletion. To reduce stress on water supplies and prevent marine intrusion, in 2003 the Integrated Water Management Plan for the Maneadero aquifer was drawn up by CONAGUA, which is the Mexican government agency in charge of national water management, through application of the National Water Law. Although the plan was never made official, it suggested the use of reclaimed water (RW) and recommended the construction of infrastructure for agricultural irrigation and artificial (not incidental) aquifer recharge (Gilabert-Alarcón et al. 2018a).

Unlike other valleys in Mexico, since 2014 in Maneadero water users are already using RW coming from the city of Ensenada. This RW for agricultural irrigation and incidental discharges on a riverbed have been applied in Maneadero Valley, resulting in a mix of dissolved salts attributed to natural geochemical processes, anthropogenic-derived processes (urban solid waste, wastewater and animal waste) and marine intrusion (Gilabert-Alarcón et al. 2018a) that have also affected the recharge of the aquifer. The main wastewater treatment plant in Ensenada is called "El Naranjo" and has a treatment capacity of 500 L s^{-1}. El Naranjo is located approximately 13 km north of the Maneadero aquifer. It was in June 29th of 2014 when the implementation of the pilot project ($6 million U.S. dollars) for the reuse of RW started with 152 ha of irrigated lands with 200 L s^{-1}. Then, 80 L s^{-1} were used to reactivate agriculture production of 200 ha in areas that were mostly abandoned and 120 L s^{-1} directly to the Las Animas creek, as an incidental aquifer recharge (Mendoza-Espinosa and Daesslé 2018).

In summary, the water subsystem has changed historically due to some factors in the human subsystem, disturbing the quality and the quantity of the Maneadero aquifer and conditioning the groundwater supply. Then, the coupled human–water system or DSES in Maneadero Valley considers the following components and variables (Table 5.2).

Table 5.2 Components and variables of a SES in Maneadero Valley

Component	Variables

Water subsystem (biophysical environment) influenced by human subsystem	• Climate: extreme temperature variations and low precipitation (rainfall) rate
	• Groundwater depletion levels: caused by sustained groundwater pumping (over-extraction)
	• Natural biogeochemical processes: sorption, ion exchange, filtration, precipitation and biodegradation that cause soil desalinization
	• Groundwater pollution
	  – Leachate infiltration from agricultural (fertilizers, herbicides, pesticides) and domestic activities (wastewater from septic tanks)
	  – Infiltration of RW from agricultural irrigation
	  – Infiltration of RW from incidental discharges to creeks
	  – Marine intrusion, which caused the increase of the TDS that exceed the maximum permissible limit of 1 g L^{-1}

(continued)

Table 5.2   (continued)

Component	Variables

Human (socio-economic, governance) subsystem influenced by the water subsystem	• Population growth and immigration in Ensenada: demand of water, sanitation and public health services
	• Economic activities and land-use changes: agricultural, public-urban, services, domestic, livestock and industrial uses
	• Markets and prices
	• Competing users
	• Ineffective water policies: regulations of water quantity and quality without follow-up, excessive number of concessions, absence of an integrated water management plan
	• Inadequate institutional arrangements
	• Outdated legal framework to reduce groundwater variations
	• Alternative water sources: RW reuse in foraging and flower crops, groundwater recharge through incidental discharges on riverbeds and private use of desalination plants
	• No capacity of water users to influence decision-making, even though COTAS
	• Variety of social actors
	  – COTAS
	  – State agencies (CONAGUA, SEDAGRO, CESPE)
	  – Local level administrative staff
	  – Water users (farmers), landowners, producers
	  – Academics, NGOs, farmworkers

Groundwater Users' Participation and Perception in Maneadero Valley

In this section we analyse the results of a survey carried out between August 14th and November 9th on 2017 with a group of farmers (groundwater users with a concession) in Maneadero Valley ($n = 52$). The objective of the survey was to document aspects regarding socioeconomic attributes, attitudes towards public participation and agricultural and water management practices, and then, to propose recommendations to improve groundwater management and governance.

The instrument consisted of 27 questions divided into three sections of variables: (1) sociodemographic, (2) public participation in water management and governance and (3) agricultural practices and implemented solutions to groundwater scarcity and pollution. The questionnaire was designed as a result of prior research, interviews with experts and communication with the technical manager of COTAS Maneadero. To carry out the surveys we use the office of the COTAS. The surveys were conducted in person and individually with farmers.

Previously, we have been acting as active observers in water meetings and forums in Baja California where Maneadero Valley is mentioned, in order to gather information on the context, stakeholders and how the decisions on quantity and quality of water supply are taken. We also have done field work that provides a general overview of the study area, especially having close meetings with people related to the environmental and socioeconomic issues in Maneadero Valley. Even when there are more than 300 water users registered in the COTAS (according to the REPDA database), just 52 of them were available to participate in the survey. This was due to the fact that: (1) most of them only rent the land, (2) they no longer live in Maneadero or (3) they are elderly people whose age and medical condition did not allow them to participate in the survey.

Consistency of data input was ensured by creating a formatted Excel spreadsheet. Quality control for data input was ensured through checks on data input from the original paper questionnaires into the spreadsheet and through direct clarification of specific translated answers to avoid misinterpretation. We gave careful consideration to data handling processes from data input to translation and interpretation (coding, classification and statistics with Stata software).

Sociodemographic Profile of Surveyed Farmers

Seventy-three percent of the sample is formed by men and the other 27% are women. The age range with greater frequency was 50–69 years, with an average age of 62.7 years. The level of studies of 44.2% of the respondents corresponds to the elementary school level, 23.1% with high school level, 19.2% undergraduate level and 7.7% have equivalent studies to PhD and masters. The rest (5.7%) have no school education.

The main economic activity of the respondents is agriculture (80.8%). The rest of the respondents are renting their land (9.6%), and in some cases, they also have another employment (7.7%) or are retired (1.9%). The monthly economic income with more frequency was from 2000 to 10,000 pesos (48.1%), and the next one was 11,000–20,000 pesos (26.9%). Almost 80% of the participants live in the same land they own and the rest have an independent house.

Farmers Participation in Water Management and Governance

We explored the rationales for farmers' participation: the perceptions of the current groundwater management, the problem and the proposed solutions; water management responsibilities and the usefulness of the participatory spaces provided by CONAGUA, through COTAS meetings. In doing so, we found that 90.4% of the respondents assist to the COTAS meetings, because they are members and to stay informed on water management in Maneadero Valley. Farmers attend meetings every third Sunday of the month organized by COTAS, but monitored by CONAGUA, who is the authority that makes the final decisions.

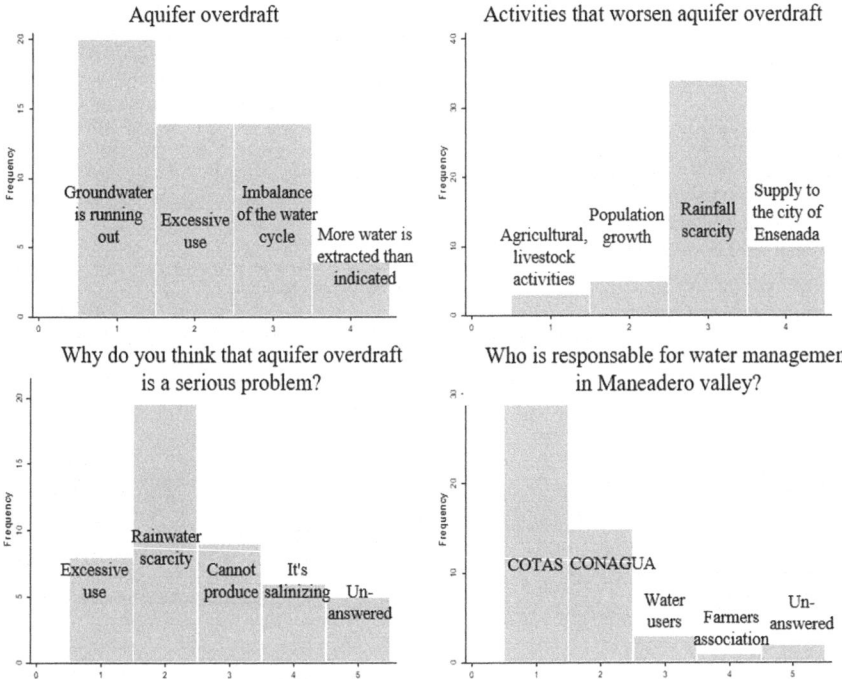

Fig. 5.2 Farmers' perception on the Maneadero aquifer condition

Farmers' perception of the aquifer condition is mainly attributed to rainfall scarcity, excessive water supply to Ensenada city and population growth, causing conflicts, especially in their productive activities. Although it is perceived as a problem that affects everyone, when they are questioned about who is responsible for water management, most of them respond that it is a task of official authorities and few respond that should be in charge of all water users (Fig. 5.2).

Regarding farmer's participation on water management, as a factor to contribute in the water governance, the existence and effectiveness of dialogue spaces with training and access to information on the use of RW were identified with a statistically significant association. Even when the interviewees perceive that scientific information is freely discussed, only 65% of them consider that their opinion on water issues is taken into account by authorities and scientists. That affects the inclusion of local knowledge, the trust among participants and usefulness of COTAS meetings as a learning and participatory space. The main obstacles identified were attributed to the lack of communication and information, in addition to the disinterest of the authorities, considering COTAS as informative spaces and without the possibility of intervening in decision-making (Table 5.3).

Table 5.3 Main results on the questions about attitudes towards public participation

Participatory spaces	Attendance (0.775)		Your opinion is taking into account (0.564)		The scientific data is freely discussed (0.218)		Training and access to information on RW[a] (0.047)	
	Yes	No	Yes	No	Yes	No	Yes	No
The spaces exist and are effective (%)	88	12	63	37	88	12	83	17
It does not exist/they are not effective (%)	91	9	73	27	73	27	55	45

Statistics of probability in parentheses. Chi-square test for independence
[a]Significance level = 0.05

Table 5.4 Users' knowledge on desalination plant operation and impacts

Know	Advantages and risk						
	(1)	(2)	(3)	(4)	(5)	(6)	Total
Yes	1	5	8	3	0	3	20
No	0	1	15	4	4	8	32
Total	1	6	23	7	4	11	52

(1) = Reactivate agriculture, (2) = High productivity, (3) = Reduce groundwater salinity, (4) = Short-term solution, (5) = Negative environmental impact, high costs, (6) = Unanswered

Implemented Technical Solutions to Groundwater Scarcity and Pollution

Two strategies were explored through the survey: the use of private desalination plants and the use of RW from treated wastewater plants. Private desalination plants have been perceived as a good alternative to maintain agricultural production by 90% of respondents. Some of farmers are already using desalination plants, but they do not have the appropriate permits for their use. Table 5.4 shows the relationship between knowledge of the operation of a desalination plant and the advantages and risks perceived.

Most respondents consider that desalination plants are useful because of local problematic conditions, and those who have less knowledge of its operation have more uncertainty about the environmental impact. The concern for the environmental impact is due to the lack of knowledge of where the waste is disposed and the possibility of affecting the subsoil and the aquifer again. There are other solutions to reduce the aquifer overdraft, as considered by farmers, and the main one is the use of RW (31%). They perceive advantages on this technological solution, but they also have some concerns about it. Figures 5.3 and 5.4 show how advantages and risk are distributed according to the attendance to COTAS meetings.

Attending to meetings has played an important role in informing the advantages, risks and knowledge of treated wastewater; among the advantages are reactivating farmland and encouraging the cultivation of flowers and forage. Some of the main

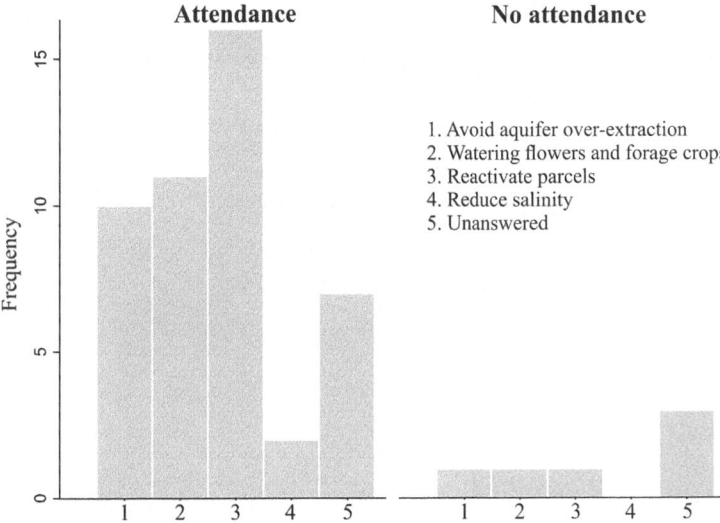

Fig. 5.3 Farmers' perception on advantages of reclaimed water (RW) reuse

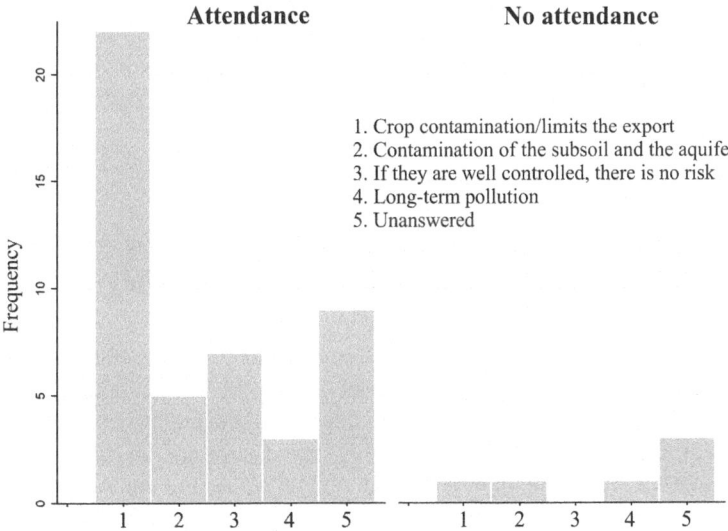

Fig. 5.4 Farmers' risk perception on reclaimed water (RW) reuse

concerns are of economic nature and public health. Finally, on the alternative use of RW for the crops irrigation or the incidental recharge of the aquifer, 98% of the farmers perceive that it has been beneficial. The only constant observation among the respondents is that the RW discharged into the Las Animas creek has produced mosquito and fly pests. Figure 5.5 shows the reasons associated with these opinions.

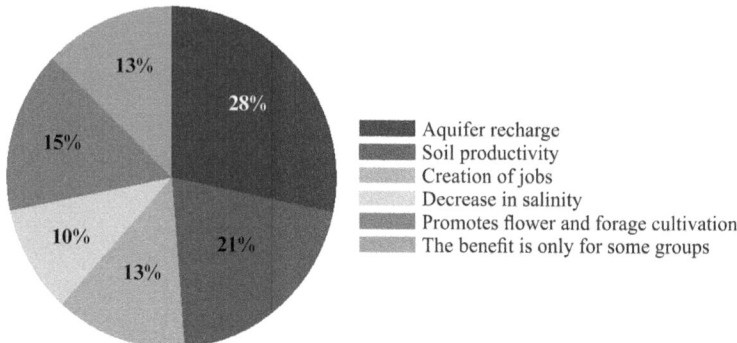

Fig. 5.5 Perceived benefits of the use of reclaimed water (RW) strategy

Most of the farmers perceive the recharge of the aquifer as the main benefit, followed by soil productivity. However, some of the farmers manifest discontent because the benefits have been targeted only to certain powerful groups. In addition to seeking that participatory spaces become accessible to different groups, it is necessary to show how the benefits of these strategies are extended in groups of the population with different sociodemographic characteristics and knowledge levels. Therefore, it is important to promote and empower the groups which least perceive these benefits to increase their participation and influence in the participation spaces through COTAS meetings.

Discussions: The Proposal of a Strategy to Improve the Groundwater Management in Maneadero Valley

The current state of the Maneadero aquifer conditions the subsistence system and the sustainability of the communities located in the Maneadero valley, especially the farmers. Their participation and leadership is fundamental to regulate water use, monitor groundwater depletion and quality, make informed and conscious decisions, as well as co-create knowledge with official authorities and scientific knowledge brokers to improve water management and governance. But first, it was imperative to understand the current water management and governance, especially the participation and perceptions of a group whose well-being is linked to the water subsystem.

De Vente et al. (2016) find that in participatory processes on drylands social-ecological systems certain design principles must be fulfilled. We use their seven recommendations to apply in the empirical case of Maneadero Valley and to improve water management and governance (Table 5.5).

Table 5.5 Strategy to improve the groundwater management in Maneadero Valley (modified from De Vente et al. 2016)

Recommendations	Description
1. Select participants carefully	In addition to official authorities, water users or farmers, include not just natural but social scientists from local universities and research institutions that can provide data and information to improve COTAS meetings, at least as guests
2. Make participation attractive and easy	Technical manager of COTAS Maneadero, farmers' leaders and social scientists can identify the problems of and with water users, establish new communication mechanisms and relationships between actors and consider emerging solutions to implement an integrated groundwater and RW management
3. Foster trust	Take advantage of existing networks and technologies for communication between actors, as in the case of COTAS meetings. Respect the knowledge of all participants by critically evaluating both scientific and local knowledge
4. Provide participants with information and decision-making power	Through COTAS meetings, offer participants the real power of decision-making that affects them and provide high quality, accessible and impartial information inviting research institutions to communicate their academic results in a simple way
5. Use professional independent facilitation and structured methods of information aggregation	Use a competent independent facilitator who can manage power imbalances, stimulate the active participation of all actors and ensure fair participation and deliberation with equal opportunities
6. Promote long-term commitment	Create a long-term commitment for all participants and realistic economic support for the implementation of solutions not only with CONAGUA support, but also farmers' associations
7. Adapt language	Use accessible language and informative resources adapted to the educational level of the participants

Conclusions

In this chapter, we have established the basic conceptual framework of the drylands socio-ecological systems through the elements of a coupled human–water system, in order to reach an integrated groundwater management in an agricultural dryland in Baja California based not only on sound science, but also on public perception and participation of water users. In this regard, we consider that integrated groundwater management in drylands should engage the wider community and relevant stakeholders in decision-making. We recognize that it is important to determine how public participation occurs in decisions regarding groundwater management and what perception water users have of the water issues and technical solutions implemented to reduce the impact of aquifer overexploitation and increase agricultural irrigation through the use of reclaimed water and desalinization plants. In doing so, we can identify the key aspects influencing public participation, agricultural practices and implemented solutions to groundwater scarcity and pollution.

Acknowledgements We thank the support from CONACYT BMBF, UNAM-UABC 155 Evaluation of the bio-economic risks due to aquifer over-exploitation in arid and coastal regions, urban and agriculture lands.

References

Bautista S, Llovet J, Ocampo Melgar A, Vilagrosa A, Mayor ÁG, Murias C, Vallejo VR, Orr BJ (2017) Integrating knowledge exchange and the assessment of dryland management alternatives – A learning-centred participatory approach. J Environ Manage 195(Part I):35–45

Berkes F, Folke C (1998) Linking social and ecological systems: management practices and social mechanisms for building resilience. Cambridge University Press, New York

CONAGUA (2018) Estadísticas del agua en Mexico. Secretaría de Medio Ambiente y Recursos Naturales, Comisión Nacional del Agua, Mexico

De Vente J, Reed MS, Stringer LS, Valente S, Newig J (2016) How does the context and design of participatory decision making processes affect their outcomes? Evidence from sustainable land management in global drylands. Ecol Soc 21(2):24

DOF (1965) DECRETO por el que se establece veda para el alumbramiento de aguas del subsuelo en el Estado de Baja California, Mexico

DOF (2015) ACUERDO por el que se actualiza la disponibilidad media anual de agua subterránea de los 653 acuíferos de los Estados Unidos Mexicanos, mismos que forman parte de las regiones hidrológico-administrativas que se indican, Mexico

DOF (2018) ACUERDO por el que se actualiza la disponibilidad media anual de agua subterránea de los 653 acuíferos de los Estados Unidos Mexicanos, mismos que forman parte de las Regiones Hidrológico-Administrativas que se indican, Mexico

Foster S, Ait-Kadi M (2012) Integrated Water Resources Management (IWRM): how does groundwater fit in? Hydrogeol J 20(3):415–418

Garduño E, Navarro A, Ovalle P, Mata C (2011) Caracterización socioeconómica y cultural de las mujeres indígenas migrantes en los valles de Maneadero y San Quintín, Baja California, Mexico. Boletín de Antropología 25:57–83

Gilabert-Alarcón C, Daessléa LW, Salgado-Méndeza SO, Pérez-Flores MA, Knöller K, Kretzschmar TG, Stumpp C (2018a) Effects of reclaimed water discharge in the Maneadero coastal aquifer, Baja California, Mexico. Appl Geochem 92:121–139

Gilabert-Alarcón C, Daesslé LW, Mendoza-Espinosa LG, Villada-Canela M (2018b) Regulatory challenges for the use of reclaimed water in Mexico: a case study in Baja California. Water 10:1432–1454

GWP (2000) Integrated water resources management. Tack background papers no. 4. Global Water Partnership, Stockholm, Sweden

Herrera-Franco G, Alvarado-Macancela N, Gavín-Quinchuela T, Carrión-Mero P (2018) Participatory socio-ecological system: Manglaralto-Santa Elena, Ecuador. Geol Ecol Landsc 2(4):303–310

Holzer J, Carmon N, Orenstein D (2018) A methodology for evaluating transdisciplinary research on coupled socio-ecological systems. Ecol Indic 85:808–819

Huber-Sannwald E, Maestre F, Herrick J, Reynolds J (2006) Ecohydrological feedbacks and linkages associated with land degradation: a case study from Mexico. Hydrol Process 20:3395–3411

Huber-Sannwald E, Ribeiro Palacios M, Arredondo JT, Braasch M, Martínez Peña M, García de Alba J, Monzalvo K (2012) Navigating challenges and opportunities of land degradation and sustainable livelihood development. Philos Trans R Soc B 367:3158–3177

Jakeman A, Barreteau O, Hunt RJ, Rinaudo JD, Ross A (2016) Integrated groundwater management: an overview of concepts. In: Integrated groundwater management: concepts, approaches and challenges. Springer, Cham, pp 3–20

King C, Salem B (2012) A socio-ecological investigation of options to manage groundwater degradation in the Western Desert, Egypt. Ambio 41(5):490–503

Lopez-Porras G, Stringer L, Quinn C (2018) Unravelling stakeholder perceptions to enable adaptive water governance in dryland systems. Water Resour Manag 32:3285–3301

McGinnis M, Ostrom E (2014) Social-ecological system framework: initial changes and continuing challenges. Ecol Soc 19(2):30

Medellin-Azuara J, Mendoza-Espinosa L, Pells C, Lund J (2013) Pre-feasibility assessment of a water fund for the Ensenada region: infrastructure and stakeholder analyses. The Nature Conservancy. Center for Watershed Sciences, UC Davis, Davis, CA

Mendoza-Espinosa L, Daesslé L (2018) Consolidating the use of reclaimed water for irrigation and infiltration in a semi-arid agricultural valley in Mexico: water management experiences and results. J Water Sanit Hyg Dev 8:679–687

Millennium Ecosystem Assessment (2005) Ecosystems and human well-being: synthesis. Island Press, Washington, DC

Onduru DD, Du Preez C (2008) Farmers' knowledge and perceptions in assessing tropical dryland agricultural sustainability: experiences from Mbeere District, Eastern Kenya. Int J Sust Dev World 52(2):145–152

Ostrom E (2009) A general framework for analyzing sustainability of social-ecological systems. Science 24:419–422

Ouessar M, Gabriels D, Tsunekawa A, Evett S et al (2017) Water and land security in drylands: response to climate change. Springer, Cham

Rica M, Petit O, López-Gunn E (2018) Understanding groundwater governance through a social ecological system framework – relevance and limits. In: Advances in groundwater governance. CRC Press, Leiden, pp 55–72

SEDAGRO (2015) Panorama General de Maneadero, Baja California, Baja California. Secretaría de Desarrollo Agropecuario, Mexico

Sivapalan M, Hubert H, Blöschl G (2011) Socio-hydrology: a new science of people and water. Hydrol Process 26:1270–1276

Thomas R (2008) Opportunities to reduce the vulnerability of dryland farmers in Central and West Asia and North Africa to climate change. Agric Ecosyst Environ 126(1-2):36–45

UN (2015) Transforming our world: the 2030 agenda for sustainable development, A/RES/70/L.1. Resolution adopted by the General Assembly. United Nations, New York

UNCCD (2009) UNCCD 1st scientific conference: synthesis and recommendations. ICCD/COP(9)/CST/INF.3. UNCCD, Buenos Aires

Virapongse A et al (2016) A social-ecological systems approach for environmental management. J Environ Manage 178:83–91

Whitfield S, Reed M (2012) Participatory environmental assessment in drylands: introducing a new approach. J Arid Environ 77:1–10

Chapter 6
Forced Modernization in Drylands: Socio-Ecological System Disruption in the Altiplano of San Luis Potosí, Mexico

L. Ortega and J. Morán

Abstract The United Nations Convention to Combat Desertification (UNCCD) recognizes desertification as a serious global problem in arid and semi-arid zones. It is necessary to conduct research on, and inform stakeholders of, the processes related to the degradation of these drylands before it is too late. The purpose of this chapter is to describe some threats caused by disruptive socio-ecological system and landscape transformations in the drylands of San Luis Potosí (SLP), a state in north-central Mexico. This is due to a process of forced modernization, which crystallizes from a vision aimed at increasing productivity through the incorporation of disruptive technologies and overproduction of renewable energies. The problem of this modernization model is that it disregards the integrity and functions of the socio-ecological system itself. In the case of the central plateau of SLP, known as the "altiplano," the state government and transnational entrepreneurs appear to view the issue of modernization through the logic of bringing progress and development to a space that they view as otherwise empty, unoccupied, and with low economic value. Therefore, implementation of projects is promoted, aiming to turn these "empty spaces" into productive areas, but also culturally and ecologically decontextualizing the traditional livelihoods of these drylands.

Keywords Drylands · Socio-ecological system · Disruptive technologies · Renewable energies

Introduction

According to the United Nations Convention to Combat Desertification (UNCCD) desertification is the degradation of land in arid, semi-arid, and sub-humid areas, known as drylands, as a result of complex interactions between physical, biological, political, social, cultural, and economic factors. The impact these areas undergo is

L. Ortega (✉) · J. Morán
El Colegio de San Luis, San Luis Potosí, Mexico
e-mail: laura.ortega@colsan.edu.mx

© Springer Nature Switzerland AG 2020
S. Lucatello et al. (eds.), *Stewardship of Future Drylands and Climate Change in the Global South*, Springer Climate,
https://doi.org/10.1007/978-3-030-22464-6_6

concerning because they are the habitat and the source of sustenance for a large part of the world's population (UNCCD n.d.).

Drylands occupy about 60% of Mexico's territory (Verbist et al. 2010; CONAFOR-UAch 2013). A large portion of it is known as the "Altiplano Mexicano," which covers approximately a quarter of the country's expanse, consisting of an uninterrupted sequence of highlands, bounded by the Sierra Madre Occidental, the Sierra Madre Oriental, and the Trans-Mexican Volcanic Belt. Throughout most of this dryland, rainfall is in the order of 200–500 mm/year (Rzedowski 2006). Within the "Altiplano," from the state of San Luis Potosí (SLP) to a strip in southern USA, is the ecoregion known as the *Chihuahuan Desert*, which is considered the largest in North America and has the second greatest biodiversity worldwide (WWF 2018). Particularly, the area occupied by SLP stands out because it concentrates the greatest diversity of cacti in the ecoregion (Hernández 2006).

The "altiplano" in San Luis Potosí extends to the northwest of the state and includes only a portion of the drylands, since its delimitation owes to a political-administrative division. It covers almost half of the state (48.1%) and includes 15 municipalities, which have the lowest population density in the state (11 inhabitants per km^2) and is referred to as a low-level development region (SDSR 2016). It has two natural protected areas: (1) *Real de Guadalcázar* and (2) *Sitio Sagrado Natural de Wirikuta,* promoted as a World Heritage Site by the UNESCO because of its biological importance and spiritual significance to indigenous people (Giménez et al. 2018).

The aim of this chapter is to analyze a process of forced modernization in San Luis Potosí's "altiplano" through observed consequences in socio-ecological system and landscape transformations. The implementation of disruptive technologies, both in the case of high-productivity horticulture and renewable energies production, constituted "pressure points" that favor cultural reorientations in the traditional forms of space appropriation. Conventional relationships between rationality of modernity and other cultural identities are questioned from the approaches of interculturality (Alsina 2004).

Relationship Between Drylands as Socio-Ecological Systems (SES) and Landscape

Drylands are analyzed as socio-ecological systems (SES) due to the complex network of interactions that sustain (a) the biotic communities and (b) the ecosystem services. Characteristically, they present non-linear dynamics and maintain a high level of response to the interactions between different scales, as well as self-organization capabilities and emergent properties (Huber-Sannwald et al. 2012). In accordance, the analytical framework called *Drylands Development Paradigm* recognizes that ecological and social issues, along with options for livelihood support and ecological management, are interwoven (Reynolds et al. 2007). Therefore, it proposes the simultaneous exploration of biophysical and socio-economic dimensions on different spatial and temporal scales through a variety of methods including recovery of local knowledge (Reed et al. 2011).

The proposal to contribute to the analysis of drylands as a socio-ecological system in the altiplano potosino, complemented by the approximation of the theoretical framework of the landscape, sets forth not only the conjunction of the natural, social, and cultural components, but the search for interdisciplinary dialogues that point to a holistic understanding of complex processes. In this way, the interest in combining the socio-ecological system and the landscape is to establish a connection that is largely unaddressed, if not nonexistent, in which the landscape becomes a communication bridge between socio-ecological system and fields of knowledge such as history, anthropology, geography, ecology, architecture. These and other disciplines that focus on the study of space, culture, forms of appropriation, exploitation, and use that social groups make of resources, in a dialectical connection that conditions certain dynamics, explain the characteristics of an environment and give meaning to an area, zone, or region that differentiates it from others.

Thus, the existence of certain biotic and abiotic elements, their availability, their function, appropriation, and meanings construct a landscape and socio-ecological system that require different tools for their understanding and study. The link between the socio-ecological system and landscape is therefore possible because environmental, geological, ecological, social, economic, and cultural factors are visualized within both of them, all of which account for the sociohistorical construction it experiences, through a process of transformation of space, nature, and society. In addition, the analysis scale can acquire diverse dimensions, from different attributes or elements that support a geophysical, morphological, biotic, historical, symbolic composition with the development of certain collective practices (from their specific linguistic, ethical, and esthetic appropriation); forms of interaction between social groups with their environment, with each other and in the way they appropriate space; as well as through the material modification they exercise over the natural environment (Vázquez and León 2015; Ruiz and Roque 2017; Martínez and Soto n.d.).

Given the above, in this chapter, we analyze the forced transformation processes that the "altiplano" in San Luis Potosí has undergone. The altiplano dryland is a substantially valuable region which has been coveted and historically reconfigured over the course of many years, starting from the confluence of assorted extractive, developmentalist, and modernization processes.

Forced Modernization

There is a paradox in productive transformation processes, according to Tudela (1989). One the one hand, there is a multiplication in investment, a wide use of technological resources, increased productivity, and, in general, increased creation of wealth. But, at the same time, they bring about processes of social and ecological decline, generating setbacks in the quality of life.

In the case of San Luis Potosí, the state government detected in 2015 that the "altiplano" only contributed to 6.4% of the state's gross domestic product, since most of the population engages in temporary agriculture focused on subsistence farming (SDSR 2016). Under this logic, agroindustrial development was guided by adopting technologies, genetic improvement, and an increase in the installed capacity of crop production units, under a system of contract farming (GESLP 2018a). Thus, the reproduction of a trend towards standardizing and transnationalizing technological policies is observed. In this process, the government becomes an advocate for the adoption of these technological packages, ignoring the socio-environmental circumstances that prevail in the regions (Tudela 1989).

According to official data, between 2015 and 2018, protected agriculture enabled the creation of 2238 jobs and allowed the installation of greenhouses for green chile and tomato production. The state became the second largest producer of the latter nationwide in 2018 (GESLP 2018a). These figures seem to justify the promotion of a new form of production, which is compared with the apparent inefficiency of traditional forms of production. This new model promotes employment, training, the reduction of uncertainty for producers, and increased income across the population (GESLP 2018b). However, these events do not guarantee a better quality of life and restrict the type of dryland adapted production, mainly benefiting the market economy through outsourcing.[1]

Likewise, the adoption of renewable energies was prompted by the altiplano's natural features (GESLP 2018a). San Luis Potosí is located in the region called "the solar belt," where capturing 5% of the irradiation received in the state would sustain energy consumption for the entire country (GESLP 2018b). Theoretically, this climate resource would contribute to regional socio-economic development because it would help to achieve the goal of generating 35% of the country's energy through renewable sources by the year 2024.

In this context, the installation of a solar energy plant in the municipality of Villa de Arriaga was projected at the end of 2018, and another solar farm would be added in Villa de Ramos in 2019. In addition, two wind farms in Charcas and another in Santo Domingo municipalities have been built. According to the technical study, the installation of the electric transmission line for the Dominica project alone, in Charcas, would entail a loss of vegetation coverage of 65,149 ha, of which 10,024 ha would experience a permanent loss. In addition, the removal of 134.3 m^3 of soil and 3072 m^3 of earth movement is calculated (MIA n.d.).

---

[1] According to the third state government report, for 2018, the production of barley, soybean, and sunflower crops under the contract farming system has been promoted by guaranteeing the sale of the harvest at a set price and by backing it up with different government support such as technical assistance, technology packages, and production incentives. In the spring–summer 2017 cycle, 12,489 ha of barley was harvested in Villa de Arriaga, with a value of 47.9 million pesos for the benefit of 1336 suppliers to Grupo Modelo México. Under the same model, producers in the municipalities of Cerritos and Guadalcázar cultivated 786 ha of sunflower under contract with the company AAK México S.A. of C.V. at a price of 7100 pesos per ton, for a total of 4.8 million pesos (GESLP 2018a).

Although San Luis Potosí has registered an increase in investment in technified agriculture, renewable energy, volume of production, and wealth generation, this form of production and appropriation of resources deteriorates the socio-ecological system and landscape, and therefore modifies the livelihoods that are consubstantial to them. For this reason, according to Tudela (1989), it is fundamental to gain knowledge of each one of the mechanisms of transformation and link them to the processes of deterioration, as well to analyze the specific modalities of economic activities that trigger their action, to critically analyze and rethink if there are actually development processes.

Livelihoods in the Altiplano Potosino

The socio-ecological system of this region was shaped by the occupation and forms of natural resource appropriation carried out by hunter and gatherer groups during the Late Postclassic period. These groups took advantage of the distribution and phenological patterns of the wide diversity of species that have evolved in conditions of chronic droughts, infrequent rainfall, and even excessive thermal variation on the surface of the soil, affecting the subsistence of plants and animals (Hernández 2006; Rodriguez-Loubet 2016).

From the sixteenth century, the productive activities of these places were suddenly reoriented, historically forming a landscape, and thus a socio-ecological system, that has resulted from juxtapositions and confluences between new productive activities and diverse cultural interactions. The influence of the new modes of appropriation of nature side-by-side with traditional ones accounts for the processes of socio-ecological system transformation:

(a) Mining. The region is located in a metallogenic province abundant in silver, lead, and gold, in addition to zinc, mercury, fluorite, and arsenic (SGM 2018). The foundational origin of the settlements during the conquest was mining extraction, which began in 1574 (SGM 2017). Later, new mining districts were incorporated, some of which have mines in operation to this day. At the end of the nineteenth century, a transition from artisanal mining to larger-scale mining began, using dynamite for the first time, mechanical perforations with air, and the movement of large volumes with winches, associated with the first power plant (SGM 2017). The endemic presence of metallic and non-metallic minerals and their accumulation in soil and water because of extractive activities have created risk scenarios for human health and for other exposed organisms (Espinosa-Reyes et al. 2014; Jasso-Pineda et al. 2007; Monzalvo-Santos et al. 2016; Razo et al. 2004), whose magnitude of effects has not yet been sufficiently measured.

(b) Salt production. The "harvest of salt" in the saline lagoons initially depended on knowledge of seasonal cycles and weather conditions. Afterwards, ponds were built, in which the salt water extracted from the subsoil was deposited, accelerating the volume of extraction (Vázquez 2014). The main purpose of this salt was use in silver amalgamation, followed by animal feed. Salt production complexes were preserved as a historical trace of the transformation of the landscape, warehouses, salt factories, "norias," and watercourses, as well as communication routes (Vázquez 2014). Some operation sites are still active, though on a smaller scale.

(c) Livestock-agriculture-hunting-collection. The introduction of small cattle in the "altiplano" was made possible by the construction of traditional hydraulic engineering techniques of temporary water reservoirs (Ruiz 2014). A cyclical production system called "trashumancia" was implemented, consisting in an extensive grazing system based on the mobility of herds between summer pastures, in lower altitude areas where there is also farming[2] and winter pastures in areas of higher altitude (Mora 2012). In this way, soil compaction degree is reduced, increasing its fertilization and contributing to biodiversity conservation and fire prevention (Mora 2011; UPA 2009). This activity is sustained through the "peasant economy" based on family participation (Barrera et al. 2018), creating income through the commercialization of meat, milk, cheese, and offering products for self-consumption, valuable in terms of food quality (Grünwaldt et al. 2016). It also articulates rainfed agriculture, hunting, and harvesting (Maisterrena and Mora 2000). In some arid zones, these systems are threatened by overgrazing (Guzmán 2004; Negrete-Sánchez et al. 2016), leading to desertification processes (soil degradation and loss of productive capacity) that are usually evident when is too late to stop them (Manzano et al. 2000).

Disruptive Technologies in High-Productivity Horticulture

Disruptive innovations do not seek to offer better products, but rather the introduction of products and services that, although not as effective as the existing ones, are simpler, more convenient, and sometimes cheaper (Bower and Christensen 1995; Smith 2003).

In the agricultural sector, disruptive technologies seek to maximize production levels continuously, developing greater automation and mechanization of the process (Kelly et al. 2017). High-productivity horticulture, which consists of large-scale production in smaller areas, generated large changes in agricultural production, using improved procedures implemented in greenhouses (Roel 1998). These technologies have been disruptive to the market in which they are introduced and, in a

---

[2] Is essential for these farming systems the use of adapted "criollo" seeds to water stress conditions and rapid growth, as well as the design of plots to the optimization of rainwater and soil control (Guzmán 2004).

broader sense, they can become disruptive to both the landscape and the social structures where they are implemented.

Peasant production systems, especially those of family structure, of a cyclical and seasonal nature, reflect a historical process of adaptation to the ecological conditions and of adaptation of productive modes: an answer to the conditions of aridity. The opposite model is one based on maximization of sustained productivity, aimed at international market demands. So far, in this region both models of food production coexist. In the "altiplano," the development of the industrial agricultural model, where intensive production and monoculture practices predominate, was enhanced owing to the guidance of businessmen from other regions, electrical energy subsidies, and increased road interconnection (Maisterrena and Mora 2000).

The valuation of an area of "conjuncture agricultural" implies its complete appropriation to maximize productivity in the short term and, after the depletion of local resources, the abandonment of lands (Macías-Macías 2008). Excessive pressure for irrigation sowing and the use of greenhouses results in a combination of effects due to high energy consumption (Salazar et al. 2012), overexploited aquifers (CONAFOR-UAch 2013), and abandonment of infrastructure, which accumulates as scrap. Altogether, this affects the forms of production of family nuclei, and even the alteration of the hydrological cycle of the area (Maisterrena and Mora 2000).

Energy Overproduction Projects

The intensive use of renewable resources such as solar and wind energy leads to the erroneous impression of the absence of limitations for their implementation as a source of energy production. Hidden social and environmental costs in (1) its productive chain and (2) energy overproduction are blurred.

Although the operation of alternative energy technological devices does not generate greenhouse gas (GHG) emissions, it is necessary to consider the energy and material extraction costs for their production. In a particular way, the manufacture of photovoltaic solar cells (PVSC) uses various chemical compounds and elements (cadmium, lead, arsenic, and nickel) cataloged as toxic. Used on a massive scale, they are highly unhealthy for the local habitat, and they have even provoked dissatisfaction with the establishment of production factories (Aman et al. 2015; EJATLAS n.d.). In addition, the production of batteries in PVSC requires a large amount of energy and raw materials (Tsoutsos et al. 2005).

The average lifetime of a solar panel is between 20 and 40 years. Final disposal has harmful implications if the necessary precautions are not taken. Thus, this electronic waste may represent a challenge. Another drawback is maintaining solar panels free of dust and dirt generated by cattle grazing or wind, which requires a variety of dust suppressants, considered dangerous, which are a risk factor for polluting surrounding and underground waters (Aman et al. 2015).

On the other hand, it has been documented that the impact of wind turbines in the life of birds and bats varies widely. They are highly dependent on the characteristics of the environment (weather conditions, topography, and seasonality), the species of organisms under study (mobility, density, and behavior), and wind park design (size, alignment, and rotor speed) (Drewitt and Langston 2006; NWCC 2010). It is necessary to highlight the particularities of the contexts where they have been observed, rather than focusing on all those cases that have not shown significant impacts. The main direct causes of death are collisions and, in the case of bats, pulmonary barotrauma (Baerwald et al. 2008). The displacement of birds due to visual interference and disturbance of habitats, the interruption of connectivity between areas of feeding, resting, and distant reproduction, can represent an effective loss of habitat (Drewitt and Langston 2006). Species with a long lifespan and low reproduction rates, migratory behaviors, those under some protection status, or even those we know very little about[3] may be particularly vulnerable.

Both turbines and PVSC require the clearing of vegetation cover and the removal of soils, as well as the opening of roads for their construction and the installation of electric conduction lines, which negatively affect the native vegetation and the wildlife. This is because it contributes to displacement and loss of habitats, increased soil erosion, and the loss of nutrients contained therein, disturbs natural drainage and irrigation systems, and causes the fragmentation of the landscape, creating barriers for the flow of organisms (Aman et al. 2015; Zaldúa 2012; Hernández et al. 2014; Santos et al. 2018).

If environmental costs are difficult to measure, more so are those social and cultural impacts, many of which also have no monetary value and are not susceptible to mitigation. The social dimension implies all those impacts for the well-being of people: their environment, the land, the community, the livelihoods, and the culture (Huesca-Pérez et al. 2018). It is evident that the implementation of these systems in an area of natural beauty or biocultural value will generate visual impacts on the landscape (Tsoutsos et al. 2005).

Regarding the second point, despite the increase in the contribution of renewable sources in the production of electricity in Mexico, mainly wind in the last decade,[4] the total supply of primary energy, including sources such as coal, natural gas, and oil, has also sustained an increase (Fig. 6.1). The final balance translates into a net increase in energy sources and, consequently, a greater consumption of total energy and its corresponding emissions. Under a scenario of forced modernization, the energy harvest that is carried out contributes to a model of economic development that increases its productive capacity, based on the increase in energy appropriation.

---

[3] García-Morales and Gordillo-Chávez (2011) report that knowledge about bats in the arid and semi-arid zones of San Luis Potosí is scarce, due to the lack of continuous and systematic studies.

[4] The participation of the generation of electric power for the wind in 2016 was 3.2% and the solar energy of 9.6%, with respect to the total sources; but if only renewables are considered, their contribution is 16.7% and 5.7%, respectively (IEA n.d.).

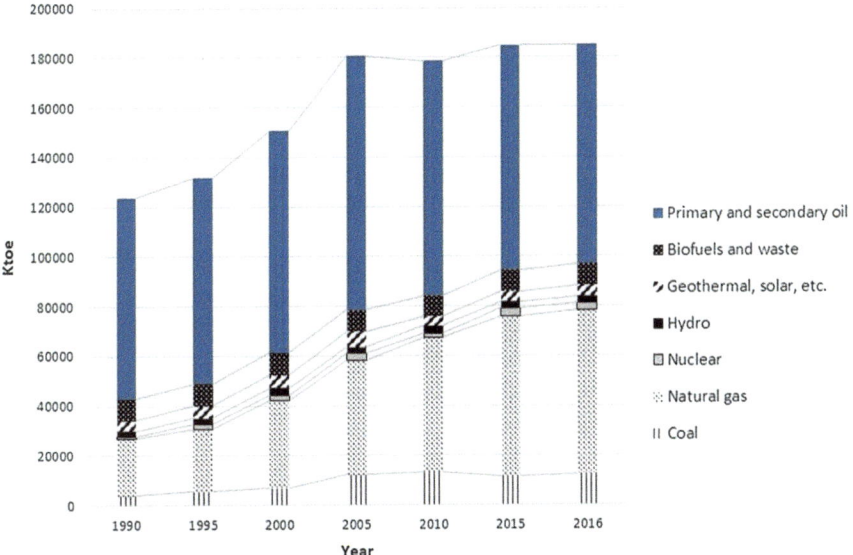

Fig. 6.1 Total supply of primary energy according to source. Mexico 1990–2006. Source: Prepared by the authors based on International Energy Agency (n.d.)

Final Considerations

Starting from the fact that the lack of a systemic vision in the past has led to serious and unforeseen environmental and social effects, we consider the following:

Energy production projects and high-productivity horticulture are introduced into a landscape whose socio-ecological system has faced intense socio-economic pressures in recent decades, leading to multiple processes of degradation of the socio-ecological system of the "altiplano," pointed out here through the different landscape disruptions. Then, it generates a marginalized periphery of small producers, affects the biotic and abiotic resources of the place, and denies the possibility of maintaining alternative forms of production where, under the dominant economic discourse, resources are wasted.

It is important to act in response to the indicators that the region shows: scarcity and contamination of water sources and soils, saline intrusion, and reduction of grazing conditions. This complex scenario integrates new factors such as the implementation of devices to produce renewable energies, which requires a careful balance since, from a sustainable development perspective, it should not undermine the environmental or biocultural heritage conditions that the communities harbor or the future enjoyment thereof.

Systemic complexity of this landscape requires studies that allow us to approach the possible impacts, which will not necessarily be linear and cumulative, where the transformations, beyond being the sum of several individual development projects, imply a disruption of the landscape and the socio-ecological system at different levels.

However, there are aspects that cannot be technically or scientifically evaluated (Martínez 2006). Therefore, decisions on the transformation of the landscape (functional, esthetic, and symbolic changes) cannot be taken in the margin of the communities that historically have shaped the existing socio-ecological system. In this sense, intercultural communication can provide tools for better representation in decision-making.

In short, we propose an articulation of two analytical components for the study of drylands, landscape, and socio-ecological systems, to gain a greater understanding of the complexity of these systems and the transformation processes that they undergo. Thus, this work is a preliminary approach of a more extensive research. Then, we identify the need to develop further research that analyzes these issues in greater depth and exposes the impacts that occur at different scales and, mainly, highlights the biological and cultural loss triggered by the so-called modernizing processes.

Finally, the interest of this chapter is to outline some of the needs of a research agenda that crosses different spheres of knowledge, and integrates their components in an analysis of the consequences facing the transformation of drylands and the alterations they generate in the landscape and the socio-ecological system. Despite the discourses in favor of modernization that does not necessarily mean improvement for the inhabitants of the impacted areas, the main issue here is to focus on changes which can be irreversible in the ecosystem and its cultural components.

References

Alsina M (2004) Cuestionamientos, características y miradas de la interculturalidad. Sphera Pública 4:53–68

Aman M, Solangi K, Hossain M, Badarudin A, Jasmon G, Mokhlis H, Bakar A, Kazi S (2015) A review of Safety, Health and Environmental (SHE) issues of solar energy system. Renew Sustain Energy Rev 41:1190–1204

Baerwald E, D'Amours G, Klug B, Barclay R (2008) Barotrauma is a significant cause of bat fatalities at wind turbines. Curr Biol 18(16):R695–R696

Barrera O, Sagarnaga L, Salas J, Leos J, Santos R (2018) Viabilidad económica y financiera de la ganadería caprina extensiva en San Luis Potosí, México. Mundo Agrario 19(40):e077

Bower J, Christensen C (1995) Disruptive technologies: catching the wave. Harv Bus Rev 73:43–53

CONAFOR-UACh (2013) Línea base nacional de degradación de tierras y desertificación. Informe final. Universidad Autónoma Chapingo

Drewitt A, Langston R (2006) Assessing the impacts of wind farms on birds. Ibis 148(Suppl. 1):29–42

Environmental Justice Atlas (EJATLAS) (n.d.) Antipollution protests against solar panel manufacturers, Quanzhou, Fujian, China. https://ejatlas.org/conflict/antipollution-protests-against-local-solar-panel-manufacturers-quanzhou-fujian-china. Entry: 01 Dec 2018

Espinosa-Reyes G, González-Mille D, Ilizaliturri-Hernández C, Mejía-Saavedra J, Cilia-López G, Costilla-Salazar R, Díaz-Barriga F (2014) Effect of mining activities in biotic communities of Villa de la Paz, San Luis Potosi, Mexico. Biomed Res Int 2014:1–13

García-Morales R, Gordillo-Chávez E (2011) Murciélagos del estado de San Luis Potosí, México: revisión de su conocimiento actual. Therya 2(2):183–192

Giménez J, Fernández H, Candelario T, Lira R, Llano M (2018) Diagnosis cultural y natural de la Ruta Huichol a Huiricuta: Criterios para su inclusión en la Lista del Patrimonio Mundial. Inv Geogr (96):1–18. https://doi.org/10.14350/rig.59604

Gobierno del Estado de San Luis Potosí (GESLP) (2018a) Tercer Informe de Gobierno. Gobierno del Estado de San Luis Potosí, México

Gobierno del Estado de San Luis Potosí (GESLP) (2018b) Tercer Informe de Gobierno, síntesis. Gobierno del Estado de San Luis Potosí, México

Grünwaldt J, Castellaro G, Flores E, Morales-Nieto C, Valdez-Cepeda R, Guevara J, Grünwaldt E (2016) Pastoralismo en zonas áridas de Latinoamérica: Argentina, Chile, México y Perú. Rev Sci Tech L'office Int Des Epizooties 35(2):543–551

Guzmán M (2004) Ecología humana y nuevas territorialidades en el altiplano potosino. Rev El Colegio de San Luis 6(17):49–74

Hernández H (2006) La vida en los desiertos mexicanos. Fondo de Cultura Económica, Mexico

Hernández R, Easter S, Murphy-Mariscal M, Maestre F, Tavassoli M, Allen E, Barrows C, Belnap J, Ochoa-Hueso R, Ravi S, Allen M (2014) Environmental impacts of utility-scale solar energy. Renew Sustain Energy Rev 29:766–779

Huber-Sannwald E, Ribeiro M, Arredondo J, Braasch M, Martínez R, García de Alba J, Monzalvo K (2012) Navigating challenges and opportunities of land degradation and sustainable livelihood development in dryland social-ecological systems: a case study from Mexico. Philos Trans R Soc B 367:3158–3177

Huesca-Pérez M, Sheinbaum-Pardo C, Köppel J (2018) From global to local: impact assessment and social implications related to wind energy projects in Oaxaca, Mexico. Impact Assess Project Appraisal 36(6):479–493

International Energy Agency (IEA) (n.d.) Statistics. Global energy data at your fingertips. https://www.iea.org/statistics/?country=MEXICO&year=2016&category=Keyindicators&indicator=TPESbySource&mode=chart&categoryBrowse=false&dataTable=BALANCES&showDataTable=false. Entry: 20 Nov 2018

Jasso-Pineda Y, Espinosa-Reyes G, González-Mille D, Razo-Soto I, Carrizales L, Torres-Dosal A, Mejía-Saavedra J, Monroy M, Ize A, Yarto M, Díaz-Barriga F (2007) An integrated health risk assessment approach to the study of mining sites contaminated with arsenic and lead. Integr Environ Assess Manag 3(3):344–350

Kelly S, Bensemann J, Bhide V, Eweje G, Imbeau J, Scott J, Lockhart J, Taskin N, Warren L (2017). Disruptive technology in the agri-food sector. An examination of current and future influence on sustainability, bio-security and business effectiveness (final report)

Macías-Macías A (2008) Costos ambientales en zonas de coyuntura agrícola. La horticultura en Sayula (México). Agroalimentaria 26:103–118

Maisterrena J, Mora I (2000) Oasis y espejismo. Proceso e impacto de la agroindustria de jitomate en el valle de Arista S.LP. SIHGO/El Colegio de San Luis/Secretaría de Ecología del Estado de San Luis Potosí, San Luis Potosí

Manifestación de Impacto Ambiental (MIA) (n.d.) "Descripción del proyecto", Línea de Transmisión Eléctrica Dominica–Charcas. http://sinat.semarnat.gob.mx/dgiraDocs/documentos/slp/estudios/2011/24SL2011E0012.pdf. Entry: 06 Nov 2018

Manzano M, Návar J, Pando-Moreno M, Martínez A (2000) Overgrazing and desertification in northern Mexico: highlights on north-eastern region. Ann Arid Zone 39(3):285–304

Martínez J (2006) Agua y sostenibilidad: algunas claves desde los sistemas áridos. Polis 14:1–13

Martínez F, Soto J (n.d.) El barrio de la Banda, paisaje y valor histórico. Universidad Autónoma Metropolitana, Mexico

Monzalvo-Santos K, Alfaro-De la Torre M, Chapa-Vargas L, Castro-Larragoitia J, Rodríguez-Estrella R (2016) Arsenic and lead contamination in soil and in feathers of three resident passerine species in a semi-arid mining region of the Mexican plateau. J Environ Sci Health A Tox Hazard Subst Environ Eng 51(10):825–832

Mora I (2011) "Vámonos con todo y chivas". Sistemas de supervivencia en las culturas ganaderas del norte de San Luis Potosí. Rev El Colegio de San Luis 1(1):48–66

Mora I (2012) Estrategias de reproducción en el desierto chihuahuense: ganadería trashumante y diversificación en el altiplano potosino. In: Fábregas-Puig A, Nájera M, Valdés C (eds) Dinámica y transformación de la región chichimeca. Universidad Autónoma de Coahuila, Universidad Autónoma de Zacatecas, Universidad Autónoma de Aguascalientes, Universidad Autónoma de Nayarit, El Colegio de San Luis, El Colegio de Michoacán, El Colegio de Jalisco, Universidad Estatal de California L.B., U, Mexico, pp 67–83

National Wind Coordinating Collaborative (NWCC) (2010) Wind turbine interactions with birds, bats and their habitats: a summary of research results and priority questions. https://www1.eere.energy.gov/wind/pdfs/birds_and_bats_fact_sheet.pdf

Negrete-Sánchez L, Aguirre-Rivera J, Pinos-Rodríguez J, Reyes-Hernández H (2016) Beneficio de la parcelación de los agostaderos comunales del ejido "El Castañón", municipio Catorce, San Luis Potosí: 1993-2013. Agrociencia 50(4):511–532

Razo I, Carrizales-Yáñez L, Castro-Larragoitia J, Díaz-Barriga F, Monroy M (2004) Arsenic and heavy metal pollution of soil, water and sediments in a semi-arid climate mining area in Mexico. Water Air Soil Pollut 152(1):129–152

Reed M, Buenemann M, Atlhopheng J, Akhtar-Schuster M, Bachmann F, Bastin G, Bigas H, Chanda R, Dougill A, Verzandvoort S (2011) Cross-scale monitoring and assessment of land degradation and sustainable land management: a methodological framework for knowledge management. Land Degrad Dev 22(2):261–271

Reynolds J, Stafford M, Lambin E, Turner B II, Mortimore M, Batterbury S, Downing T, Dowlatabadi H, Fernández R, Herrick J, Huber-Sannwald E, Jiang H, Leemans R, Lynam T, Maestre F, Ayarza M, Walker B (2007) Global desertification: building a science for dryland development. Science 316(5826):847–851

Rodriguez-Loubet F (2016) San Luis Potosí y Gran Tunal en el Chichimecatlán del México Antigüo. Arqueología y etnohistoria. El Colegio de San Luis/Fomento Cultural del Norte Potosino A.C, Mexico

Roel V (1998) La tercera revolución industrial y la era del conocimiento. UNMSM Fondo Editorial, Lima

Ruiz C (2014) Paisajes culturales y territorio en Real de Catorce, 1772–1800. In: Ruiz C, Roque C, Coronado L (eds) Paisajes culturales y patrimonio en el centro-norte de México, siglos XVII al XX. El Colegio de San Luis, San Luis Potosí, pp 49–95

Ruiz C, Roque C (2017) Introducción. In: Ruiz C, Roque C (Coords) La dimensión histórica y social del paisaje cultural y el patrimonio en México. El Colegio de San Luis, San Luis Potosí, pp 11–24

Rzedowski J (2006) Vegetación de México. Mexico: Comisión Nacional para el Conocimiento y Uso de la Biodiversidad.

Salazar R, Cruz P, Rojano A (2012) Eficiencia del uso de la energía en invernaderos mexicanos. Rev Mex Cienc Agrícolas 4:736–742

Santos J, Marques J, Neves T, Marques T, Ramalho R, Mascarenhas M (2018) Environmental impact assessment methods: an overview of the process for wind farms' different phases— from pre-construction to operation. In: Mascarenhas M, Marques A, Ramalho R, Santos D, Bernardino J, Fonseca C (eds) Biodiversity and wind farms in Portugal. Springer, New York, pp 35–86

Secretaria de Desarrollo Social y Regional (SDSR) (2016) Diagnóstico mircrorregional Altiplano Oeste. Gobierno del Estado de San Luis Potosí, San Luis Potosí

Servicio Geológico Mexicano (SGM) (2017) Panorama Minero del Estado de San Luis Potosí. Secretaría de Economía, San Luis Potosí

Servicio Geológico Mexicano (SGM) (2018) GeoInfoMex. Sistema de consulta de información geocientífica. https://www.sgm.gob.mx/GeoInfoMexGobMx/

Smith R (2003) The innovator's solution. Creating and sustaining successful growth. Soundview Execuitve Book Summaries, Concordville 25(11):1–8.

Tsoutsos T, Frantzeskaki N, Gekas V (2005) Environmental impacts from the solar energy technologies. Energy Policy 33(3):289–296

Tudela F (Coord) (1989) La modernización forzada del trópico: El caso de Tabasco. Proyecto integrado del Golfo. Federación Internacional de Institutos de Estudios Avanzados/Instituto de Investigaciones de las Naciones Unidas para el Desarrollo Social/Centro de Investigación y de Estudios Avanzados del Instituto Politécnico Nacional/El Colegio de México, Mexico

Unión de Pequeños Agricultores y Ganaderos (UPA) (2009) Informe UPA. La trashumancia en España. La Tierra 213:49–56

United Nations Convention to Combat Desertification (UNCCD) (n.d.) Convención de las Naciones Unidas de Lucha ontra la Desertificación en los países afectados por sequía grave o desertificación, en particular en África. https://www.unccd.int/sites/default/files/relevant-links/2017-08/UNCCD_Convention_text_SPA.pdf

Vázquez D (2014) El complejo salinero del altiplano potosino y el noreste de Zacatecas: hacia su valoración patrimonial paisajística y cultural. In: Ruiz C, Roque C, Coronado L (eds) Paisajes culturales y patrimonio en el centro-norte de México, siglos XVII al XX. El Colegio de San Luis, San Luis Potosí, pp 123–168

Vázquez A, León M (2015) Guía metodológica para el paisaje cultural ecuatoriano. Instituto Nacional de Patrimonio Cultural y Universidad Autónoma de Querétaro, Quito

Verbist K, Santibañez F, Gabriels D, Soto G (2010). Atlas de zonas áridas de América Latina y el Caribe. Documento técnico No 25. UNESCO

World Wild Found (WWF) (2018) Desierto Chihuahuense. http://www.wwf.org.mx/que_hacemos/programas/desierto_chihuahuense/. Entry: 01 Dec 2018

Zaldúa N (2012) Principales impactos del desarrollo eólico sobre la avifauna: Síntesis de la revisión de bibliografía internacional de referencia. PNUD Uruguay

Part II
Transdisciplinarity in Drylands

Chapter 7
Public Participation Approaches for a New Era in Dryland Science and Stewardship in the Global South

D. L. Coppock

Abstract The drylands of the Global South are facing challenges from human population growth, unsustainable land-management practices, and climate change. Such problems are complex and can no longer be adequately addressed using traditional, top-down means. Rather, increased reliance on public participation is needed to better identify key research questions and interventions that promote positive change. Major actors in these approaches would include communities, applied researchers, outreach agents, policy makers, and planners working in tandem. The process involves embracing "engaged research" within a framework of an "innovation system"—referred to here by the acronym ERIS. Such concepts have recently gained acceptance, and now is the time to implement them. Project donors are key to this transformation of applied research and professional practice because they can provide incentives. ERIS projects must be long term with a diversity of pragmatic and transdisciplinary scholarly achievements. An ERIS platform for interproject coordination could be part of a comprehensive Agadir platform. The main option to institutionalize ERIS approaches is via policies that alter how government agencies and their stakeholders work together.

Keywords Action research · Engaged research · Extension · Community-based research · Innovation systems · Outreach · Participatory rural appraisal · Participatory policy making · Participatory planning

Introduction

The objective of this chapter is to provide a concise overview of participatory perspectives as they pertain to the science, management, policy making, or planning for improved stewardship of dryland socio-ecological systems (SESs) in the Global

D. L. Coppock (✉)
Department of Environment and Society, Utah State University, Logan, UT, USA
e-mail: layne.coppock@usu.edu

© Springer Nature Switzerland AG 2020
S. Lucatello et al. (eds.), *Stewardship of Future Drylands and Climate Change in the Global South*, Springer Climate,
https://doi.org/10.1007/978-3-030-22464-6_7

South. The term "participation" primarily refers to the active inclusion of dryland inhabitants (e.g., community members) into processes that seek to improve their lives. This inclusion is long overdue. The past has been characterized by too many "top-down" efforts where little input is obtained from resource users or project beneficiaries. This situation has discounted the immense value from knowledge and experiences of local people. Heavy reliance on top-down approaches has often led to the wrong research questions being addressed, wasting of limited resources, loss of valuable time, and lack of trust-building between communities and change agents. The irony is that applied scientists, outreach agents, and policy makers sincerely hope their work will lead to positive and practical impacts on humanity, but the previous ways of working reduce the likelihood this will occur. Only by embracing participatory approaches will such odds improve. The chapter begins with a brief overview of participatory methods and how these have evolved over time. The chapter concludes with some thoughts as to how participation might help shape a new era of dryland science and stewardship.

Why does the science and stewardship of drylands in the Global South need to change? The answer is simply that the inhabitants of drylands are being increasingly marginalized and the sustainable use of dryland landscapes is increasingly under threat (MEA 2005). The drylands are also becoming warmer and drier due to climate change (Huang et al. 2017). Dryland systems are thus ripe for destabilization via increased resource competition, social conflict, and decline in traditional governance (Coppock et al. 2017). One scenario is that a rural exodus from the drylands will occur, overburdening infrastructure elsewhere that is ill-equipped to cope.

Background of Participatory Thought and Methods

The Land-Grant Model in the US: Using Extension to Connect Research with the Public

Public tertiary education in the United States of America (US) offers a template to understand the historical interplay between research, outreach, and the public. This template has evolved over the past 160 years. Current trends suggest that research and outreach will become increasingly connected as universities strive to enhance their relevance to society by addressing a growing diversity of issues. The land-grant university system in the US started in 1862 as a way to make college education more accessible to American citizens.[1] Today there are over 100 land-grant schools, and they embrace a mission focused on teaching, research, and extension (e.g., public outreach). The extension component was intended to connect applied research with societal needs. Special attention was given to improving agriculture in rural areas. The extension agent would identify problems with producers, and then serve

---

[1] https://www.nap.edu/read/4980/chapter/2.

Fig. 7.1 Diagram of
connections among applied
researchers, the extension
service, and the public in a
traditional land-grant model
(Source: D. L. Coppock)

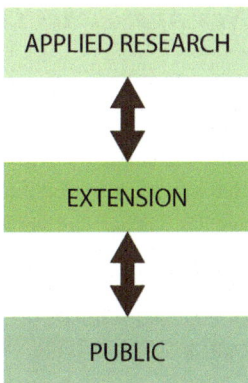

as a link to researchers back on campus who would develop technologies or management practices to solve problems. Extension would complete the loop by taking new innovations back to producers for local testing and adoption (Fig. 7.1).

While input from agricultural producers was always essential to this process, it is fair to say that the extension system, in general, could drift toward a "top-down" orientation. Experts would provide tools and techniques to improve productivity on farms and ranches, and producers were often viewed as somewhat passive participants eager to adopt innovations (Rogers 2003). The process was not always smooth, however. Adoption of seemingly beneficial new technologies could be difficult. Numerous barriers to rapid uptake were identified, ranging from the cultural to economic or environmental. An excellent example is the long time it took to achieve widespread use of improved corn varieties in Iowa starting in the 1940s; this and other situations are well-summarized by Rogers (2003). Despite numerous challenges, the fundamental structure and function of the land-grant extension system appeared sound, however, as evidenced by the positive performance of American agricultural systems.[2]

Land-grant schools have evolved by diversifying their core mandates, with funding sources shared by state and federal sources. Academic curricula have been greatly expanded and faculty now compete for a growing array of research opportunities via contracts and grants. The traditional applied research/extension component has become a smaller part of university activities overall, although still politically important given the need for schools to demonstrate connectivity with rural constituencies. And by the 1990s priorities for extension began to fundamentally change as client populations became more urban and problems shifted from agricultural productivity to a wider spectrum of social and economic issues.[3] Extension also began to actively critique itself, noting the difficulties in demonstrating impact among diverse target groups and contemplating new ways of working.[4]

---

[2] https://www.usda.gov/sites/default/files/documents/history-american-agriculture.pdf.

[3] https://nifa.usda.gov/cooperative-extension-history.

[4] https://www.nap.edu/read/5133/chapter/7.

New Approaches Linking Research with the Public in the USA

There is growing awareness at American colleges and universities that public engage-ment matters in ways that go beyond the traditional roles of teaching or extension. In essence, academic leadership is questioning how universities can regain public trust that has been eroded due to perceptions that campuses have become insular environ-ments. As a case in point, schools across the US have recently embraced the concept of the "engaged university," whereby researchers and educators partner with local communities in processes of mutual learning and problem-solving. The emergence of engaged scholarship recognizes that academic expertise can be valuable for soci-etal problem-solving (Whitmer et al. 2010). As an offshoot of this momentum, "engaged research" is recognized as an advanced expression of community-based involvement (Fig. 7.2). As with the highly respected Carnegie systems for classifying universities with reference to teaching or research prowess,[5] Carnegie is now pursu-ing classifications based on public engagement metrics.[6] Making public engagement a pervasive reality for universities, however, requires that academics are rewarded for efforts that go beyond traditional aspects of teaching, research, and professional ser-vice.[7] Transaction costs of working in transdisciplinary teams, and confronting the risk exposure inherent with public involvement,[8] are added burdens that must be recognized and compensated for if public engagement is to thrive (Whitmer et al. 2010). Evidence suggests that knowledge created through engaged research is more likely to be socially acceptable, policy relevant, and influence favorable environmen-tal outcomes (Overdevest et al. 2004).

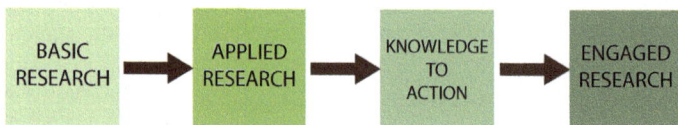

Fig. 7.2 A research continuum that represents stages of immediate relevance to society. Basic and much of applied research can occur with minimal public engagement. Knowledge-to-action research deals with fine-tuning a process of generating innovations where public input is focused on problem identification. Engaged research, in contrast, is where public input constitutes more of a continual partnership from problem identification to full implementation of a solution (adapted from concepts in Whitmer et al. 2010)

---

[5] http://carnegieclassifications.iu.edu/definitions.php.

[6] https://compact.org/initiatives/carnegie-community-engagement-classification/.

[7] https://www.chronicle.com/article/But-Does-It-Count-/147199.

[8] Uncertainty of outcomes—a fact of real life—is a major risk of public engagement with regard to the number or types of research products generated. The traditional research model is one where specific publication outputs are essentially promised by grant recipients. In contrast, researchers must be more flexible when conducting engaged research because the content or types of key publications may not be accurately predicted. The key is to build added flexibility into grant pro-posal outputs.

Examples as to how academia is moving toward greater public engagement come from diverse sources. These include public sociology (Feagin et al. 2015) and public history (Cauvin 2016) that advocate study in pursuit of social justice. Connecting the public with biophysical research includes citizen science, whereby laypeople are recruited to collect data on research projects in a form of social inclusion and empowerment (Cavalier and Kennedy 2016).

Linking Research with the Public in the Global South

Given the massive challenges limiting the growth of crop and animal agriculture in the Global South, it is ironic that strong, integrated systems of applied research with extension have still not emerged to a significant extent. A general lack of financial resources and the need for institutional reforms are two reasons why this situation has occurred.[9] Applied research has typically been conducted via government agencies or at international centers,[10] while extension has often been delegated either to other governmental or non-governmental organizations (GOs, NGOs). Applied research conducted by government is often restricted to field stations where experiments can be conducted, but replication of on-farm conditions is limited. Universities typically do not play a role in fostering agricultural extension services in the Global South, as few (if any) systems similar to the US land-grant model have been created.

In any case, when they exist, extension agents in the Global South may struggle in terms of effectiveness and having sustained, positive impacts on producer communities.[11] Lack of budget to regularly interact with producers is a common problem, negatively affecting transportation and other operational needs; this is especially evident in the under-resourced drylands. Extension agents often follow a "top-down" approach where input from beneficiaries is typically minimal in shaping the extension agenda. Beneficiaries are often seen as passive recipients of extension technology or information. In some countries government extension can even be politicized, with extension agents spending time indoctrinating clients on political matters in addition to filling their role as providers of technical advice. The NGOs—both national and international—usually have more financial resources than the GOs, but the former also have mixed records of success. The development agenda pursued by NGOs can also be "top down" in that they advocate for the priorities of their international donors who may have little knowledge of on-the-ground realities (Derr 2018). In addition, NGOs may have few (if any) open lines of

---

[9] http://www.fao.org/3/a-y2709e.pdf.

[10] Such as the Consortium of International Agricultural Research Centers (CGIAR). https://www.cgiar.org/.

[11] D. L. Coppock, personal observations.

communication with the research-based programs of government or the international centers.[12]

Farming Systems Research and Extension   One response to this research/extension gap in the Global South began in the 1970s. It was called Farming Systems Research & Extension (FSR&E). The FSR&E approach was somewhat of a re-creation of the traditional land-grant extension model, although more attention was given to tighter, collaborative feedbacks among researchers, extension agents, and producers. This occurred due to the recognition that expatriate personnel were typically less familiar with the details of small-farm agricultural problems in foreign lands. Manuals were generated summarizing FSR&E approaches (Shaner et al. 1982). While the FSR&E perspective once supported peer-reviewed journals and annual meetings, the general trend has been for FSR&E to evolve into other interrelated forms of inquiry. A history is provided by David Norman.[13] The FSR&E perspective has been very interdisciplinary, as is required for real-world problem-solving. This was groundbreaking for its time.

Participatory Rural Appraisal (PRA)   Perhaps the best known critiques of the challenges associated with "top-down" diagnosis of rural problems in the Global South were forwarded by Robert Chambers (1983). Chambers illuminated many disconnects between the world of "development experts" and the needs of the rural poor. He noted that the knowledge held by local communities was largely ignored by experts, contributing to widespread failures of development projects (Fig. 7.3). Chambers is also recognized as the author of one of the most influential synthesis papers describing participatory rural appraisal (PRA; Chambers 1994). The PRA process is an approach that challenges the top-down model by providing mechanisms for sharing power and experiences between experts and community members in the co-production of new knowledge. As with FSR&E, the PRA approach was a consolidation of diverse perspectives that responded to the "extension gap" evident for developing nations. Procedures for PRA and related techniques are summarized in various field manuals[14] and scholarly texts (Narayanasamy 2009).[15] A concise review emphasizing research applications for PRA is provided by Coppock (2016).

The PRA process is generic in the sense that it can be effectively used to guide development interventions or identify priority research questions. The main focus

---

[12] There are exceptions to this rather bleak picture, however. There are some cases where, in the course of annual performance reviews, government researchers may receive tangible incentives when their research has demonstrably led to impact on communities. This can foster stronger interactions among researchers, extension agents, and producers (D. L. Coppock, personal observations).

[13] https://www.researchgate.net/publication/251791709_The_farming_systems_approach_A_historical_perspective.

[14] https://financiamentointernacional.files.wordpress.com/2013/12/manual-de-formac3a7c3a3o-de-formadores-em-gestc3a3o-de-projectos.pdf.

[15] PRA has also subsequently morphed into several other, but closely related, forms of community-based analysis and intervention.

Fig. 7.3 Development experts or scientists (standing, enlarged) impose their vision of change over local community members (sitting, reduced) illustrating an imbalance of power. Because the knowledge or circumstances of the locals is often ignored, the resulting project can have limited impact (Source: D. L. Coppock)

of PRA is to identify the key, solvable problems in a community and then devise a community action plan (CAP) to pursue sustainable solutions. The CAP lays out the pathway for change and identifies the human, technical, and financial resources needed to move forward.

While the PRA approach has been adapted to fit a variety of circumstances, the core toolkit involves about a dozen elements. A thorough PRA can be typically conducted in about five days. The toolkit elements include analyzing: (1) Maps of local landscapes and key resources; (2) timelines of important historical events; (3) seasonal activity calendars; (4) wealth class dynamics; (5) gendered activities; (6) sustainability of livelihoods; and (7) institutions (formal and informal) that affect community well-being. The cornerstone of the process is a community-based ranking of priority, solvable problems using matrices. Matrices compare each major problem with every other major problem in terms of perceived costs and benefits.[16] The time needed for problem identification through solution implementation can take months or years (Coppock 2016).

Group dynamics are diverse and can be tailored to each PRA situation. Methods include the use of plenary opening and closing gatherings that involve as many community members as possible; this helps ensure that awareness is raised and surprises are avoided. Smaller breakout groups can be used to focus on details of problems or solutions; they can also be conduits for input from marginalized people unwilling to speak in larger assemblies (Fig. 7.4). In-depth interviews of key informants can also supplement the process, as needed. In general, change agents—whether they be researchers, outreach staff, policy makers, or planners—are only observers. They can provide expert feedback when the priority problems or CAP

---

[16] While any solvable problem can emerge as a priority from a PRA, recent experiences in Africa and Asia regarding community adaptation to climate change often end up focused on water issues (D. L. Coppock, personal observations).

Fig. 7.4 Community focus group held as part of a participatory rural appraisal (PRA) process in Nepal. The project concerned helping communities adapt to climate change (Divakar Duwal in Coppock 2016)

details are under debate. It is best if an independent third party—with skills managing group dynamics—provides the facilitating functions. Community members or others suspected of wanting to hijack the process for personal benefit can be sidelined in creative ways (Coppock 2016).

Participatory Action Research (PAR)   Another movement involving participatory approaches began in the 1970s and 1980s. One thread emanated from Latin America (Bonfil 1970; Friere 1970; De Silva et al. 1979; Fals Borda 1979), while another occurred in the US (Whyte 1989). The emergence of PAR was in response to the need for problem-solving in diverse sectors including public education, public health, agricultural development, and industrial management. Like FSR&E or PRA, PAR is a process whereby researchers or other change agents work closely with project beneficiaries. The PAR approach in general involves a series of iterative steps shown in Fig. 7.5. The process continues until final interventions are implemented. Some of the voices advocating for increased use of PAR have been American sociologists seeking new frontiers and intellectual challenges for academic social science (Greenwood and Levine 2007). Both PRA and PAR can be combined. A PRA provides a problem diagnosis, while PAR can provide the details (sometimes via conventional research) that support creation of new technology, management systems, or policy interventions needed to solve a problem (Derr 2018).

Innovation Systems (IS)   A more recent area of participatory inquiry is called "innovation systems" (Röling 2009; Waters-Bayer et al. 2009). An IS can be defined as a multiple-stakeholder, problem-solving partnership. While this sounds similar to PRA or PAR, an IS differs because it is larger and more diverse in scope (Table 7.1). Creation

Fig. 7.5 The iterative nature of participatory action research (PAR) (Coppock 2016)

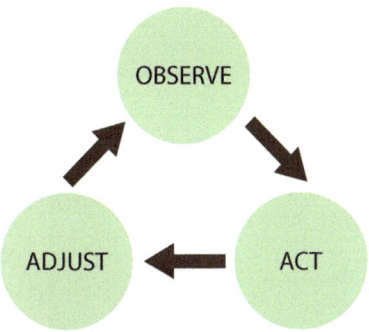

Table 7.1 Comparisons of participatory rural appraisal (PRA), participatory action research (PAR), and innovation systems (IS) in the context of research and development projects in the southern Ethiopian rangelands, 1997–2015 (Coppock et al. 2009; Coppock 2016)

Scale	PRA	PAR	IS
Organizational	Several communities, change agents, and institutions	Several communities, change agents, and institutions	Dozens of communities, scores of change agents, and scores of institutions
Spatial	Local	Local	Local, regional, international
Temporal	Up to 1 year	Up to 3 years	Up to 10 years

of a successful IS requires centralized leadership and vision. Use of IS in rural, agricultural contexts has been limited (Sanginga et al. 2009). More long-term field testing and team learning must occur before best practices can be verified. As with PRA or PAR, advocates for IS affirm that creative solutions for problem-solving require co-production of knowledge from among diverse stakeholders (Waters-Bayer et al. 2009). A good dryland example of using an IS can be found in Coppock et al. (2009).

Problem solutions discovered using an IS may not be rooted in new technology, but rather in policies or processes that empower people by building human or social capital. An example of a change process from southern Ethiopia illustrates how pastoral women were empowered to improve household resilience by reducing poverty (Fig. 7.6). Each step required a different disciplinary lens as well as a different suite of facilitating IS actors. An IS can thus promote transdisciplinary work.

A successful IS should be a long-term commitment because trust and learning pathways need to be established among stakeholders. While what constitutes "long term" is open to debate, it certainly is longer than the three-year grant periods that typically characterize donor programs. An IS will have a life cycle (i.e., birth, growth, death; Coppock et al. 2009). Stakeholder incentives are needed to maintain an IS.[17]

---

[17] An IS supporting a pastoral women's empowerment program in southern Ethiopia existed for 10 years. The IS grew from a few members in 1999 to a peak of 70 by 2006, and then declined as activities wrapped up by 2009 (D. L. Coppock, personal observations). Once project funding ended after 2009 the IS quickly disappeared. Collaboration thus depended on funding incentives. The good news is that the women's empowerment program endured. The IS thus played an essential—but temporary—role in this capacity-building process. It is also noteworthy that for this project it was never originally planned to create an IS, it simply happened out of necessity to create a development pathway. Each actor in the IS was needed to help create the final product.

Fig. 7.6 Example of a stepwise capacity-building approach for pastoral women's groups in southern Ethiopia. Each step required different disciplinary perspectives. From the bottom up: Sociology (diffusion of innovations), non-formal education, micro-finance, small-business management, livestock production/livestock marketing (Coppock 2016)

A successful IS also requires an operating platform for coordination and communication (Sanginga et al. 2009). A platform would organize relevant actors to facilitate sharing of information and provide frameworks for reporting, analysis, and planning. Hubs could be nested according to regional locations, and stakeholders could be networked according to local, regional, national, or international domains.

Public Participation, Policy Making, and Planning

In a new era for dryland science and stewardship, engaged research leads to technical, management, or policy interventions that are climate-smart and help make communities more sustainable. This could be at local or regional scales. Outreach or extension are processes that directly inform research and make research products more useful and relevant. It is thus obvious that engaged research requires strong connections to outreach or extension.

What is less obvious, however, is that policy making and planning should also be part of an integrated process from the start. For example, integrated research and outreach can identify needs for policy changes that govern the use of innovations. Policy recommendations can therefore emerge from engaged research. The challenge is to do research or outreach that is relevant to policy and planning; this happens by bringing policy makers and planners into a project loop at the beginning. Researchers, outreach practitioners, or community members may otherwise only be dimly aware as to how national policies or regional plans affect current or future circumstances for their projects.

The formulation of public policy is an essential government function to address societal challenges (Birkland 2016). Policy-making steps include: (1) Identifying priority problems; (2) shaping potential policy solutions to solve priority problems; (3) legitimizing policy prescriptions via public input or legislative processes leading to laws or regulations; (4) implementation of laws or regulations; (5) re-evaluation of implemented laws or regulations; and (6) maintenance, succession, or termination of policies as needed. In wealthy democracies recent trends suggest that policy makers are more actively engaging the public today—this also helps to educate policy makers (Gramberger 2001; Priscoli 2004).

Public planning includes environmental, land use, regional, urban, and other infrastructural domains (Kelly 2010). Planning consists of processes that integrate activities or investments required to achieve desired public policy goals. Mainstream planning has been evolving in the US to conduct research and be more inclusive in obtaining diverse sources of public input, also helping to educate planners (Umemoto 2001; Forsyth 2012). Planning is related to forecasting, where alternative future scenarios for a society can be compared with respect to goal achievement. In the context of the drylands, planning plays vital roles with respect to transportation networks, water resources, marketing infrastructure, telecommunications, energy use, or managing human populations.[18]

In wealthy democracies with well-educated citizens, both policy making and planning have significant avenues for public input. This includes voting to elect policy makers or use of face-to-face meetings or electronic means to solicit stakeholder input.[19] In some nations of the Global South, however, democratic processes may be muted and wide communication gaps between government and citizens may occur. Policy making and planning can thus largely be top-down processes in such situations.[20] There are exceptions, however. Valencia-Sandoval et al. (2010) used participatory methods to plan for sustainable community landscapes in Mexico.

---

[18] Eastern Africa provides good examples of the potential transformation of dryland economies with respect to recent investments to improve road and rail networks (https://thediplomat.com/2019/01/the-chinese-railways-remolding-east-africa/). Such infrastructure will promote many aspects of pastoral development. In addition, Ethiopian planners are preparing for an anticipated exodus of farmers to urban locales as a result of rural overpopulation and climate change. Plans include designing intermediate settlements to attract and retain migrants before they reach congested, big-city destinations where many public services are already overwhelmed (http://www.spiegel.de/international/tomorrow/ethiopia-plans-to-build-8000-new-cities-in-countryside-a-1197153.html).

[19] While planning is often associated with urban settings, US federal agencies have long incorporated participatory public planning into rural conservation and management efforts involving federal lands or federal funds. Examples of federal planning efforts include the National Environmental Policy Act (NEPA; https://www.epa.gov/laws-regulations/summary-national-environmental-policy-act). The Environmental Protection Agency (EPA) has established best practices for public participation (https://www.epa.gov/international-cooperation/public-participation-guide-introduction-public-participation).

[20] D. L. Coppock, personal observations.

Synthesis

A United Approach

ERIS In general, the advantages of integrating more public participation with applied research, outreach, policy making, and planning are self-evident. A heightened utility of project outcomes or impacts should improve dryland stewardship. This unified approach can be consolidated as "*E*ngaged *R*esearch within an *I*nnovation *S*ystem," or ERIS for short.

Process Comprehensive approaches like ERIS to integrate the various actors—from community members to researchers, outreach agents, development specialists, policy makers, and planners—should be a key feature of a new way of working in the dry lands. All of these people (and associated institutions) should be part of a multi-layered process when a project begins (Fig. 7.7). Operational steps can then include:

- Basic or applied research (Fig. 7.2) supplies background information on important systemic features such as climate change forecasts or population projections;
- Inventories of key policies and planning efforts provide additional systemic background;

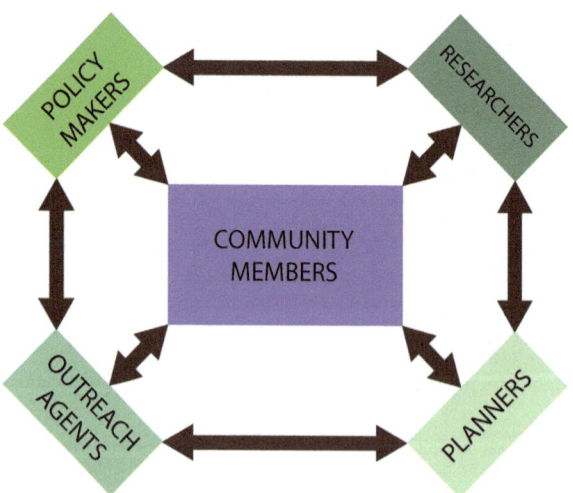

Fig. 7.7 Diagram of major actors in an engaged research/innovation system (ERIS). Such a framework can potentially have several more layers of actors to achieve a complex goal (D. L. Coppock)

- An initial membership of an IS can be identified;
- Community engagement begins with PRAs that are conducted across multiple locations; this can reveal patterns in priority problems and generate hypotheses. Refinement of potential problem solutions can be assessed across multiple locations using PAR. Once major issues are clarified, an expanded membership of an IS can be proposed;
- The research process can then morph into other forms of hypothesis testing, followed by a knowledge-to-action mode and/or an engaged research mode (Fig. 7.2). Feedback loops are solidified to better connect the evolving roster of the IS;
- Local or regional partnerships within the IS can be expanded depending on resources and emerging needs; and
- Multiple IS across regions or nations can be connected via a central coordinating platform; this expands the possibilities for supra-regional or international partnerships, further enlarging the scope for information sharing—and hence impact—within any given IS.

Potential Pitfalls

While the process of creating an innovation system appears straightforward, in reality it can be difficult (Coppock et al. 2009). First, there usually are no traditions for the various actors to work closely together, and the transaction costs of doing so can be a deterrent. Second, lack of continuity among participants is a common challenge. Rapid turnover among key personnel leads to a revolving door; the project then incurs added costs to recruit and educate new participants. Third, the aforementioned short funding cycles rarely afford enough time to get sufficient momentum to achieve impact. Fourth, inadequate funding levels can undermine collaborative partnerships given that incentives matter. Some of the most acute trade-offs would be felt by scientists given that research traditions typically have not involved the public (Table 7.2). Benefits to researchers from public involvement could increase if institutional reward systems recognize the value of engaged research (Whitmer et al. 2010).

Table 7.2 Differences between participatory and conventional research methods in the context of community-based problem-solving (Coppock 2016)

Topic	Participatory research	Conventional research
Research innovation	Higher	Lower
Publication output	Lower	Higher
Problem-solving	Higher	Lower
Funding required	Higher	Lower
Transaction costs	Higher	Lower

Conclusions

If we are to achieve improved project performance in the context of a new era in dryland science and stewardship, use of an approach like ERIS is needed. The ERIS perspective can also make a large contribution to promoting transdisciplinary work to foster problem-solving for dryland social-ecological systems. Donors are key to such efforts because ERIS-type projects must be longer in duration and broader in scope than what has typically been supported in the past. This requires greater donor patience and larger funding commitments to incentivize and sustain the ERIS process. Similarly, institutions that contribute personnel to ERIS-type efforts must recognize the value of community-based work and reward the adherents accordingly. An ERIS platform for inter-project coordination could easily be part of a comprehensive Agadir platform (see Chapter 13, this volume). Successful implementation of a multi-national ERIS requires visionary leadership and commitment to detail. The main way to institutionalize ERIS-type approaches starts at regional or national levels with policies supporting changes in how government agencies and their stakeholders work together.

Acknowledgements Special thanks to Elisabeth Huber-Sannwald and the editorial team for the invitation to submit this chapter. Several illustrations (i.e., Table 7.2 and Figs. 7.4, 7.5, and 7.6) have been reprinted from a paper by Coppock (2016) that has open-access status with Elsevier. Brian Kartchner (Utah State University) assisted with producing the figures.

References

Birkland T (2016) An introduction to the policy process: theories, concepts, and models of public policy making, 4th edn. Routledge, New York
Bonfil G (1970) La antropología social en México: Ensayo sobre sus nuevas perspectivas. An Antropol 7:24–39
Cauvin T (2016) Public history: a textbook of practice, 1st edn. Routledge, New York
Cavalier D, Kennedy E (2016) The rightful place of science: citizen science. Consortium for Science, Policy & Outcomes, Tempe, AZ
Chambers R (1983) Rural development: putting the last first. Longman, Harlow
Chambers R (1994) The origins and practice of participatory rural appraisal. World Dev 22(7): 953–969
Coppock DL (2016) Cast off the shackles of academia! Use participatory approaches to tackle real-world problems with underserved populations. Rangelands 38(1):5–13
Coppock DL, Desta S, Tezera S, Gebru G (2009) An innovation system in the rangelands: using collective action to diversify livelihoods among settled pastoralists in Ethiopia. In: Sanginga P, Waters-Bayer A, Kaaria S, Njuki J, Wettasinha C (eds) Innovation Africa: enriching farmer's livelihoods. Earthscan, London, pp 104–119
Coppock DL, Fernández-Giménez M, Hiernaux P, Huber-Sannwald E, Schloeder C, Valdivia C, Arredondo JT, Jacobs M, Turin C, Turner M (2017) Rangelands in developing nations: conceptual advances and societal implications. In: Briske D (ed) Rangeland systems: processes, management, and challenges. Springer, Berlin, pp 569–641
De Silva G, Wignaraja P, Mehta N, Rahman M (1979) Bhoomi Sena, a struggle for people's power. Dev Dialogue 2:3–70

Derr T (2018) Climate change perceptions and adaptation among small-scale farmers in Uganda: a community-based approach. M.Sc. thesis, Department of Environment and Society, Utah State University, Logan, UT

Fals Borda O (1979) The problem of investigating reality in order to transform it: the Colombian experience. Dialect Anthropol 4(1):33–55

Feagin JR, Vera H, Ducey K (2015) Liberation Sociology, 3rd edn. Paradigm Publishers, Boulder

Forsyth A (2012) Commentary: alternative cultures in planning research—from extending scientific frontiers to exploring enduring questions. J Plan Educ Res 32(2):160–168

Friere P (1970) Pedagogy of the oppressed. Continuum, New York

Gramberger M (2001) Citizens as partners: OECD handbook on information, consultation and public participation in policy making. Organisation for Economic Co-operation and Development, Paris

Greenwood D, Levin M (2007) Introduction to action research: social research for social change, 2nd edn. Sage Publications, Thousand Oaks

Huang J, Li Y, Fu C et al (2017) Dryland climate change: recent progress and challenges. Rev Geophys 55(3):719–778

Kelly E (2010) Community planning: an introduction to the comprehensive plan, 2nd edn. Island Press, Washington, DC

MEA (Millennium Ecosystem Assessment) (2005) Ecosystems and human well-being: synthesis. Island Press, Washington, DC

Narayanasamy N (2009) Participatory rural appraisal: principles, methods, and application. Sage Publications, New Delhi

Overdevest C, Huyck Orr C, Stepenuck K (2004) Volunteer stream monitoring and local participation in natural resource issues. Human Ecol Rev 11:177–185

Priscoli J (2004) What is public participation in water resources management and why is it important? Water Int 29(2):221–227

Rogers E (2003) Diffusion of innovations, 5th edn. Free Press, New York

Röling N (2009) Conceptual and methodological developments in innovation. In: Sanginga P, Waters-Bayer A, Kaaria S, Njuki J, Wettasinha C (eds) Innovation Africa: enriching farmers' livelihoods. Earthscan, London, pp 9–34

Sanginga PA, Waters-Bayer A, Kaaria S, Njuki J, Wettasinha C (2009) Innovation Africa: beyond rhetoric to praxis. In: Sanginga P, Waters-Bayer A, Kaaria S, Njuki J, Wettasinha C (eds) Innovation Africa: enriching farmers' livelihoods. Earthscan, London, pp 374–386

Shaner W, Phillipp P, Schmehl W (1982) Farming systems research and development: guidelines for developing countries. Consortium for International Development, Tucson

Umemoto K (2001) Walking in another's shoes: epistemological challenges in participatory planning. J Plan Educ Res 21:17–31

Valencia-Sandoval C, Flanders D, Kozak R (2010) Participatory landscape planning and sustainable community development: methodological observations from a case study in rural Mexico. Landsc Urban Plan 94:63–70

Waters-Bayer A, Sanginga P, Kaaria S, Njuki J, Wettasinha C (2009) Innovation Africa: an introduction. In: Sanginga P, Waters-Bayer A, Kaaria S, Njuki J, Wettasinha C (eds) Innovation Africa: enriching farmers' livelihoods. Earthscan, London. pp 1–5 (chapter 1)

Whitmer A, Ogden L, Lawton J, Sturner P, Groffman P, Schneider L, Hart D, Halpern B, Schlesinger W, Raciti S, Bettez N, Ortega S, Rustad L, Pickett S, Killilea M (2010) The engaged university: providing a platform for research that transforms society. Front Ecol Environ 8(6):314–321

Whyte W (1989) Advancing scientific knowledge through participatory action research. Sociol Forum 4(3):367–385

Chapter 8
Spatial and Temporal Analysis of Precipitation and Drought Trends Using the Climate Forecast System Reanalysis (CFSR)

D. A. Martinez-Cruz, M. Gutiérrez, and M. T. Alarcón-Herrera

Abstract Determination of spatial and temporal patterns of precipitation and drought remains a challenge in many arid and semi-arid regions. Generally, in these areas, precipitation measurements obtained from field stations are few or unreliable. Reanalysis precipitation data has opened up new large-scale hydrological application in data-sparse or ungauged catchments. This study presents a method based on GIS, statistical tests, and satellite-based precipitation for the analysis of spatial and temporal precipitation and drought trends. As a case study, we used 35 years (1979–2013) of precipitation data from the climate forecast system reanalysis (CFSR) for the Mexican state of Durango. For trend detection, we used the statistical test of Mann–Kendall at 5% standard precipitation index timescale of 12 months (SPI-12) to analyze the spatiotemporal trends of drought, and principal component analysis (PCA) was applied to characterize the spatial patterns. In many of the grid locations analyzed, the precipitation trends were not statistically significant. However, the months within winter and spring, except for June, showed a decreasing trend in precipitation, while the months of July, August, and September showed an increasing trend. This variation in rainfall intensity agrees with the climate pattern reported for this region. The frequency of drought events for each month during the period of analysis was mapped. The incidence of drought events was higher in June and September. Drought events are present all around the state of Durango, but they are more frequent in the northern part, in the Sierra Madre Occidental, Mapimi, and Laguna regions.

D. A. Martinez-Cruz · M. T. Alarcón-Herrera (✉)
Centro de Investigación en Materiales Avanzados, S.C. Durango, Durango, Mexico
e-mail: teresa.alarcon@cimav.edu.mx

M. Gutiérrez
Department of Geography, Geology and Planning, Missouri State University,
Springfield, MO, USA

© Springer Nature Switzerland AG 2020
S. Lucatello et al. (eds.), *Stewardship of Future Drylands and Climate Change in the Global South*, Springer Climate,
https://doi.org/10.1007/978-3-030-22464-6_8

Keywords Spatial–temporal · Trends · Precipitation · Drought · Satellite-based precipitation

Introduction

Detailed temporal and spatial trend analysis of precipitation is essential to supply useful and reliable information to water managers. This analysis provides detail on how much rain falls, and also its intensity, temporality, and variability. This information can be applied in agricultural planning, flood frequency analysis, flood hazard, hydrological modeling, and water resource assessment (Gallego et al. 2011). Precipitation can be measured by gauge observations, weather radar observation, and remotely sensed observation (Ashouri et al. 2016). From these, gauge stations are subject to major limitations, including sparse network observations, complex terrains, interrupted/limited data monitoring records, instrumental error, and, especially in the case of developing countries, the lack of equipment and limited funds (Vu et al. 2018).

An alternate way to obtain an estimation of precipitation in areas where rain gauge data are scarce or uncertain is radar observations. Hydrological studies based on radar data solve some of the problems associated with gauge measurements by discriminating different forms of precipitation such as hail, snow, and rainfall and by determining the appropriate relationship between radar reflectivity and rain rate, storm, and rainfall. They are however available only in limited land regions (Fuka et al. 2014).

Satellite remote sensing and reanalysis techniques have gained more attention recently, because of their ability to provide continuous and high-resolution precipitation estimates at a quasi-global scale, and by not being limited by topography (Gao et al. 2012). Reanalysis data are an important source of high quality data for analyses of precipitation of the current situation and past conditions (Chen and Han 2016). The reanalysis dataset is obtained by combining advanced forecast models and data assimilation systems to create global datasets of the atmosphere–land surface and ocean. The climate forecast system reanalysis and reforecast (CFSR) products are available hourly and with a resolution of up to 20 s, horizontal of 0.5° latitude × 0.5° longitude (Saha et al. 2010). The National Prediction Center (NCEP), The National Center for Atmospheric Research (NCAR), and the National Oceanic and Atmospheric Administration/Climate Diagnostics Center (NOAA/CDC) developed data from the numerical model NCEP/NCAR Reanalysis. These are available at the levels of 2.5° latitude by 2.5° longitude (278.3 km by 278.3 km) (Kistler et al. 2001). The Japanese Meteorological Agency (JMA) also has its reanalysis databases: the JRA-25 (Japanese 25-year Reanalysis)—JCDAS (JMA Climate Data Assimilation System) and the JRA-55 (Japanese 55-year Reanalysis) (Kobayashi et al. 2015).

In the state of Durango, Mexico, an arid to semi-arid climate predominates and drought events are common. This climate contributes to the scarcity of water resources and compromises the sustainability of the socio-ecological systems of the region. The analysis of documented droughts could help provide an early diagnostic, identify the zones facing a higher risk of drought, and provide the information needed by management programs to cope with its adverse effects. Currently, geographic information systems (GIS) and remote sensing (RS) have been fundamental in studying different types of hazards, either natural or anthropogenic. This study emphasizes the application of RS and GIS in the field of drought risk evaluation to better understand the spatial and temporal variability of rainfall and drought trends over these predominantly arid and semi-arid zones with otherwise limited precipitation (gauging) data.

Materials and Methods

Standardized Precipitation Index (SPI)

McKee et al. (1993) developed the standardized precipitation index (SPI) to define and monitor droughts (Ghosh and Mujumdar 2007). In the last decade, the SPI became the most popular drought index, based on its theoretical development, robustness, and versatility for drought analysis (Rivera and Penalba 2014). The SPI represents the number of standard deviations from which a precipitation value is above or below the climatological average of a particular location. For the calculation of the SPI, the accumulated rainfall series on different timescales are divided into 12 monthly series, which are adjusted to a theoretical probability distribution that represents the variations of rainfall in the study region. In this study, the two-parameter gamma distribution was used because of its known capability for modeling the variability in precipitation of long-term rainfall series in semi-arid zones of Mexico (Mosiño Alemán and Garcia 1981). The gamma probability distribution of two parameters is described by Comtois (2000) as:

$$g(x) = \frac{x^{\alpha-1} e^{\frac{-x}{\beta}}}{\beta^{\alpha} \Gamma(\alpha)} \quad x, \alpha, \beta > 0 \tag{8.1}$$

where $\Gamma(\alpha)$ is the gamma function; and the two parameters α, β are, respectively, scale and shape parameters of the space under consideration. The calculation of the parameters was obtained for each location, for a 12-month scale. In order to estimate the parameters, probability weighted moments were used (Greenwood et al. 1979). The cumulative gamma probability distribution was then transformed to a normal distribution in order to obtain the SPI. This process was repeated in each of the 117 locations. Then, the R Package created by Beguería and Vicente-Serrano (2017) was used to calculate SPI. This program, available at http://sac.csic.es/spei, standardizes a variable following a gamma distribution function.

SPI class	Drought category
2.0+	Extremely wet
1.5 to 1.99	Very wet
1.0 to 1.49	Moderately wet
−0.99 to 0.99	Near normal
−1.0 to −1.49	Moderately dry
−1.5 to −1.99	Severely dry
−2 and less	Extremely dry

Table 8.1 Standardized precipitation index (SPI) values (McKee et al. 1993)

The values acquired by the SPI represent the current hydric condition concerning the historical series. It is classified according to Table 8.1.

Statistical Tests for Trend Detection: Mann–Kendall (MK) Test

The nonparametric Mann–Kendall (MK) test (Khaliq et al. 2009) is a widely applied technique for the detection of trends in climatic and hydrologic time series. This test is computed for assessing if there is a monotonic upward or downward trend of precipitation over time. Although it is a nonparametric (distribution-free) test, there is a necessary assumption for no correlation to assure the power of the test. The assumption of independence requires that the time between samples be sufficiently large so that there is no correlation between measurements collected at different times. In such cases, the existence of serial correlation will affect the ability of the MK test to assess the significance of the trend, and the MK and the Theil–Sen slope would be unable to consider the AR process of the time series (Hamed and Rao 1998).

The nonparametric MK test is used for studying patterns to identify trends in a time series, where N is known as the total number of data in the time series, and the equation calculates statistic S:

$$S = \sum_{i=1}^{N-1}\sum_{j=i+1}^{N} \operatorname{sgn}\left(Yj - Yi\right) \tag{8.2}$$

where Yj is the value of the jth data, n is the number of data, and $\operatorname{sgn}(\theta)$ is the sign function:

$$\operatorname{sgn}(\theta) = \{+1 \text{ if } \theta = Yj - Yi > 0 \quad 0 \quad \text{if } \theta = Yj - Yi = 0 - 1 \quad \text{if } \theta = Yj - Yi < 0 \tag{8.3}$$

When statistic S exhibits a positive value the trend is upward, while a negative value of S indicates the opposite. The S has a normal distribution when $N \geq 8$, and its mean and variance are calculated using the below equations:

$$E[S] = 0 \qquad (8.4)$$

$$\mathrm{var}(S) = \frac{\left[N(N-1)(2N+5) - \sum_{i=1}^{n} t_i i(i-1)(2i+5)\right]}{18} \qquad (8.5)$$

where t_i is the number of data in the ith tied set. The test statistics Z is computed by Eq. (8.6):

$$Z = \{(S-1)/\sqrt{\mathrm{Var}(S)} \quad S>0 \quad 0 \quad S = 0(S+1)/\sqrt{\mathrm{Var}(S)} \quad S<0 \qquad (8.6)$$

The value of Z is used to detect significant trends, at a significance level of $\alpha = 0.05$. A positive value of Z greater than 1.96 confirms an increasing trend, while a negative value smaller than 1.96 reveals a decreasing trend.

Autocorrelation Function

The k-order autocorrelation coefficient of a stationary stochastic process measures the degree of linear association between two random process variables separated k periods:

$$p_k = \frac{\mathrm{cov}(Y_t Y_{t+k})}{\sqrt{V(Y_t)V(Y_{t+K})}} \qquad (8.7)$$

As this is a correlation coefficient, it does not depend on units. The autocorrelation function of a stationary stochastic process is a k function that collects all the autocorrelation coefficients of the process and is denoted by ρk, $k = 0, 1, 2, 3, \ldots$ The autocorrelation function is usually represented graphically using a bar graph called a correlogram.

Theil–Sen's Estimator

The Theil-Sen estimator (Theil 1950; Sen 1968) was used to compute the magnitude of the trend in rainfall time series, which is calculated as follows:

$$\beta = \mathrm{Median}\left[Yi - Yj / i - j\right] \qquad \forall j < i \qquad (8.8)$$

where $1 < j < i < n$. If N indicates all combinations of record pairs for the complete dataset, and the value of the slope is calculated as $n = N (N - 1)/2$ and β is considered to be the median of these n values.

Principal Component Analysis (PCA)

The algorithm for the calculation of the PCA can be described in a summarized form using matrix algebra (Jolliffe and Cadima 2016), as a matrix with N rows as intervals of time and n as points in the space. The matrix resultant will be $N \times n$, where i is the index time and j the space, in this case j represents each location with information of precipitation. The value in any point in time and space is denoted then by x_{ij}. The purpose of the PCA is to find a coefficient matrix (E) that modifies x by lineal operations to get a new group of variables Z. These new variables must represent the majority of variance in terms of n points in the space, in a smallest number possible, where $p \leq n$

$$Z = XE \tag{8.9}$$

$$Z\_ij = \Sigma\_(k=1)^p x\_(i,k) e\_(k,j) \qquad i = 1,...,N; \qquad j = 1,...,n \ \ p \leq n \tag{8.10}$$

The columns of the matrix E are called vectors or orthogonal empirical functions, and the columns of the matrix Z are called principal components.

We applied PCA in a drought index (SPI-12) time series at 117 locations of gridded point precipitation. The loadings of main rotated PCA for each location across Durango were mapped. The number of leading components retained for rotation was evaluated using the sampling errors of eigenvalues associated with the principal components (O'Donnell et al. 2011). PCA results were used to divide the subregions according to drought characteristics. Since the topographical and climatic features are complex, recognizing the spatial patterns of drought characteristics is useful to regional planners.

Case Study

Study Area and Data

The state of Durango is located to the northwest of the central part of the Mexican Republic, between parallels 22°17′ and 26°50′ north latitude and between meridians 102°30′ and 107°09′ west longitude (Fig. 8.1). Eight physiographic sub-provinces characterize the state (Pedroza Sandoval et al. 2014): (1) Chihuahua's Great Plateau and Canyons, (2) Durango's Great Plateau and Canyons, (3) Southern Plateaus and Canyons, (4) mountains and plains of Durango, (5) Mapimi Basin,

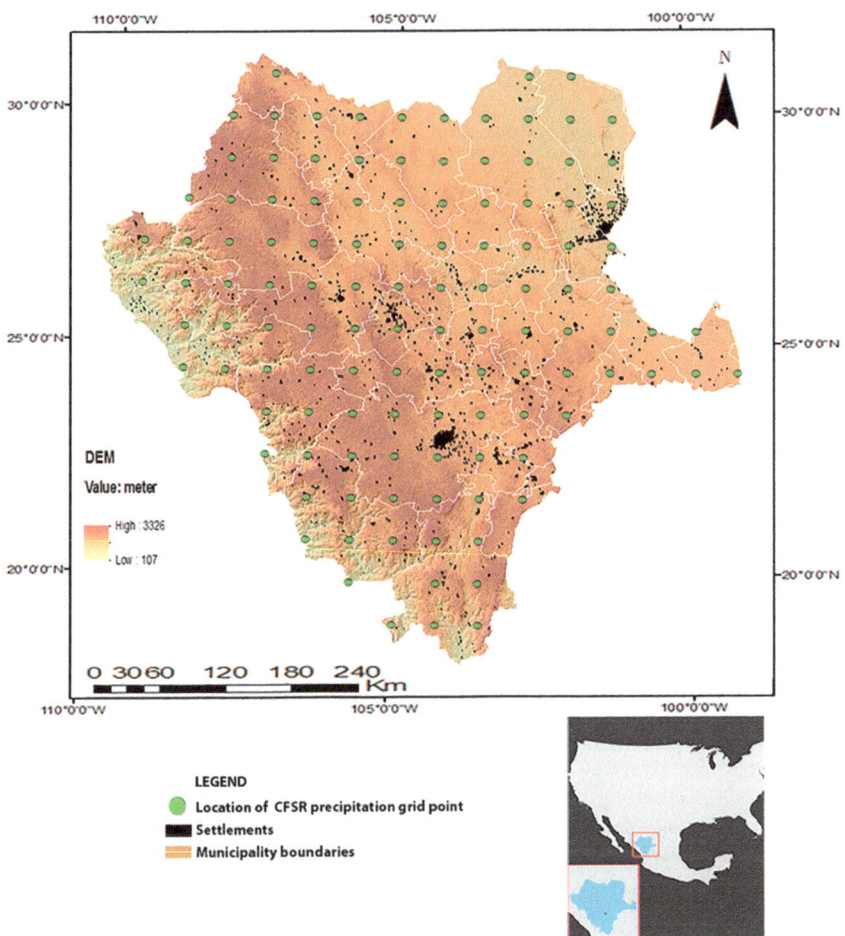

Fig. 8.1 Geographical distribution of the municipality boundaries and topographical characteristics of the CFSR precipitation grid point locations in the study area

(6) Transverse Ranges, (7) mountains and hills of Aldama and Rio Grande, (8) mountains and plains of the North (Fig. 8.2).

Precipitation Data

The CFSR is the product of a global climate reanalysis and is provided as gridded data generated by the NCEP Global Forecast System (Saha et al. 2010). The CFSR weather data are a product based on habitual ground observation data, satellite remote sensing data, and other advanced technology and represent a global,

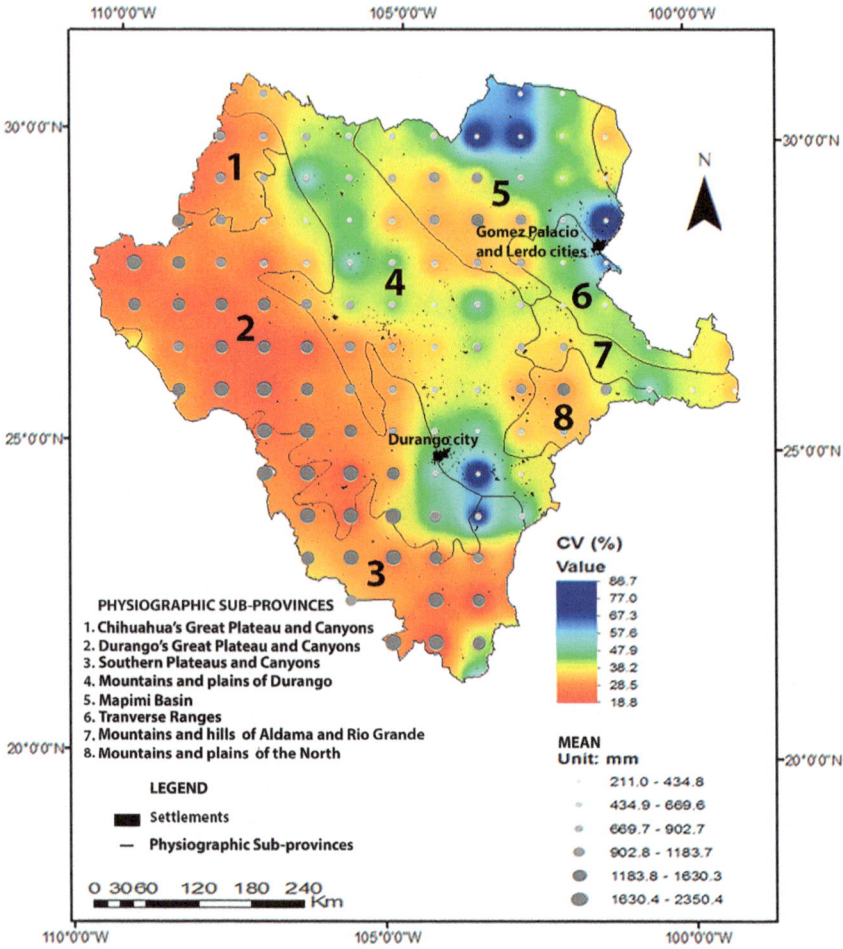

Fig. 8.2 Mean and coefficient of variation (CV %) for annual precipitation (*n* = 35) at 117 CFSR precipitation grid point locations

high-resolution coupled atmosphere–sea ice–ocean–land system. The CFSR global atmosphere data have a spatial resolution of approximately 38 km, and the data are available since 1979. The CFSR weather data for the study area (bounding box: latitude: 27.21°–22.16°N and longitude: 102.18°–107.63°W) were downloaded from the SWAT model's official website (http://globaweather.tamu.edu/). The data includes daily rainfall, maximum and minimum temperature, as well as wind speed, and solar radiation for 117 locations (Fig. 8.1).

Results and Discussion

Spatial Variability of Annual and Monthly Precipitation

The spatial patterns of mean and coefficient of variation (CV %) for the annual precipitation data from January 1979 to December 2013 at 117 locations are shown in Fig. 8.2. The lowest values of mean annual precipitation (MAP), less than 434 mm, occur in lowland areas (less than 800 m of altitude) of the state of Durango. These areas belong to five physiographic sub-provinces: mountains and plains of Durango, Mapimi Basin, Transverse Ranges, mountains and hills of Aldama and Rio Grande, and mountains and plains of the North. The highest MAP values in the state, greater than 1630 mm, occur at highlands (more than 2400 m of altitude). These areas correspond to two of the physiographic sub-provinces: Durango's Great Plateau and Canyons and Southern Plateaus and Canyons. These results show that elevation affects significantly the precipitation in the study area. A similar distribution of annual precipitation in Durango was reported by Návar (2014) using 80 climatic ground stations.

The degree of rainfall variability can be classified according to CV% (Hare 1983). When CV% < 20 it is considered to have low variability, 30 > CV% > 20 moderately variability, and CV% > 30 is highly variable. Areas with CV > 30% are also considered vulnerable to drought. Approximately three quarters of the territory of Durango state have a CV > 30%. Only in the highlands the CV reaches 18–20%. The five physiographic sub-provinces with the lowest values of MAP also showed higher variability (CV > 30%). Analysis of the variability of precipitation revealed that within the arid and semi-arid regions a high interannual variation is characteristic of the arid zones (Kiros et al. 2017). Previous studies established that a CV > 30% implies a high variation in the rainfall amount and distribution patterns (Ayanlade et al. 2018). It should be noted that the most populated urban centers in the study area, Victoria de Durango and Gomez Palacio-Lerdo, are located in areas with high interannual rainfall variability (CV > 47% and CV > 67%, respectively). These variability levels are similar to variability record levels reported for the arid regions with greater climatic variability in the world, such as the northern area of Australia and Somalia (Van Etten 2009). Interannual rainfall variability is a crucial component of climate, and can even be a more critical climatic feature than the long-term average (Thornton et al. 2014). Climate variability has direct impacts on biophysical and biological systems and can have important repercussions on social phenomena. Further, to make possible the sustainable utilization of arid lands, land managers need to be aware of connection between rainfall variability and biotic/landform responses; for instance, when making decisions about fire management and surface water storage (Ffolliott 2012; Pandey 2005; Waswa 2012).

Figure 8.3 shows the average of accumulated rainfall per month in the state of Durango, for the years 1979–2013. The area is characterized by the strong mid to late summer precipitation maximum in July, August, and September with considerably less precipitation during the rest of the year. Most of the state, except the

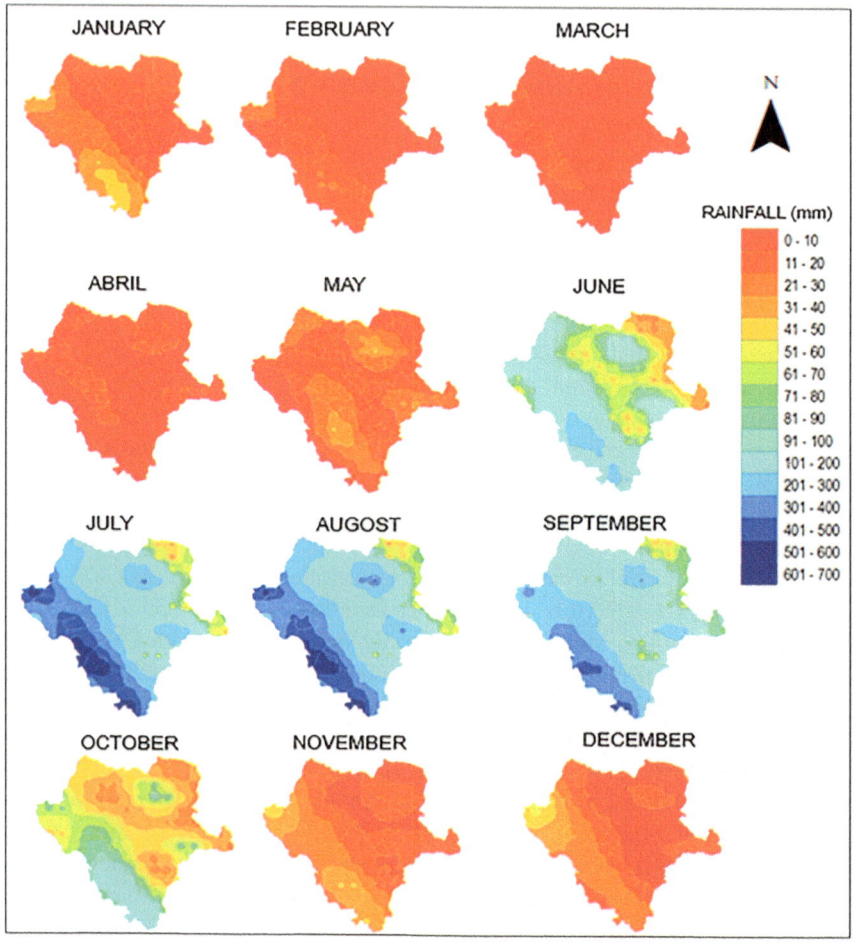

Fig. 8.3 Spatial distribution of accumulated rainfall for each month in the state of Durango (average for the period from 1979 to 2013)

eastern-most tip, lies within the region called the monsoon region, which stretches along the Sierra Madre Occidental in northwestern Mexico and follows the continental divide through eastern Arizona, most of New Mexico and extends into southern Colorado in the USA (Comrie and Glenn 1998). Seasonal monsoon mainly controls precipitation in the north of Mexico and is partially responsible for the seasonal and, potentially, interannual variations and oscillations in precipitation (Comrie and Glenn 1998; Higgins et al. 1999). Several climatic trends and oscillations contribute to the monsoonal variability in the region, including the El Niño–Southern Oscillation (ENSO), the Pacific Decadal Oscillation (PDO), and the North Atlantic Oscillation (NAO) (Návar 2014).

Spatial Patterns of Temporal Trends in Precipitation

Positive or negative autocorrelation affects the variance (S). Positive autocorrelation increases the variance (S) and negative autocorrelation decreases the variance (S). Thereby, positive autocorrelation overestimates the significance of both positive and negative trends, and negative autocorrelation underestimates the significance (Dinpashoh et al. 2013). It is important to check the autocorrelation in a given series and to adjust the test if necessary, as often the trend of precipitation autocorrelation receives little attention.

We calculated the coefficient of autocorrelation lag-1 as significant (positive or negative) autocorrelation coefficient ($r1$) for the 12 months of the year, in 117 CFSR grid point locations. Table 8.2 lists the results of autocorrelation lag-1 in the monthly time series. The autocorrelation close to zero is observed in the time series January–April and October–December. For April, May, June, July, and September, a few locations showed a positive correlation. This pattern can be due to the shaping effect of the monsoon season in the occurrence of precipitation in the state of Durango. In contrast, during winter the unpredictable phenomena like ENSO and PDO influence precipitation in the northwest of Mexico (Magaña et al. 2003). The scarce autocorrelation found in winter and part of spring for the analyzed locations suggests an increased randomness in the occurrence of the precipitation during those months.

Temporal trends in monthly and annual precipitation and their respective change values (mm/year) for each 117 localizations were obtained by the modified Mann–Kendall test (modMK1) (Hamed and Rao 1998) and the Theil–Sen's estimator (Fig. 8.4). The temporal trends were mapped to identify their spatial distribution across all months of the year. The standard test statistic Z was used as the criterion to identify temporal trends; $Z < -1.96$ indicated a significant decreasing trend, $-1.96 < Z < 0$ indicated a no significant decreasing trend, $0 < Z < 1.96$ indicated a not significant increased trend, $Z > 1.96$ indicated a significant increased trend.

Table 8.2 Number of locations showing significant and not-significant lag-1 autocorrelation and significant and not significant trends throughout the state of Durango

	+	–		Increasing	Decreasing	Increasing	Decreasing
January	0	0	117	0	2	0	115
February	0	0	117	0	0	18	98
March	0	2	115	0	0	33	84
April	9	0	108	7	110	0	0
May	37	0	80	2	0	10	101
June	6	0	111	1	4	38	60
July	27	0	90	3	1	114	7
August	8	0	109	0	14	26	86
September	14	0	103	4	2	82	23
October	2	2	113	0	0	52	67
November	0	2	115	0	1	21	90
December	0	3	114	0	42	0	64

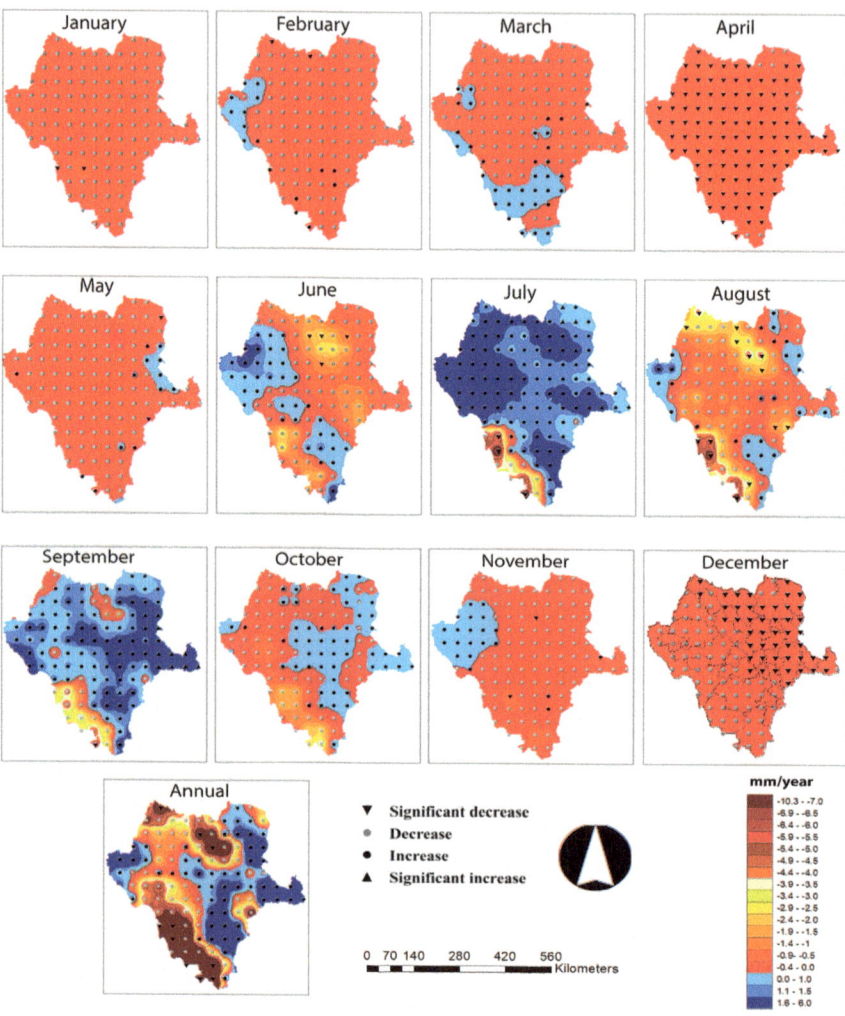

Fig. 8.4 Maps showing the spatial patterns of Sen Slope (mm/year) and Z statistic using mod MK1 at 5% significance level for monthly and annual precipitation in Durango

Figure 8.4 shows the spatial distribution of the positive (increasing) and negative (decreasing) trends of each one of 117 locations. In the vast majority of the 117 locations in Durango, the precipitation trends were not statically significant at the 95% (Table 8.2). The drier months (January, February, March, April, May, November, and December) showed a decreasing tendency, but were not statistically significant. The wetter months (July, September, and October), except August, showed an increasing trend although they were also not statistically significant.

Spatial Patterns of Drought Frequency

Since droughts vary in time and space, the regional distribution of drought using drought events with SPI-12 values less than −1.0 was interpolated and mapped for 12 months in a year. This value of SPI is typically chosen as an indicator of drought. In some regions, a good rainfall in one particular month can create the impression that the drought is over, but until the SPIs are not above a particular value at all scales (typically −1), a drought will still affect a region in one way or another (Patel et al. 2007).

We calculated the drought event for each location in the state of Durango and interpolated the spatial distribution of the drought events (Fig. 8.5). More than half of the state experienced 53 or more droughts during the period of analysis of 35 years. SPI field station calculations in the ground station (Rivera del Río et al. 2007) recorded an average of 15 dry periods in 14 years. There is a certain proportion of SPI drought events detected with ground field measurements and with satellite precipitation. The station (−104.38, 22.63) in the municipality of Mezquital registered fewer drought events (14), whereas the location (−105.63, 25.13) in Tepehuanes presented the most (74) drought events. The average of drought events in all stations was 54.

The spatial distribution of frequency droughts events for the first 5 months was similar. At this time of year, there are slightly more droughts in the east and a small portion of the northeast of the study area. In the east, these areas include much of Chihuahua's and Durango's Great Plateau and Canyons and the northern portion of Southern Plateaus and Canyons physiographic sub-provinces. And in the north part, some areas of Mapimí Basin. By June, an increase in the frequency of drought events throughout the state of Durango was noticeable. At the beginning of the rainy season (July), the frequency drought events decrease in the study area and in this month the droughts are more frequents in the east part. This area covers the following physiographic sub-provinces, the eastern part of Mapimí Basin, the huge portion of Transverse Ranges, mountains and hills of Aldama and Rio Grande, mountains and plains of the North, and small portion in the south of the mountains and plains of Durango. However, at the end of the rainy season (September) the frequency of drought events increased throughout the state. October, November, and December have a similar spatial trend in the frequency of droughts, and during these months there are more incidences towards Mapimi basin, Transverse Ranges, mountains and hills of Aldama and Rio Grande and central part of Durango's Great Plateau and Canyons and mountains plains of Durango.

It is important to note that in the previous section, June and September showed many areas with increases in precipitation. Instead, this section indicates these months have large regions of frequents droughts, which seems to be contradictory. It is because the SPI for a particular month is calculated considering the accumulated rainfall from previous months, depending on the timescale used. In this case, although September and June have shown increases in precipitation, the index contemplates the decrease in rainfall that has occurred in previous months which has more intensity than the increases.

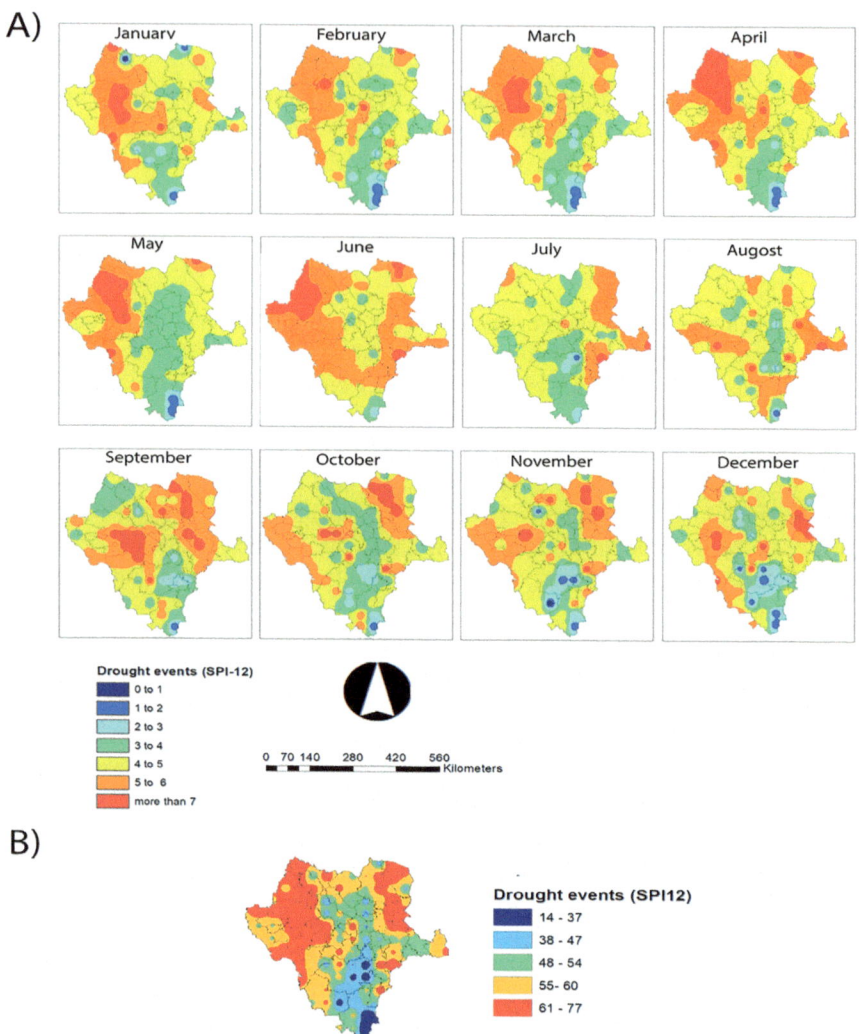

Fig. 8.5 (**a**) Spatial distribution of drought events identified by SP12 detailed for each month. (**b**) Spatial distribution of total drought events identified by SP12 since 1979–2013

Spatial Variability of Drought Using PCA

PCA on 12-month SPI was calculated for the 117 locations to identify spatial similitudes of drought over 420 months from 1979 to 2013 in the study area. The first three rotated components (RPCs) explained about 87.9% of total variance, and each principal component explained 68.6%, 13.9%, and 5.4% of the variance, respectively. The rotated loading for three RPCs for the drought time series was computed and interpolated for the 117 locations to classify and characterize the spatial

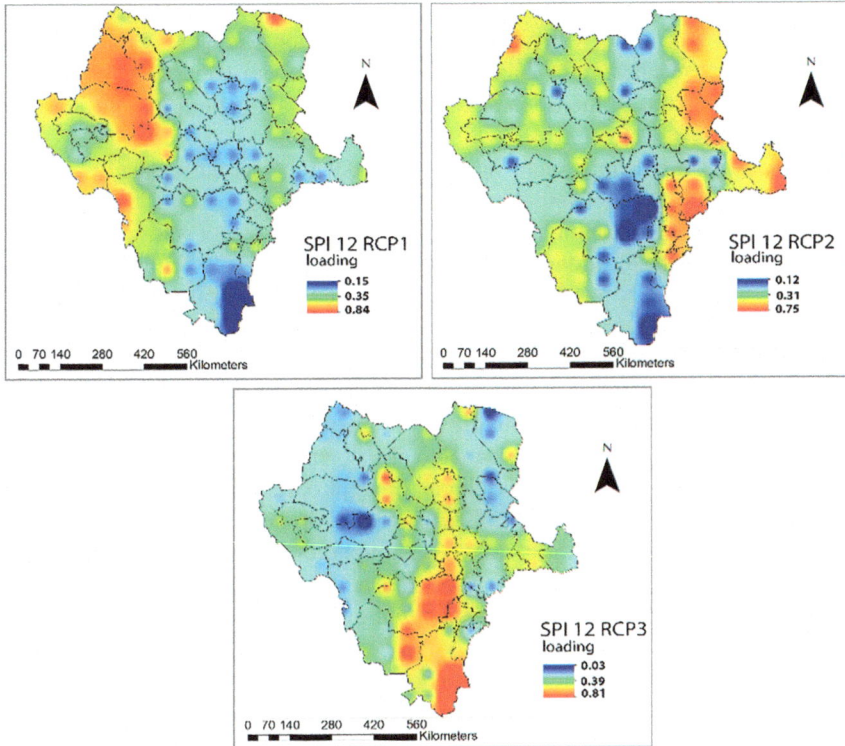

Fig. 8.6 Spatial variability of the loading for three rotated principal components obtained for the analysis of drought time series by SPI12

variability of droughts in the area of study. The PCA allows us to find regions with a similar occurrence of droughts detected by SPI-12 time series (Fig. 8.6). Considering the analysis in the previous section, the spatial distribution as principal component 1 (RPC1) indicates a high positive loading along the northern part of the state. In this area, droughts are more frequent. The second component (RPC2) indicates regions where drought events are also frequent but at a lower extent than in the area reported in the first component. The third (RPC3) shows an area where the drought events are less frequent. Although three RPCs were unable to extract the total variation of drought across the region, this operation was successful in finding the three major kinds of drought variabilities present in the area of study.

Conclusions

This study effectively identified the spatial and temporal variations and trends in rainfall and drought after 35 years of data in a semi-arid region of Northern Mexico. In much of the state of Durango, 70% or more of the annual rainfall occurs in July,

August, and September. The results also highlight the wide range of annual precipitation in the state, with places where precipitation averages 200 mm/year and areas where it reaches up to 2400 mm/year. The analysis confirms the dominant effect of the American monsoon season on the temporal pattern of rainfall in the area of study.

A decrease in precipitation to values of 0.5–0.9 mm/year during the dry months was observed in most of the area in contrast with an increase in values of 1.0–1.5 mm/year in the wet months. The fact that this trend was not statistically significant for many of the locations could be attributed to the effects of climate variability factors, such as El Niño–Southern Oscillation (ENSO) and the Pacific Decadal Oscillation (PDO).

The complexity of drought as a dangerous phenomenon and its interactions with socio-ecological systems has limited the progress of generating effective drought management and mitigation plans. However, quantifying the spatial and temporal occurrence of drought is a valuable tool to decision makers towards reducing social and ecological vulnerability at the time a drought occurs. Using the SPI 12 as an indicator of drought, more than 53 episodes of drought were identified in the 35 years of data, as well as the areas where droughts events are most frequent. These areas cover more than half of the state of Durango, mainly in its northwestern, northern, and eastern regions. These drought events remained relatively constant throughout the year, although a slight increase was observed in June and September.

Satellite observations offer a great opportunity for the operational monitoring of droughts and rainfall. Current and future satellite missions could be used to develop composite and multi-indicator drought models to improve early drought warning. While there are tremendous opportunities, there are also enormous challenges, including data continuity, unquantified uncertainty, sensor changes, and community acceptability. One of the significant limitations of many of the currently available satellite observations is their short temporal resolution. Many satellite missions and relevant sensors provide only a decade of data, which may not be sufficient to study droughts and precipitation trends from a climate perspective.

References

Ashouri H, Nguyen P, Thorstensen A, Hsu K-L, Sorooshian S, Braithwaite D (2016) Assessing the efficacy of high-resolution satellite-based PERSIANN-CDR precipitation product in simulating streamflow. J Hydrometeorol 17(7):2061–2076

Ayanlade A, Radeny M, Morton JF, Muchaba T (2018) Rainfall variability and drought characteristics in two agro-climatic zones: an assessment of climate change challenges in Africa. Sci Total Environ 630:728–737

Beguería S, Vicente-Serrano SM, Beguería MS (2017) Package SPEI, Calculation of the Standardised Precipitation-vapotranspiration, version 1.7. http://sac.csic.es/spei

Chen Y, Han D (2016) Big data and hydroinformatics. J Hydroinf 18(4):599–614

Comrie AC, Glenn EC (1998) Principal components-based regionalization of precipitation regimes across the southwest United States and northern Mexico, with an application to monsoon precipitation variability. Clim Res 10(3):201–215

Comtois P (2000) The gamma distribution as the true aerobiological probability density function. Aerobiologia 16(2):171–176

Dinpashoh Y, Mirabbasi R, Jhajharia D, Abianeh HZ, Mostafaeipour AJ (2013) Effect of short-term and long-term persistence on identification of temporal trends. J Hydrol Eng 19(3): 617–625

Ffolliott PF (2012) Sustainable use of natural resources of dryland regions in controlling of environmental degradation and desertification. In: Sustainable natural resources management. InTech, London

Fuka DR, Walter MT, MacAlister C, Degaetano AT, Steenhuis TS, Easton ZM (2014) Using the climate forecast system reanalysis as weather input data for watershed models. Hydrol Process 28(22):5613–5623

Gallego M, Trigo R, Vaquero J, Brunet M, García J, Sigró J, Valente MJ (2011) Trends in frequency indices of daily precipitation over the Iberian Peninsula during the last century. J Geophys Res Atmos 116(D2):7280

Gao Y, Liu MJH, Discussions ESS (2012) Evaluation of high-resolution satellite precipitation products using rain gauge observations over the Tibetan Plateau. Hydrol Earth Syst Sci 9(8):23

Ghosh S, Mujumdar P (2007) Nonparametric methods for modeling GCM and scenario uncertainty in drought assessment. Water Resour Res 43(7):W07405

Greenwood JA, Landwehr JM, Matalas NC, Wallis JR (1979) Probability weighted moments: definition and relation to parameters of several distributions expressable in inverse form. Water Resour Res 15(5):1049–1054

Hamed KH, Rao ARJ (1998) A modified Mann-Kendall trend test for autocorrelated data. J Hydrol 204(1–4):182–196

Hare FJ (1983) Climate and desertification. A revised analysis. World Meteorological Organization, WCP-44. World Climate Program, Geneva

Higgins RW, Chen Y, Douglas AV (1999) Interannual variability of the North American warm season precipitation regime. J Clim 12(3):653–680

Jolliffe IT, Cadima J (2016) Principal component analysis: a review and recent developments. Philos Trans A Math Phys Eng Sci 374(2065):20150202

Khaliq MN, Ouarda TB, Gachon P, Sushama L, St-Hilaire A (2009) Identification of hydrological trends in the presence of serial and cross correlations: a review of selected methods and their application to annual flow regimes of Canadian rivers. J Hydrol 368(1–4):117–130

Kiros G, Shetty A, Nandagiri L (2017) Extreme rainfall signatures under changing climate in semi-arid northern highlands of Ethiopia. Cogent Geosci 3(1):1353719

Kistler R, Kalnay E, Collins W, Saha S, White G, Woollen J, Chelliah M, Ebisuzaki W, Kanamitsu M, Kousky V, Van den Dool H, Jenne R, Fiorino M (2001) The NCEP–NCAR 50-year reanalysis: monthly means CD-ROM and documentation. Bull Am Meteorol Soc 82(2):247–268

Kobayashi S, Ota Y, Harada Y, Ebita A, Moriya M, Onoda H, Onogi K, Kamahori H, Kobayashi C, Endo H (2015) The JRA-55 reanalysis: general specifications and basic characteristics. J Meteorol Soc Jpn 93(1):5–48

Magaña VO, Vázquez JL, Pérez JL, JBJGi P (2003) Impact of El Niño on precipitation in Mexico. Geofis Int 42(3):313–330

McKee TB, Doesken NJ, Kleist J (1993) The relationship of drought frequency and duration to time scales. In: Preprints, eighth conference on applied climatology. American Meteorological Society, Anaheim, pp 179–184

Mosiño Alemán PA, Garcia E (1981) The variability of rainfall in Mexico and its determination by means of the gamma distribution. Geogr Ann Ser B 63(1–2):1–10

Návar J (2014) Spatial and temporal hydro-climatic variability in Durango, Mexico. Tecnol Cienc Agua 1:103–123

O'Donnell R, Lewis N, McIntyre S, Condon JJJC (2011) Improved methods for PCA-based reconstructions: case study using the Steig et al. (2009) Antarctic temperature reconstruction. J Clim 24(8):2099–2115

Pandey N (2005) Societal adaptation to abrupt climate change and monsoon variability: implications for sustainable livelihoods of rural communities. Winrock International–India, New Delhi

Patel NR, Chopra P, Dadhwal VK (2007) Analyzing spatial patterns of meteorological drought using standardized precipitation index. Meteorol Appl 14(4):329–336. https://doi.org/10.1002/met.33

Pedroza Sandoval A, Sánchez Cohen I, Becerra López J, Ramos Cortez E, Reyes Bernabé C, Rosales Palacios L, Vargas Piedra G (2014) Regionalización de zonas con escaso régimen pluvial: Estudio de caso zona Centro-Norte del estado de Durango, México. Rev Chapingo Ser Zonas Aridas 13(2):71–85

Rivera del Río R, Crespo Pichardo G, Arteaga Ramírez R, Quevedo Nolasco A (2007) Comportamiento espacio temporal de la sequía en el estado de Durango, México. Terra Latinoam 25(4):383–392

Rivera J, Penalba OJC (2014) Trends and spatial patterns of drought affected area in Southern South America. Climate 2(4):264–278

Saha S, Moorthi S, Pan H-L, Wu X, Wang J, Nadiga S, Tripp P, Kistler R, Woollen J, Behringer D (2010) The NCEP climate forecast system reanalysis. Bull Am Meteorol Soc 91(8):1015–1058

Sen PK (1968) Estimates of the regression coefficient based on Kendall's tau. J Am Stat Assoc 63(324):1379–1389

Theil H (1950) A rank-invariant method of linear and polynomial regression analysis. In: Proceedings of the Koninklijke Nederlandse Akademie van Wetenschappen, pp 1397–1412

Thornton PK, Ericksen PJ, Herrero M, Challinor A (2014) Climate variability and vulnerability to climate change: a review. Glob Chang Biol 20(11):3313–3328

Van Etten J (2009) Inter-annual rainfall variability of arid Australia: greater than elsewhere? Aust Geogr 40(1):109–120

Vu TM, Raghavan SV, Liong SY, Mishra A (2018) Uncertainties of gridded precipitation observations in characterizing spatio-temporal drought and wetness over Vietnam. Int J Climatol 38(4):2067–2081

Waswa BS (2012) Assessment of land degradation patterns in Western Kenya-implications for restoration and rehabilitation. Doctoral dissertation, Landwirtschaftlichen Fakultät Rheinischen Friedrich-Wilhelms-Universität, Bonn

Chapter 9
The Construction and Sabotage of Successful Agricultural Lands in Semiarid Lands: A Case Study of Vitivinicultural Areas in Northern México

I. Espejel, G. Arámburo, N. Badan, L. Carreño, A. Cota, G. Gutiérrez, L. Ibarra, C. Leyva, T. Moreno-Zulueta, L. Ojeda-Revah, L. Pedrín, C. Uscanga, M. Reyes-Orta, J. C. Ramírez, P. Rojas, J. Sandoval, C. Turrent, Á. Vela, and I. Vaillard

Abstract The land-use planning effort for a semiarid watershed where wine is the staple economic product was evaluated. Over the last 30 years, three top-down local public policies have been in effect with the shared objective of building a sustainable valley where rural settlements, Mediterranean-type crops, and tourism can thrive. The evaluation result was that most of the proposed guidelines were not implemented. The questions addressed in this chapter are: what are the main reasons behind the non-compliance of regulations in land-use planning and what kind of alternatives can be proposed? To answer these questions, documental analysis of land-use planning publications, reports, and laws was analyzed. All findings were interpreted by a transdisciplinary team (the coauthors: local producers, government stakeholders, ecosystems management researchers, and graduate students). The main results behind the unsuccessfulness of the evaluated public policy instruments are: (1) unclear land tenure, (2) ambiguous laws and lack of by-laws, (3) corruption between buyers and government agents, and (4) lack of awareness by users. Our alternative proposal is a bottom-up transdisciplinary plan with five main strategies: (a) a participatory observatory as a

I. Espejel (✉)
Facultad de Ciencias, Universidad Autónoma de Baja California UABC,
Ensenada, Baja California, Mexico
e-mail: ileana.espejel@uabc.edu.mx

G. Arámburo
Facultad de Ciencias Administrativas y Sociales, Universidad Autónoma de Baja California UABC, Ensenada, Mexico

N. Badan · P. Rojas
Rancho El Mogor, Ensenada, Mexico

© Springer Nature Switzerland AG 2020   147
S. Lucatello et al. (eds.), *Stewardship of Future Drylands and Climate Change in the Global South*, Springer Climate,
https://doi.org/10.1007/978-3-030-22464-6_9

science communication tool, (b) green infrastructure projects and guidelines, (c) identification and valuation of available ecosystem and environmental services, (d) diffusion of innovative wise water management techniques for the region, and lastly (e) a new larger scale land-use plan for the watershed that uses environmental units with landscape values for the local people. Our proposal might help others to change paradigms and design transdisciplinary regional plans.

Keywords Regional planning · Territorial ordinances · Watersheds

Introduction

The planning of any region is a long-term process since emergent issues are continuously arising. Socio-ecological system-based plans use the watershed as the basic territorial unit. Especially, in semiarid watersheds where the aquifer is the limiting factor and the axis of any planning. The ecosystem-based logic not always coincides with the logic followed in cultural, administrative, legal, economic, or social setting. Therefore, the land-use planning of a semiarid watershed has to be constructed in a process where environmental education plays a key role. As well, it must be a participative and creative process since the lack of data concerning subterranean water quantity, quality, and dynamics remains unmeasured and is a constant limitation. In consequence, the planning of land-use requires a transdisciplinary approach since land-use decisions are need to be taken by the government officials with the participation of local inhabitants.

L. Carreño · T. Moreno-Zulueta
Programa de Doctorado en Medio Ambiente y Desarrollo, Universidad Autónoma de Baja California UABC, Ensenada, Mexico

A. Cota · L. Ibarra · L. Pedrín · C. Uscanga
Programa de Maestría en Manejo de Ecosistemas de Zonas Áridas, Universidad Autónoma de Baja California UABC, Ensenada, Mexico

C. Leyva
Facultad de Ciencias, Universidad Autónoma de Baja California UABC, Ensenada, Mexico

G. Gutiérrez
Secretaria de Protección al Ambiente de Baja California, Ensenada, Mexico

L. Ojeda-Revah
Colegio de la Frontera Norte, Tijuana, Mexico

M. Reyes-Orta
Facultad de Turismo y Mercadotecnia, Universidad Autónoma de Baja California, Ensenada, Mexico

J. C. Ramírez · J. Sandoval · Á. Vela
Instituto Municipal de Investigación y Planeación de Ensenada, Ensenada, Mexico

C. Turrent
Taller de Arquitectura Contextual ClaCla, Ensenada, Mexico

I. Vaillard
Rancho Tres Mujeres, Ensenada, Mexico

Major Conceptual Advances in Land-Use Planning and Broader Applications

Modern land-use planning is based on bottom-up strategies (Sabatier 1986), where society stakeholders, policy makers, and academics interact and negotiate land-use priorities. The generation of knowledge, and co-design and management of natural resources emerge as key issues for socio-ecological system-based regional planning (Schuttenberg and Guth 2015). For successful land-use planning, sustainable development is a fundamental concept that involves the continuous dialogue between a team of scientist, inhabitants (at all sectors), and decision makers (at all levels) with the main objective of integrating ecological and social knowledge for the generation and development of proper and flexible public policies through a constant evaluation process.

The description and function of a watershed as a socio-ecological system has been presented by several authors (Márquez and Valenzuela 2008; Parkes et al. 2010; Rathwell and Peterson 2012; Fischer et al. 2015). Still, Fischer et al. (2015) mention that some issues need to be explored, such as the development of a stronger science–society interface that could help enhance understanding and improve the interactions involved in ecosystem stewardship. Watersheds have been considered as exploratory laboratories because they offer the ideal settings for long-term research of dynamic natural resources, such as water (Schmidt and Friede 1996; Slaughter and Richardson 2000; Malhi et al. 2010). Another useful concept is the development of an ecological network that incorporates multi-actor planning, change, and large-scale processes (Opdam et al. 2006); they also call for participatory planning but from the ecological point of view (Luz 2000). A new approach includes incorporating social-ecological concepts when co-designing and planning land-use projects (Guerrero et al. 2015) by transdisciplinary teams that combine top-down and bottom-up land-use modeling (Verburg and Overmars 2009). In summary, watersheds are great territories for land-use planning and even have been proposed to be administrative units such as municipalities (Espejel et al. 2004), especially in drylands where focused aquifer management planning is crucial.

Case Study: Guadalupe Valley Official Planning and New Proposals

A case study in Mexico was chosen because it has only enforced top-down regional planning public policies and the sectorial, ecological, and territorial planning overlap, causing all land-use decisions to favor economic priorities. The selected case study explains how three top-down land-use programs have been implemented, and why it has been unsuccessful of achieving its main goal. Based on the reasons behind the failure to follow regulation on land-use, we propose various bottom-up strategies as an alternative based on the development of a transdisciplinary learning

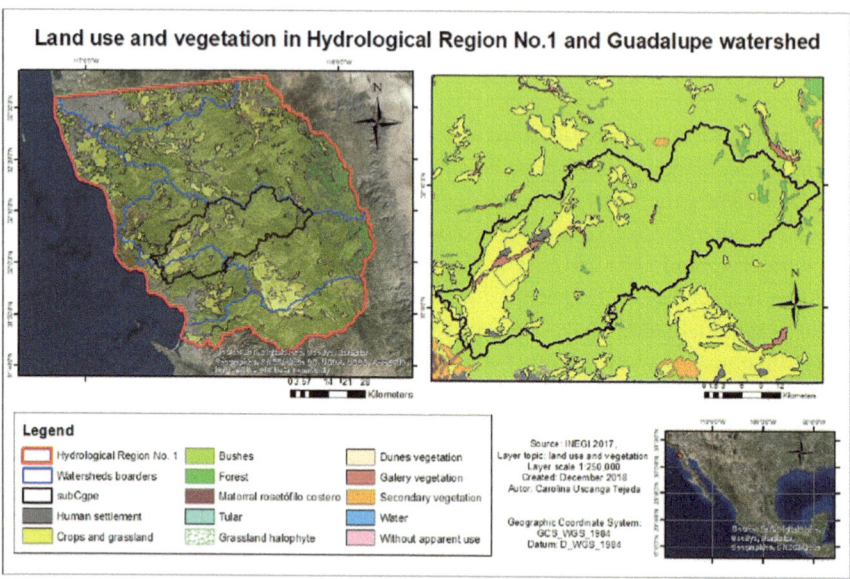

Fig. 9.1 Hydrological system of Northern Baja California, Mexico (left). The main watershed of Guadalupe and the secondary watershed of the Valley of Guadalupe (right)

community that co-manages the watershed as a long-term process, aiming to change paradigms for the future design of regional plans.

Most coauthors have been into these planning efforts since the beginning (1993) so our evaluation consisted in a chronological description of the main proposals of each program and a compliance assessment was subsequently made indicating if the strategy, goal, or action was met, if it was partially fulfilled, or if it was not carried out. To demonstrate compliance with the actions proposed by each planning program, information was sought in official documents, government reports, by field-work and satellite images (Google Earth Pro). There are many other publications and thesis on the subject, but these were especially important in the communication of the regional planning efforts and resulting programs. In the last 30 years, the watershed has been described as a socio-ecological system, since it is planned for agriculture and tourism (Fig. 9.1). To understand the complexity of land-use planning in Mexico, Fig. 9.2 shows the top-down regulations in a time line since 1930.

TU-LUP-WVE: touristic and urban sectorial land-use planning the northern zone of wine valley of Ensenada, B.C.

S-LUP-TU: state land-use planning touristic and urban

ELUP-SA-VG: environmental land-use planning San Antonio de las Minas-Valle de Guadalupe

TU-SP-WVE: touristic and urban sectorial programs of the northern zone of wine valley of Ensenada, B.C.

F-LUP: federal land-use planning

Fig. 9.2 Plans and programs related to land-use planning in Mexico since 1930–2010 (left) and documents, papers, books, and sectorial plans (right) related to the description, planning of land-use in the Guadalupe Valley

S-ELUP: state ecological land-use planning
DDPLU-GV: documents related to the description and planning of land-use in the
    Guadalupe Valley.

Figure 9.2 also shows the overlapping laws for land-use in Baja California, where the study case is located. At the national level, there are two coinciding spatial land programs. An ecological ordinance by the environmental ministry (SEMARNAT by its Spanish acronym) and a territorial ordinance with risk zone population relocation schemes implemented by the ministry of urban development (SEDUE by its Spanish acronym). At the state level, ecological, territorial, and sectorial ordinances are in effect. Although the federal ordinances have decided the ambits of implementation (urban vs. rural, contamination and nature conservation vs. human population risks), to the citizens it is unclear the reason behind the existence of two instruments regulating land-use. Moreover, in 2016 the Baja California State Environmental Agency (SPA by its Spanish acronym) published an Strategic Program (SPA, 2016) adding more issues to the valley. At the state level, there are both coincidences and contradictions among land-use laws, giving way to an ideal setting for corruption. Landowners sell land for other purposes than the law prescription, land buyers buy land in forbidden areas, and decision makers deliver land-use permits in improper lots. Breaking the law is a consequence, among many issues, of poor surveillance. In countries like Mexico, surveillance activities by the authorities lack adequate financial support. An alternative to this problem has been that residents and local users organize and help the authorities with prevention and communication tools. The first step to start self-surveillance is to understand the laws. In the case of the Guadalupe valley, there is a problem with the available law documents, all the planning laws are written in unclear phrasings or published in blurry official newspapers. A simplification of the language used in the laws and clearer maps showing the properties of localization and land-use designations are in dire need by citizens. There are four planning programs that deserve attention in terms of "translating" the information into infographics or other dissemination tools to make them understandable by most people using the laws for self-surveillance. As seen in Fig. 9.2 and Table 9.1, there are five planning programs published in the Official Newspaper of the State of Baja California, the most recent the urban ordinance published in October 2018 (POEBC 2018) which its bylaw is under review by the Guadalupe valley inhabitants (http://www.valledeguadalupe.gob.mx/site-vgpe/mt-content/uploads/2017/02/reglamento-de-zonificacion-y-control-territorial-del-municipio-de-valle-de-guadalupe.pdf).

The recent 2018 urban ordinance mentions worsening of the same problems previously stated. Besides the published regional, local, tourism and urban planning laws, difussion of the laws content was done, in a scientific publication (Espejel et al. 1999) and dissemintarion books for the general public in an e-book (Leyva and Espejel 2013), and an updated printed book (Espejel and Leyva 2017). In conclusion, studies, publications, laws, regulations, and planning programs have been found to be unsuccessful. These findings make the Guadalupe Valley a notable example of why top-down policies do not function. For example, in the three evalu-

Table 9.1 Problems mentioned in each of the five regional plans analyzed

Year of a plan/issue	Landscape	Land-use	Pollution	Soil and sand	Water	Vegetation	Urban	Treated water	Fires	Social
1993	No landscape planning	Competition between conservation and livestock land uses and between tourism	Risk of contamination by pesticides and chemical fertilizers	Erosion	Overexploitation of aquifers	Loss of native vegetation	Lack of urban infrastructure	Lack of solid waste treatment plants		
2003	Fragmentation and deterioration of the landscape	Land promotion for real estate purposes			Pressure on the water resource		Human settlements without services			
2006	Landscape deterioration	Increased competition for land uses: agriculture, commerce, restaurants, hotels, rural subdivisions	Inadequate management of solid waste	1. Decrease of soil quality 2. Sand extraction	Depletion of the water table due to overexploitation of the aquifer	Decrease in the surface of natural vegetation	1. Guidelines for infrastructure construction not followed 2. Deficit of public equipment and services	Discharges of urban water without treatment	Increase in intentional fires	

(continued)

Table 9.1 (continued)

Year of a plan/issue	Landscape	Land-use	Pollution	Soil and sand	Water	Vegetation	Urban	Treated water	Fires	Social
2010	Risk of loss of rural landscape, lack of effective urban development instruments that regulate land uses in the valley and its towns, and inexistence of urban image regulations	Disorderly growth and lack of towns image, there are no urban image regulations, lack of effectiveness in monitoring functions of urbanization actions, and irregular housing settlements isolated and without services		1. Environmental deterioration 2. Excessive extraction of sand	1. Water scarcity 2. Overexploitation of aquifers 3. Rights concession exceeding by almost 100% 4. Lack of adequate infrastructure for efficient extraction instrument for the efficient use of water significant 5. Lag in the		Limited tourism equipment, lack of road and tourist signage, limited offer and tourist equipment, and lack of alternative productive activities or sources of investment	5. Use and treatment of water and lack of a guiding		Lack of integration of the population into the economy of the valley, taxes generated by the wine industry in the region are not reflected locally

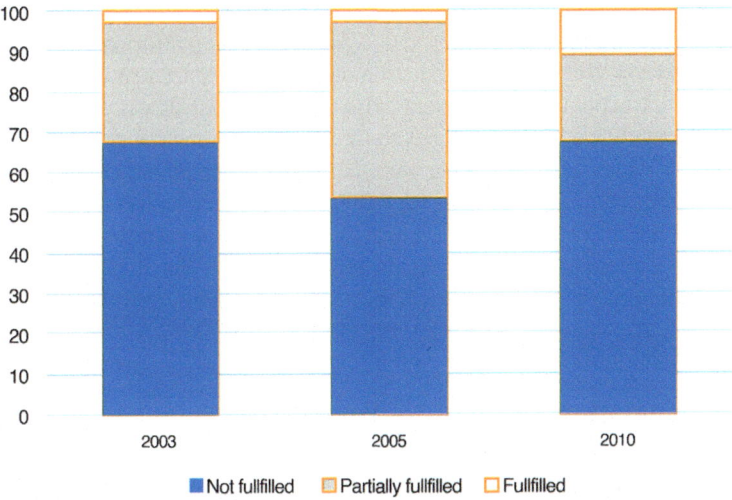

Fig. 9.3 Fulfillment of guidelines of three land-use planning laws in Guadalupe Valley, Ensenada Mexico

ated instruments, eight priority actions were identified and only one was partially carried out and corresponds to the water management plan, which has not been published, thus still lacks proper regulation. Of 38 actions, only four (7.5%) were fulfilled; as for the projects, only one (20%) was implemented (Fig. 9.3).

The best example of land-use failure is within the tourism sector. For instance, hotel rooms increased 52% between 2012 and 2016 (Reyes-Orta et al., 2016) other 60% between 2016 and 2018. Boutique hotels, villas, glamping and hostals offered 362 rooms in 2018. The number of restaurants in the valley also increased from 50 in 2016 to78 restaurants in 2018, resulting in an increase in the variation rate by 56%. It should be noted that in a region with very low water availability at least 10 hotels have swimming pools, Jacuzzi, and/or spa facilities, which are forbidden by these land-use planning laws. In addition, rapid growth of tourism has yielded an emergent necessity for country housing and support services. Thus, unplanned urban sprawl is destroying the landscape, and sustainable goals are being dismissed. For the tourism and urban sectors, an update of the land-use plan has just been published (POEBC 2018) expecting this time to be succesful because it comes with a bylaw (under review as mentioned above).

As a response to a severe drought, in 2018, wine makers promoted the transfer of treated water from the nearest metropolis, Tijuana. The project is expected to be a long-term solution for the everlasting drought situation. But the general mentality of inadequate water management and conservation still remains. Thus, more water availability will lead to increased development of vineyards, wineries, tourism, and residential areas which will push agricultural growth to expand to neighboring areas. Poor compliance of the evaluated instruments is complex, but we identified four main reasons: (1) unclear land tenure; (2) ambiguous laws and lack of by-laws, (3) corruption among all actors, the government agent, the landowner, and the buyer; and (4) lack of awareness of the planning instruments. As citizens, we can deal with anticorruption practices and design educational platforms to provide planning information. But the implementation of additional instruments for law compliance, such as economic instruments, rights market, subsidies, and no land-use policies for land owners, corresponds to the state and municipal governments.

Plenty of issues have to change with the recent urban planning law. To accompany this new planning, a citizens' surveillance committee has been proposed and a bylaw has been prepared by a heavy participatory transdisciplinary team. Nevertheless, these instruments remain top-down laws. So, it is advised to lean towards bottom-up strategies to break the recurring paradigms of the last 30 years.

Recommendations and Future Perspectives

In conclusion, the lessons learned from the unsuccessful planning of the Guadalupe Valley are that top-down strategies alone do not function and new ways of participation, such as transdisciplinary teams, might be the solution for social-ecological watersheds planning. With the mindset of up-taking new ideas in collective fashion as key to fulfill regional land-use planning a transdisciplinary team is being created to develop the following strategies:

1. *Citizen's observatory,* bringing together academics, producers, and the rural community to solve emergent problems. It consists of developing an internet portal (network platform) where specific information is available and updated constantly. The purpose of this observatory is to empower people with reliable data. It is believed to be one of the main tools that can achieve successful governance (Engelken-Jorge et al. 2014). In Mexico, the law for sustainable agriculture promotes rural observatories without success , for they are managed by the government, and government agents do not wish to be observed and/or evaluated. For this particular reason, the observatory will be participative, inclusive, managed by citizens, and co-designed with seven aquifer user groups. The key issue in designing an observatory of this nature is the negotiation of information (indicators), since it will function as a source for following projects. The results have to be uploaded as technical reports for a layman audience and infographics, since they are effective ways of presenting complex data in a visual compelling format directly useful for decision-making purposes (Otten et al. 2015).
2. *Green infrastructure projects,* built by residents, developed by architects, and coordinated with the current government urban program. The importance of green infrastructure is to develop an "interconnected network of green space that conserves natural ecosystem values and functions and provides associated benefits to human populations" (Benedict and McMahon 2002, p. 12). It includes natural, semi-natural, and artificial areas, within and around urban areas. Their quality and quantity will be based on the multifunctional role as habitat interconnectors (Tzoulas et al. 2007). The role of water is important within the green

infrastructure framework, combined they are considered as blue-green infrastructure (Ahern 2007).

The scales used for planning and implementing green infrastructure are landscape, region or city, districts or neighborhoods, and sites or parcels (Ahern 2007), each with specific strategies (Allen 2014). Green infrastructure uses mainly the landscape management approach; nevertheless, it also applies urban design in planning practices (Mell 2012). At a landscape scale, it is convenient to identify non-fragmented ecosystems that protect and improve the habitat of native species, corridors, and functional landscapes (Allen 2012). This also applies at the urban scale, where all types of green areas are part of the green infrastructure. At a site scale, urban street design elements (planters, trees, landscape strips, medians, and rain gardens) can promote ecosystem services by using water efficiency and management techniques, and by landscaping (reducing urban heat island and pollution) (Rehan 2013).

Agricultural landscape encompasses natural habitats and crops in a matrix, which provides economic output together with a range of ecosystem services. By conserving surrounding habitats and using farm and landscape-scale management techniques, the value of the ecosystem services in which vineyards and tourism economies depend is maintained (Viers et al. 2013). According to Winkler et al. (2017) and Viers et al. (2013), the main ecosystem services vineyards receive from surrounding natural ecosystems are: (1) cultivated crops; (2) filtration (promoting aquifer recharge); (3) sequestration, storage, and accumulation by the vineyards; (4) pest and disease control; (5) buffering weather (wind and temperature); (6) minimizing erosion, maintaining soil fertility through good practices; and (7) heritage, cultural, scientific services, and esthetic values.

Ahern (2011) proposed five planning strategies and urban design to build resilience capacity: multi-functionality, redundancy and modularization, diversity (biological and social), networks at various scales, and connectivity. Green infrastructure planning, participation, and interaction between potential stakeholders by means of expanding the green infrastructure range has the potential to achieve great impact in the ability of providing services by the local governance (Winkler et al. 2017). A participatory, inclusive approach facilitates local specific situations and ideas that can be incorporated in projects (Schifman et al. 2017). This approach decentralizes and extends coverage of green infrastructure over a greater area while engaging citizens. And, by means of supporting programs, these practices can expand (Winkler et al. 2017).

3. The identification and *valuation of ecosystem and environmental services* is fundamental to recognize their importance for the economy and development (Costanza et al. 2017). Their interrelationships at the ecological level will help to detect key services in Guadalupe Valley. The use of cartographic products could illustrate ecosystem services in the territory. Considering the watershed as a unit of analysis (Grizzetti et al. 2015), the maps would show the compatibility (or incompatibility) between the land uses assigned and the priority areas for the conservation or restoration of ecosystem services. Economic valuation provides useful information on the costs and benefits related to environmental impacts

during the execution of planning programs or private projects (Galán et al. 2012; CONANP 2015). These valuations will communicate the necessary information in a language understood by residents and tourists encouraging interest to maintain these services and respect spatial planning guidelines.

4. Diffusion of innovative forms of household and agricultural techniques with *wise water management* will provide a sustainable approach for Guadalupe Valley viticulture. Constructed wetlands have been used as an ecological friendly technique to treat wastewater for decades all around the world especially for small communities and remote places; however, the sustainable application of these systems is quite challenging (Wu et al. 2015). This technique if properly managed can yield a valuable conservation tool for its harnessing and cleansing properties. Constructed wetlands have been primarily utilized to treat wastewater, but they can be managed for other functions as well. They can be used to receive crop runoff, mitigate groundwater contamination from fertilizers, pesticides, heavy metals, and treated wastewater (Haberl 1999). They provide habitat for waterfowl and other animals and increase plant diversity richness. A constructed wetland surrounded by a riparian buffer zone will significantly slow down introduction of contaminants into the water and prevents erosion and crop runoff especially from agriculture plots on steep slopes, which we readily find in Guadalupe Valley viticulture practices. All of these agricultural techniques combined with conservation tillage, a type of no tillage technique, will significantly ease the daily challenges faced by reduced water resources such as poor soils, soil compaction, soil water infiltration, low soil moisture, and minimal recharge of the water table.

5. *A new larger scale for regional planning.* Agricultural land-use in the Guadalupe Valley is dwindling and eventually will be lost. Tourism and recreation for masses and dense country housing has gained and is transforming the landscape. Acceptance of such reality turns our efforts towards a larger scale plan. The watershed has other terrains apt for agriculture, and if treated water from Tijuana is going to be vast, these lands could be irrigated and become agricultural lands. But for now, the planning, with the lessons learned, could establish agriculture

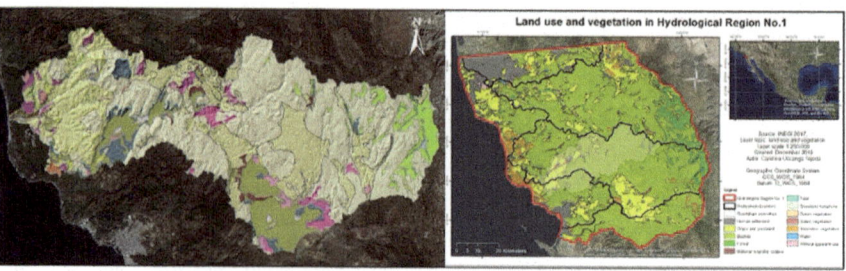

Fig. 9.4 Illustrates a map in a larger scale, where new developed country housing, eno-gastro tourism, and wine industry are shown, and the neighborhood valleys are kept as irrigated agricultural land as a result of incoming water from Tijuana's water treatment plant

reserves (as protected areas for the land-use of crops and cattle) to contain country housing sprawl. The Guadalupe Valley would become the area for tourism and country housing, hoping not to spread to the other valleys of the watersheds (Fig. 9.4). The watershed as a management unit would improve the understanding of the water origin and the role users partake in the ecosystem as proposed by Espejel et al. (2004).

In this chapter it is demonstrated that the planning of a region is a long-term process, not always successful because emergent problems continuously keep the planning in process. The plans for the study case, representative case of many of the Global South, have been socio-ecological system based and the watershed was the territory to regulate. Nevertheless, in semiarid watersheds as it is the Guadalupe basin, the aquifer is more susceptible to overexploitation by the increment of water-demanding crops, by the increase of agriculture surface, and by the growth of the resident population and water-demanding visitors. If water is a key issue worldwide, in semiarid and arid lands, it is the key limiting factor for development. Now, planning to bring water from treatment plants of the largest and nearest metropolis will change the region into a more complex socio-ecological system. Environmental education did not play the expected key role. That is the reason to propose different participative and creative processes and co-generate data concerning subterranean water quantity, quality, and dynamics, as well to use a more efficient language to transmit crucial information. A future approach has to keep the incorporation of social-ecological concepts but co-designing and planning the watershed land-use projects (Guerrero et al. 2015). The transdisciplinary teams have to combine top-down and bottom-up land-use modeling (Verburg and Overmars 2009) in order to be more successful than nowadays.

Acknowledgements The authors recognize the effort given to the Guadalupe Valley by most decision makers and residents. This research contributes to the aims of two projects, A-347 Formulation and adoption of the Guadalupe Watershed Management Plan, Ensenada, Baja California funded by Fundación Rio Arronte and by Conacyt BMBF, UNAM-UABC 155 Evaluation of the bio-economic risks due to aquifer over-exploitation un-arid and coastal regions, urban and agriculture lands.

References

Ahern J (2007) Green infrastructure for cities: the spatial dimension. In: Novotny V, Brown P (eds) Cities of the future: towards integrated sustainable water and landscape management. IWA Publishing, London, pp 267–283

Ahern J (2011) From fail-safe to safe-to-fail: sustainability and resilience in the new urban world. Landsc Urban Plan 100(4):341–343

Allen WL (2012) Environmental reviews and case studies: advancing green infrastructure at all scales: from landscape to site. Environ Pract 14(1):17–25

Allen WL (2014) A green infrastructure framework for vacant and underutilized urban land. J Conserv Plan 10:43–51

Benedict MA, McMahon ET (2002) Green infrastructure: smart conservation for the 21st century. Renew Resour J 20(3):12–17

CONANP (2015) Valoración de Servicios Ecosistémicos: Un Enfoque para Fortalecer el Manejo de las Áreas Naturales Protegidas Federales de México. Comisión Nacional de Áreas Naturales Protegidas, Secretaria de Medio Ambiente y Recursos Naturales, México

Costanza R, De Groot R, Braat L, Kubiszewski I, Fioramonti L, Sutton P, Grasso M (2017) Twenty years of ecosystem services: how far have we come and how far do we still need to go? Ecosyst Serv 28:1–16

Engelken-Jorge M, Moreno J, Keune H et al (2014) Developing citizens' observatories for environmental monitoring and citizen empowerment: challenges and future scenarios. In: Parycek P, Edelmann N (eds) Conference for e-democracy and open government, Krems, 2014

Espejel I, Leyva C (2017) Valle de Guadalupe. paisaje en tres tiempos. Universidad Autónoma de Baja California, Mexicali

Espejel I, Fischer DW, Hinojosa A et al (1999) Land-use planning for the Guadalupe Valley, Baja California, Mexico. Landsc Urban Plan 45(4):219–232

Espejel I, Hernández A, Riemann H et al (2004) Propuesta para un nuevo municipio con base en las cuencas hidrográficas. Estudio de caso: San Quintín, B.C. Gest Polít Pública 14(1):129–168

Fischer J, Gardner TA, Bennett EM et al (2015) Advancing sustainability through mainstreaming a social–ecological systems perspective. Curr Opin Environ Sustain 14:144–149

Galán C, Balvanera P, Castellarini F (2012) Políticas públicas hacia la sustentabilidad: Integrando la visión ecosistémica. CONABIO, México

Grizzetti B, Lanzanova D, Liquete C et al (2015) Cook-book for water ecosystem service assessment and valuation. Publications Office of the European Union, Luxembourg

Guerrero AM, Bodin Ö, McAllister RR, Wilson KA (2015) Achieving social-ecological fit through bottom-up collaborative governance: an empirical investigation. Ecol Soc 20(4):41

Haberl R (1999) Constructed wetlands: a chance to solve wastewater problems in developing countries. Water Sci Technol 40(3):11–17

Leyva C, Espejel I (2013) El valle de Guadalupe: conjugando tiempos. [ebook]. Universidad Autónoma de Baja California, Mexicali. http://webfc.ens.uabc.mx/documentos/El%20 Valle%20de%20Guadalupe.pdf. Accessed 5 Dec 2018

Luz F (2000) Participatory landscape ecology–a basis for acceptance and implementation. Landsc Urban Plan 50(1–3):157–166

Malhi Y, Silman M, Salinas N et al (2010) Introduction: elevation gradients in the tropics: laboratories for ecosystem ecology and global change research. Glob Chang Biol 16(12):3171–3175

Márquez G, Valenzuela E (2008) Estructura ecológica y ordenamiento territorial ambiental: aproximación conceptual y metodológica a partir del proceso de ordenación de cuencas. Gest Ambient 11(2):137–148

Mell IC (2012) Green infrastructure planning: a contemporary approach for innovative interventions in urban landscape management. J Biourban 1:29–39

Opdam P, Steingröver E, Van Rooij S (2006) Ecological networks: a spatial concept for multi-actor planning of sustainable landscapes. Landsc Urban Plan 75(3–4):322–332

Otten JJ, Cheng K, Drewnowski A (2015) Infographics and public policy: using data visualization to convey complex information. Health Aff 34(11):1901–1907

Parkes MW, Morrison KE, Bunch MJ et al (2010) Towards integrated governance for water, health and social–ecological systems: the watershed governance prism. Glob Environ Chang 20(4):693–704

POEBC Periodico Oficial del Estado de Baja California (2018) Programa sectorial de desarrollo urbano turístico de los valles vitivinícolas de la zona norte del municipio de Ensenada. https:// periodicooficial.ebajacalifornia.gob.mx/oficial/mostrarDocto.jsp?nombreArchivo=Periodico-42-CXXV-2018914-SECCI%C3%93N%20II.pdf&sistemaSolicitante=PeriodicoOficial/2018/ Septiembre

Rathwell KJ, Peterson GD (2012) Connecting social networks with ecosystem services for watershed governance: a social-ecological network perspective highlights the critical role of bridging organizations. Ecol Soc 17(2):24

Rehan RM (2013) Sustainable streetscape as an effective tool in sustainable urban design. HBRC J 9(2):173–186

Reyes-Orta M, Olague JT, Lobo MO et al (2016) Importancia y valoración de los componentes de satisfacción en la experiencia enológica en Valle de Guadalupe Ensenada, Baja California: Contribuciones al proceso de gestión sustentable. Rev Anál Turíst 22:39–55

Sabatier PA (1986) Top-down and bottom-up approaches to implementation research: a critical analysis and suggested synthesis. J Publ Policy 6(1):21–48

SAI-Sustainable Agriculture Initiative Platform (2010) Water Conservation Technical Brief: TB 9 - Use of a conservation tillage system as a way to reduce water the footprint of crops. http://www.saiplatform.org/library/sai-platform-publications. Accessed 25 Feb 2019

Schifman LA, Herrmann DL, Shuster WD et al (2017) Situating green infrastructure in context: a framework for adaptive socio-hydrology in cities. Water Resour Res 53(12):10139–10154

Schmidt WC, Friede JL (1996) Experimental forests, ranges, and watersheds in the northern Rocky Mountains: a compendium of outdoor laboratories in Utah, Idaho, and Montana. General. Technical Report INT-GTR-334. US Department of Agriculture, Forest Service, Intermountain Research Station, Ogden, p 117

Schuttenberg H, Guth H (2015) Seeking our shared wisdom: a framework for understanding knowledge coproduction and coproductive capacities. Ecol Soc 20(1):200115

Slaughter C, Richardson CW (2000) Long-term watershed research in USDA Agricultural Research Service. Water Resour IMPACT 2(4):28–31

SPA Secretaría de Protección al Ambiente (2016). Programa Ambiental Estratégico de la Región Vitivinícola del Valle de Guadalupe, Municipio de Ensenada, B.C., México. Gobierno de Baja California. http://www.spabc.gob.mx/wp-content/uploads/2018/05/PROGRAMA-AMBIENTAL-ESTRATEGICO-DE-LA-REGION-VITIVINICOLA-DEL-VALLE-DE-GUADALUPE-MUNICIPIO-ENSENADA-B.C.-2016-1.pdf

Tzoulas K, Korpela K, Venn S et al (2007) Promoting ecosystem and human health in urban areas using green infrastructure: a literature review. Landsc Urban Plan 81(3):167–178

Verburg PH, Overmars KP (2009) Combining top-down and bottom-up dynamics in land use modeling: exploring the future of abandoned farmlands in Europe with the Dyna-CLUE model. Landsc Ecol 24(9):1167–1181

Viers JH, Williams JN, Nicholas KA et al (2013) Vinecology: pairing wine with nature. Conserv Lett 6(5):287–299

Winkler KJ, Viers JH, Nicholas KA (2017) Assessing ecosystem services and multifunctionality for vineyard systems. Front Environ Sci 5:15

Wu H, Zhang J, Ngo HH, Guo W, Hu Z, Liang S, Fan J, Liu H (2015) A review on the sustainability of constructed wetlands for wastewater treatment: design and operation. Bioresour Technol 175:594–601

Chapter 10
Conservation and Development in the Biosphere Reserve of Mapimí: A Transdisciplinary and Participatory Project to Understand Climate Change Adaptation

N. Martínez-Tagüeña, E. Huber-Sannwald, R. I. Mata Páez,
V. M. Reyes Gómez, C. Villarreal Wislar, R. Cázares Reyes,
J. Urquidi Macías, and J. J. López Pardo

Abstract Mexican drylands cover over 50% of its territory. They are important socio-ecological systems (SES), like rangelands that are vital, both for the conservation of multifunctional landscapes and for human development. The UNESCO Biosphere Reserve of Mapimí (BRM) is a dryland SES that harbors an extraordinarily high level of endemism in almost all biotic kingdoms and historically holds evidence of various mobile indigenous groups; later it became an important livestock production center. Thus, the BRM has been affected by land

N. Martínez-Tagüeña
Cátedra CONACYT, Consortium for Research, Innovation and Development
of Drylands, Instituto Potosino de Investigación Científica y Tecnológica,
San Luis Potosi, San Luis Potosí, Mexico

E. Huber-Sannwald (✉)
División de Ciencias Ambientales, Instituto Potosino de Investigación Científica y
Tecnológica, San Luis Potosi, San Luis Potosí, Mexico
e-mail: ehs@ipicyt.edu.mx

R. I. Mata Páez · J. J. López Pardo
Instituto Potosino de Investigación Científica y Tecnológica, A.C., San Luis Potosí, Mexico

V. M. Reyes Gómez
Red Ambiente y Sustentabilidad, Instituto de Ecología, A.C., Chihuahua, Mexico

C. Villarreal Wislar
Comisión Nacional de Áreas Naturales Protegidas, Torreón, Mexico

R. Cázares Reyes · J. Urquidi Macías
Ejidatario de La Soledad, Reserva de la Biósfera de Mapimí, Mapimi, Mexico

© Springer Nature Switzerland AG 2020
S. Lucatello et al. (eds.), *Stewardship of Future Drylands and Climate
Change in the Global South*, Springer Climate,
https://doi.org/10.1007/978-3-030-22464-6_10

use and climate change for decades. Consequently, the BRM's management plan by the National Commission of Natural Protected Areas (CONANP) seeks both the conservation of endangered species like the desert turtle and the implementation of sustainable development programs. In parallel, while long-term ecological research has generated ample knowledge on desert ecosystems, it has remained unavailable to local inhabitants, who hold local ecological knowledge and are, and will be, the key players and decision makers at the local scale. An interdisciplinary research group has responded to the challenge of linking knowledge systems in the context of global environmental change. They formed a transdisciplinary participatory research working-team with multiple stakeholders to develop projects that reflect interests of all actors. In this chapter, we describe the process of co-designing and jointly executing this research and further present the historic and current development challenges in the RBM, alongside a discussion on the adaptive potential of local communities, their production systems, and the dominant ecosystems in response to government help programs, adverse climatic conditions, and a strong tendency of out-migration. Transdisciplinary research brings numerous benefits when tackling sustainable development challenges in complex dryland SES.

Keywords Transdisciplinary research · Rangelands · MAB Reserve · Socio-ecological Systems

Introduction

Drylands are conformed by an exceptionally high bio-, geo-, and cultural diversity that offer a broad spectrum of ecosystem services to a large proportion of their population (Safriel et al. 2005). Mexican drylands gave rise to important socio-ecological systems (SES), such as rangelands, which are vital, both for the conservation of multifunctional landscapes and for the human activity with historic, cultural, and economic relevance such as livestock production. While high variability in space and time is an inherent property of drylands to which its dwellers have adapted for millennia (Stafford Smith and Cribb 2009), reducing the vulnerability to scare resources and climate change (Davis 2016), current prolonged and repeated droughts, alongside land degradation is a complex yet true reality and risk in drylands worldwide.

In this chapter, through the example of the Biosphere Reserve of Mapimí (BRM) located in the southern area of the "Bolsón de Mapimí" (an endorheic depression typical of the mountainous areas of the dryland regions) between the Mexican states of Durango, Coahuila, and Chihuahua (Fig. 10.1), a case study is presented to illustrate a dialogue between conservation and development by various actors that share common goals but from multiple perspectives. It is the story of a group of interconnected people, who are normally physically separated because of the vastness of the region. Together they seek to consolidate themselves as a team to promote conservation programs that are developed in cooperation and collaboration with

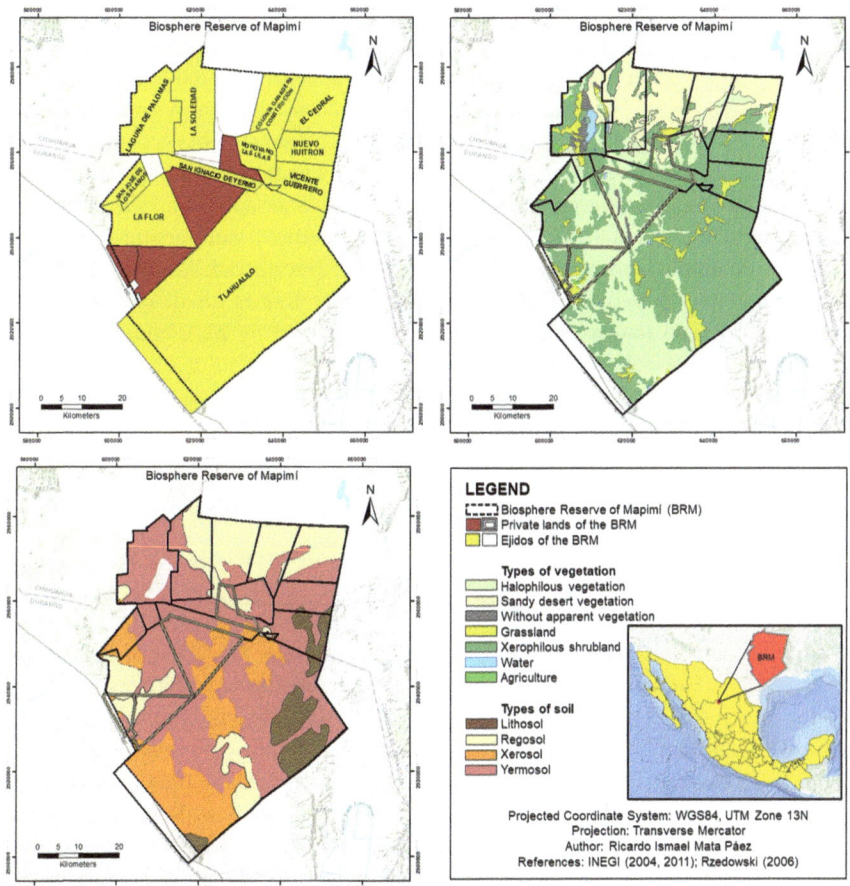

Fig. 10.1 The socio-ecological system of the Biosphere Reserve of Mapimí, Mexico including its political division composed of *ejidos* and private lands (top left), the dominant dryland vegetation (top right) (Rzedowski 2006; INEGI 2011) and soils (bottom left) types (INEGI 2004)

local inhabitants, and that incorporate the complexity of an integrated SES. As a team they are more likely to provide effective stewardship achieving long-term socio-ecological research and monitoring of global change.

Case Study: The Biosphere Reserve of Mapimí (BRM)

The Biosphere Reserve of Mapimí (BRM) was initially decreed in 1977 as an UNESCO Man and the Biosphere (MAB) Reserve founded to protect the environment and to rescue the Bolsón Tortoise from extinction (Halffter 1978; Kaus 1992). Since their origins, biosphere reserves are expected to promote conservation, research, development, and education as fundamental pillars in its management plans (di Castri et al. 1981). Thus, their establishment seeks to bridge the gap between conservation and development (Kaus 1992). Since 2002, the BRM has been managed by the National Commission of Protected Natural Areas (CONANP, by its initials in Spanish), with the dual purpose of conserving the ecosystem while promoting human development. Thus, this goal generates a complex dynamic between the conservation policies and the productive activities carried out by the local inhabitants (Kaus 1993; Toledo 2005). In the BRM there are 11 *ejidos* (the Mexican common land tenure system) and three private ranches with a total of approximately 326 people (CONANP 2006). They live in small isolated settlements; the overall low population density is explained by high emigration rates mostly due to the lack of opportunities of children education nearby. This SES is integrated by various forms of land tenure and land use activities that predate the protected area status and its official establishment.

The main socio-economic activity in the BRM is livestock production (Fig. 10.1a) that was introduced in the "Bolsón de Mapimí" at the end of the sixteenth century by the Spanish colonizers (Hernández 2001; CONANP 2006). This activity had three important historic stages in the area: (1) the boom of "haciendas" (a Spanish large estate property system) at the beginning of the nineteenth century, which prospered due to the demand for meat and traction animals (horses and mules) for agriculture and mining activities in the region; (2) the disappearance of the "haciendas" between the decade of 1910 and 1920 due to the effects of the Mexican revolution; and finally, (3) the creation of *ejidos* in the 1930s resulting from the division of the "haciendas" and distribution among the local communities. These *ejidos* continued until today; originally they raised equine livestock to satisfy the demands of traction animals in the region. It was not until the 1960s that a sharp change in cattle production was originated because of a drop in the demand for equine livestock (Kaus 1993; CONANP 2006). Another important socio-economic activity that only takes place in one *ejido* called Estación Carrillo is the production of salt through "salinas" that are saltwater wells (Fig. 10.2b).

Fig. 10.2 The main economic activities of the BRM are (**a**) livestock production and (**b**) salt production in the "Salinas"

In the "Bolsón de Mapimí" the distribution of dryland vegetation and its composition mainly depend on the slope, the water availability, the salinity, and the texture of the soil, among other features (Grunberger et al. 2004). Therefore, for functional distinctions of the region it is more adequate to divide the BRM into large landscape units characterized of dominant vegetation following a topo-sequence with a marked slope gradient reaching from the volcanic hills in the upper zone of the basin down to the sebrkra (Montaña 1988a; Grunberger et al. 2004) (Fig. 10.3). Overall, there are more than 403 plant species registered (Montaña 1988b) and the dominant vegetation types are xerophilic desert scrub communities with *Prosopis glandulosa* Torr., *Larrea tridentata* (DC.) Cav., and *Flourensia cernua* DC located mainly in the "bajada" and the "high playa," and the halophytic shrubs *Suaeda nigricans*, *Atriplex* spp., and edaphically determined halophytic grasslands (Rzendowski 1975) with *Sporobolus aeroides* and *Hilaria mutica* (Buckley) Benth often forming monospecific communities covering the "low playa" (Grunberger et al. 2004; Rzedowski 2006). Large open interspaces free of vascular plants are covered by diverse communities of soil biological crust (biocrust) composed of cyanobacteria, green algae, lichens, and bryophytes stabilizing the soil.

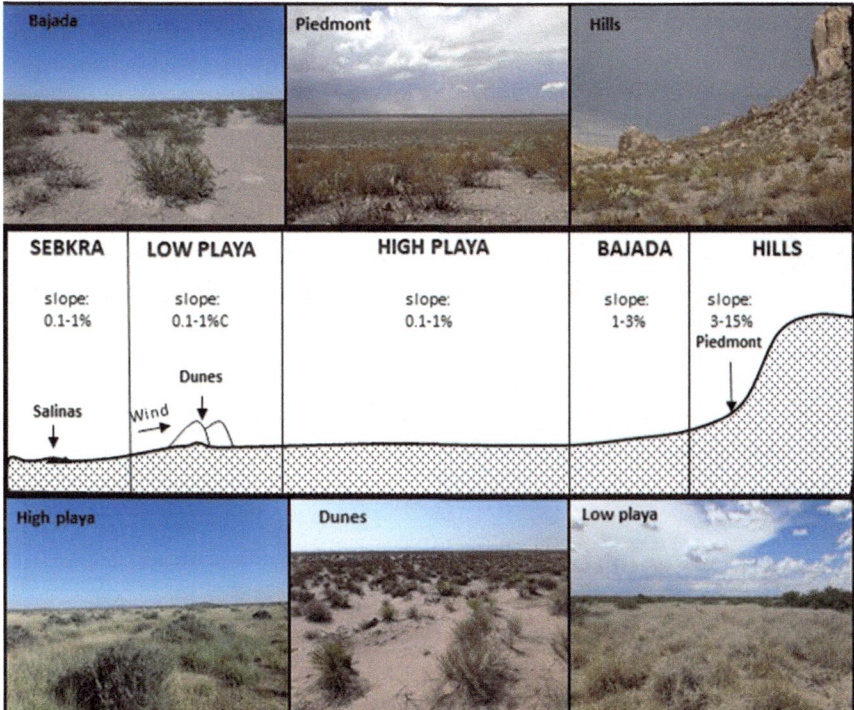

Fig. 10.3 The toposequence of the BRM with photographs of the typical vegetation types. Modified from Grunberger et al. (2004)

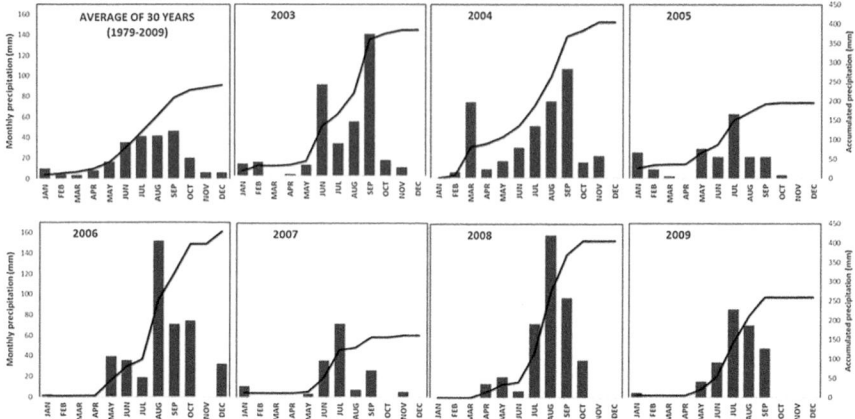

Fig. 10.4 Monthly precipitation values of the "Ceballos" meteorological station near to the BRM. The graphs represent intra and interannual variation in monthly precipitation between 2003 and 2009; the first chart (top left) shows the average precipitation of the 30-year period of 1979–2009. The available data was obtained from the Mexican National Meteorological Service (SMN 2009)

This region is a closed basin with large extensions of plains that have an average altitude of 1150 m.a.s.l. (CONANP 2006). In the RBM there are soils of alluvial and colluvial origin; the most dominant are calcareous regosols, xerosols, yermosols, and saline vertisols (Delhoume 1992). As in other closed basins, soils of the lower zones show a high degree of salinity (NaCl and KCl) and in the case of the Estación Carrillo, important saltwater wells are present and used for salt production (see Fig. 10.3). The climate of the BRM is BWhw(e) characterized as extreme arid with the main rainy season during the summer months (July–September) (Fig. 10.4 top left). For the period 1981–2010, average annual rainfall is 263 mm, of which 67.5% correspond to the summer months and 11.06% to the winter months (December–February). The dry season can last for 8–9 months with a significantly higher evaporation than precipitation rate. The average annual temperature is 20.2 °C with a normal minimum of 10.7 °C in the month of January and a normal maximum of 27.6 °C in the month of June (CONANP 2006; SMN 2010). Nevertheless, there is a great variation in monthly precipitation patterns between years (Fig. 10.4).

Research Focus in the BRM

After the official formation of the UNESCO MAB Reserve in 1975, the BRM was managed by the "Instituto de Ecología" (INECOL); it coordinated mostly basic science research activities in the area. Alongside the development of the biosphere reserve, the "Instituto de Ecología" established a research facility in a donated land called "Laboratorio del Desierto" (Desert Laboratory) which was inaugurated in 1978. At that time, a civil association (a type of NGO) was formed giving the

Reserve's management a legal status; it consisted of representatives: local inhabitants, the research institute INECOL, the National Council of Science and Technology (CONACYT), and MAB-Mexico (Halffter 1981; Kaus 1992). The first studies in the BRM were basic evaluations of vegetation and vertebrate populations, undertaken by researchers from the Instituto de Ecología, "Ecole Normale Supérieure" from France, the National Mexican Autonomous University (UNAM), and the "Instituto Nacional de Investigaciones sobre Recursos Bióticos" (INIREB) (Halffter 1978).

Soon after, an integrated research program was developed to study interrelated topics in a multidisciplinary approach. The following priorities were identified: climate, soils, vegetation, wildlife, livestock, and human population with the common objective of exploring the potential of ecosystem production that would be sustainable for conservation and human development (Montaña 1984). This project developed out of common interests among the researchers and some of the ranchers that had expressed their concerns during the reserve formation regarding limitations of water and pasture (Kaus 1992). However, this initial momentum lost force and by the time Kaus (1992) conducted her anthropological study in the late 1980s and early 1990s, research was being conducted mainly in the vicinities of the Desert Laboratory and had focused on ecological, biological, and meteorological topics with no emphasis on livestock and human population (see Kaus 1992, p. 116–123) for an ample description of this impressive and broadly conducted research.

Therefore, the role of researchers of the natural sciences had been "to coordinate the available research projects rather than actively develop programs which seeks out needed expertise, combines research results for regional application, or provides mechanisms for local participation in research activities" (Kaus 1992, p. 115–116). In occasions, local inhabitants were hired as labor or guides, while the relationship between the researchers and local inhabitants was cordial there was no collaboration. Some complaints expressed by the local inhabitants were described as lack of courtesy from the local perspective, for example, leaving trash behind, damaging fence lines, failing to close gates properly, driving fast when passing through settlements, not informing residents about research activities nor taking time to visit, or giving unsolicited advice (Kaus 1992, p. 124). Kaus's (1992) research contributed to the discussions on how local people might be included within a conservation program, how development and conservations goals can be compatible, and how conservation can benefit local human populations. She describes how the human perspective of the RBM was usually considered from an outsider perspective. Along this line, solutions to apparent problems that the inhabitants did not even know existed were imposed. Hence, conservation programs were developed that ignored alternative perceptions of the environment in which humans are not detached from nature but are an integral part of it.

Kaus (1992) recommendations were crucial then and some still remain even more pertinent and urgent today: (1) the need for obtaining baseline data and long-term studies, (2) collaboration with local residents on research projects or including residents in meetings about the activities in the protected area, (3) develop visible demonstration of the relative benefits of alternative forms of land use or non-use or the meaning of conservation, (4) develop a council that binds together the community to mediate land use and internal conflict, access to information and a voice in management practices (CONANP developed a council that will be later described in detail) (5) apply regional research to increase the incentives and opportunities for long-term management of the land and its resources, and (6) establishment of a revolving fund for small conservation/resource management loans.

Biospheres promote long-term environmental research and monitoring the effects of global change, but little attention has been given to the monitoring of social variables such as population dynamics, land tenure, or land use that directly affect other ecological processes and the conditions in and surrounding biospheres (Kaus 1992). Without describing in detail all the research that has been conducted in RBM until today, a simple search on Scopus and Web of Science databases demonstrates the emphasis of ecological studies over any other studied topic (Fig. 10.5). In search of publications, the filter item included BRM in the title, summary, or keywords. Then, each publication was reviewed by reading the summary, or if

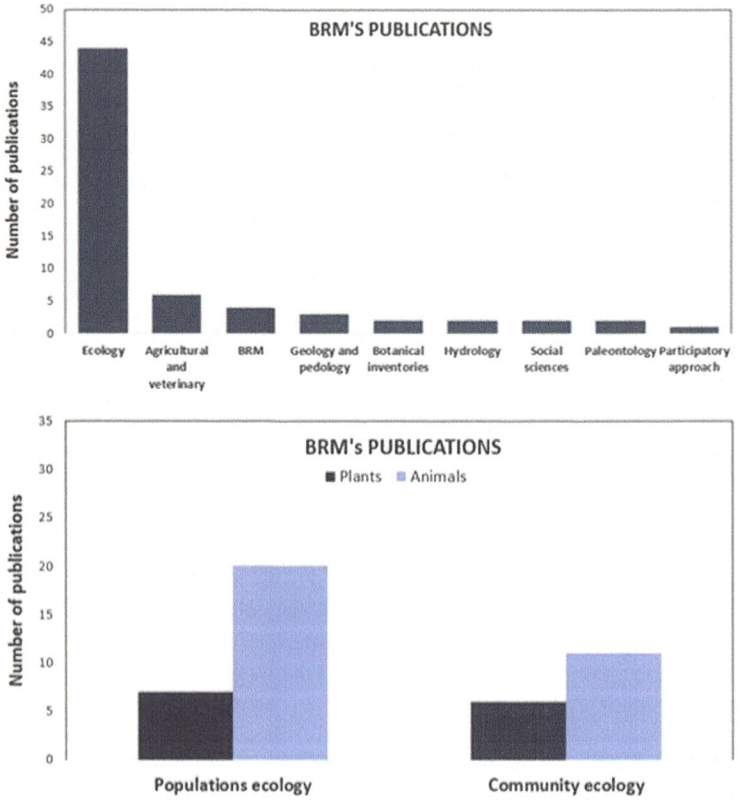

Fig. 10.5 Publications on the BRM in several disciplines indexed in the Scopus and Web of Science databases, filtered with the words "Mapimi" + "Reserve" (Scopus 2019; Web of Science 2019)

necessary, the content to corroborate that the study was conducted in the BRM. Finally, the publications were classified according to the area of knowledge to which they corresponded (Scopus 2019; Web of Science 2019). A total of 66 publications were found published between 1981 and 2019; of which, 44 articles correspond to the area of ecology. Within the ecological studies, only 13 address the ecology of plants, while 31 talk about animal ecology, with emphasis on mammals in 19 studies, on reptiles (9), followed by insects (3), and finally birds (1). Most of these studies focused on population and community questions and few included functional aspects. Thus, most of the scientific studies carried out in the BRM are related to the description of singular elements of the socio-ecological system, without taking into account the human component that has existed in the reserve since its foundation. Furthermore, it is relevant to remark that published information is unavailable to local inhabitants and its access difficult and limited to other actors outside of academia (from NGO and GO) that are currently promoting conservation and development, thus managing the RBM.

Participatory Research Approach in the BRM

Participatory research has the overall goal of solving problems in socio-ecological systems through the integration of scientific, technical, and local/traditional knowledge (Cornwall and Jewkes 1995). This is usually accomplished with the collaboration between the researcher and the local people, starting from the diagnosis of problems to the application of solutions (Coppock et al. 2004). The implementation of this approach provides an opportunity to integrate scientific and local knowledge into public policies and community actions, since it promotes learning between the different actors to flow in a bidirectional manner and between the different scales of the system (Stringer et al. 2006). In particular, in biosphere reserves, the integration of local inhabitants in research and in the generation of management programs is necessary, since they use the natural resources present in their lands and they depend on the services dryland ecosystems provide for their well-being. Besides they make decisions on management practices using intergenerational learning, local/traditional knowledge, as well as technical knowledge learned from NGOs and GOs capacitation programs. These factors influence jointly and directly the management and functioning of the ecosystems of arid zones such as those that are present in the BRM (Agrawal and Gibson 1999; Walker et al. 2009).

In order to better understand the SES of the BRM and to identify and study the socio-ecological problems present in the Reserve through the participatory approach, an interdisciplinary and inter-institutional group of researchers was formed with expertise in the areas of ecology, anthropology, geography, and hydrology. The project had three main stages: (1) exploration of general characteristics of the SES, (2) detection of needs and interests, both corresponding to the diagnosis of the system, and (3) the development of collaborative research projects. The first stage began with an interview with the director of the RBM of CONANP, to become acquainted with the interdisciplinary and inter-institutional research group and the objective to conducting participatory research in the RBM. The objective of this first encounter was twofold, first to obtain authorization to work in the reserve, and second to invite him and his team in this government institution to collaborate and to jointly tackle this new transdisciplinary endeavor. Furthermore, two exploratory visits were made in April and July 2017, led by one key *ejido* member, who facilitated the contact to local inhabitants and quick access to those *ejidos* with permanent residents. During these visits, we presented ourselves as a participatory research group, learned about the RBM from different people's perspectives, got to know the types and challenges of different production systems, and the state of the rangelands. In this stage, informal interviews were conducted to identify the main actors and key issues present in the BRM.

For the second stage of the project, in October 2017, a third visit was made to the *ejidos* of La Soledad, La Flor, Colonia Ganadera, and Los Álamos to establish collaborative partnerships with local inhabitants and to conduct 12 structured interviews with the objective to learn about the relationships between the main stakeholders and the external factors that control their decision-making processes in the RBM. In addition, information was gathered about their needs, interests and, if there was one, the problems of several local inhabitants of the RBM (Moraine et al. 2017) in order to identify future research opportunities. Although, a common participatory method called Participatory Rural Appraisal (Chambers 1983) is used to identify a community's problems and solutions, due to lack of funding and time, and also due to low population in the Reserve, this information was obtained through informal and semi-structured interviews. From the total population of permanent settlers in the RBM, 300 people belong to one *ejido* and the others range from 1 to 9 local inhabitants.

Within the RBM SES, four main groups of stakeholders are involved in the stewardship of the reserve (Fig. 10.6): the local inhabitants, CONANP as the government institution, the non-governmental organization PRONATURA, A.C., and the academic group of different institutions and organizations. The local inhabitants are the main group, consisting of people who live inside of the reserve and/or in the surrounding villages (usually by means of a dual residency, explained in detail in Kaus 1992) who however realize their socio-economic activities within the limits of the BRM. As mentioned before, the main socio-economic activities are livestock production, and in the case of the *ejido* Carrillo and La Flor, salt production and ecotourism, respectively. These activities influence the quality of ecosystem services and therefore, the quality of their lives (understood differently by the various stakeholders in the RBM). Also, the projects granted by both CONANP and PRONATURA play a role, because they provide technical knowledge and represent up to 10% of the annual income of the *ejidatarios* interviewed.

As previously mentioned, CONANP formed a consultative council in the RBM; it includes different stakeholders: field technicians from CONANP and PRONATURA, local representatives, and researchers; they make jointly important decisions such as priority of projects and allocation of budgets. This council is a local creation that does not happen in other protected areas mainly due to the

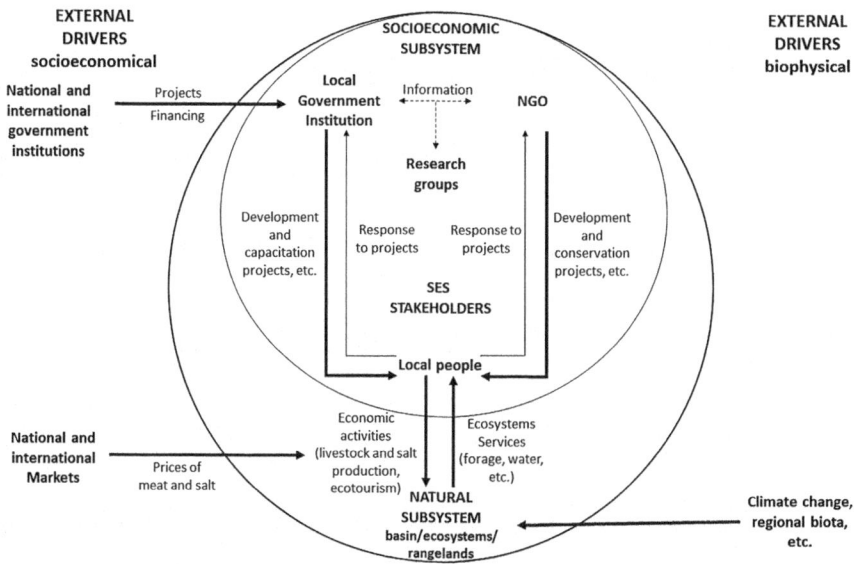

Fig. 10.6 Generalized diagram of the local socio-ecological system of the BRM including dominant external drivers and the strength of relationships among principal stakeholders in their role as stewards of the BRM. The thicker the solid arrow, the stronger the relationship in that direction. Dotted arrows refer to intermittent relations. Modified from Mata-Páez et al. (2018)

initiative of the director of the Reserve to promote strong local participation. Therefore, while the RBM represents a positive example from CONANP, overall there is still low feedback among local inhabitants and government institutions at other scales. This council is still trying to improve cohesion and seeking better local participation to mediate land use and internal conflict, access to information and a voice in the reserve management scheme. While it is not the scope of the present chapter and it is not necessarily the case in the RBM, CONANP has been criticized for its top-down management approaches and lack of adequate management plans. Protected land is widely dispersed, often remote and not connected throughout the country, and their stewardship mainly follows a neoliberal conservation model, which responds almost exclusively to economic valuation and mercantilization of nature, with a utopic fascination for sustainable development (i.e., Durand 2014; Guzmán 2016).

Another group of interest is composed of different research groups that are present intermittently in the BRM. As it was described earlier, these groups have been present ever since the foundation of the reserve and while they seek to generate scientific knowledge that can potentially serve the government, settlers, and NGOs in designing local development and/or conservation projects, there has not been a direct link between this group and others, especially with the local residents. Therefore, this situation highlights the importance of generating place-based knowledge in collaboration with the different actors involved (Agrawal and Gibson 1999). In case of BRM, the conducted interviews revealed two main topics for applied research: (1) analysis of the quantity and quality of water for human and animal use, which is the principal limiting resource in a dryland region, and (2) the condition of the rangelands inside the reserve, due to the interest of the settlers to learn about the quantity and quality of available forage for their cattle. It is worth mentioning that these two topics were stated as a priority by local inhabitants during the Reserve formation, in addition to having explicitly mentioned an interest in the possibilities of seeding the floodplains or the degraded pasture areas with native forage plants (Kaus 1992, p. 152).

As a final stage of the project, several possible research studies were identified that included various themes ranging from socio-economic and socio-environmental appraisals for the development of ecotourist activities; mapping and compilation of historic information also for ecotourism and outreach activities; socio-ecological research on organic meat production projects; monitoring of different SES attributes; social conflict resolution in one *ejido* to solve land tenure and organization issues; market study to explore different options for the salt production development, and socio-ecological evaluation of the ecotourism facility. However, the interdisciplinary participatory research team had to demarcate the scope of the research regarding (1) funding, (2) time, (3) expertise and access to students, and (4) local inhabitants interested in collaborating in the research project. Therefore, the *ejido* "La Soledad" was chosen and it was visited again in October 2017 to form transdisciplinary work teams to establish the research objectives and methodology to study (1) the evaluation of the effect of grazing intensity on the ecohydrological functionality of the rangelands (Mata-Páez 2019) and (2) the evaluation of the success of restoration projects implemented in the area (López-Pardo 2019), both investigations were done between January–June of 2018.

The overall goal of the first project was to understand the ecosystem functioning in relation to cattle ranching to provide data for better decision-making in the RBM. An ecological monitoring approach was combined with ranchers' local knowledge and perspective to determine rangeland health. Much was learned on the local vegetation and soil surface conditions, cattle preferences for forage plants and on the system's potential to recover after what was locally referred to as a good rainy season (Mata-Páez 2019) (see Kaus 1992 for a similar discussion on the recovery capacity after the rain season). Furthermore, for the second project the goal was to propose an integrated methodology that weaves together different knowledge systems to ultimately define from a transdisciplinary and participatory perspective a success index to evaluate the restoration projects, whose objectives were to retain run-off and sediments and stimulate forage production of the desert turtle. While local inhabitants have participated with CONANP and PRONATURA in the creation of these works, they were only occasionally asked about their preferences to select a location for their establishment. In this project, they participated during all research stages. The monitoring and evaluation of restoration projects is crucial, since ample resources have been invested given to this enterprise. Finally, collaboration also revealed that the toposequence and their associated vegetation types in the BRM (see Fig. 10.3) should be considered as different socio-ecological contexts when designing, implementing and monitoring the local, regional, national and internationally defined management and restoration plans.

Historic and Current Conservation and Development Challenges

The single most important condition for the ranchers' acceptance of the Reserve was land-tenure security, followed by an economic gain (Kaus 1992, p. 393). Early on, people like Nigh (1989) questioned local participation in the RBM since they were not part of the decision-making processes. As defined by Kaus (1992, p. 129) the problem was that the local residents had no visible demonstration of the relative benefits of alternative forms of land use or non-use or the meaning of conservation during the establishment of the core area in the RBM. In addition, neither researchers nor the decision makers knew about the necessary natural resources, their economic and cultural value, the social systems in place to manage or control access to these resources, and the incentives to use the land in a sustainable manner.

CONANP has employed several participatory methodologies to collaborate with local inhabitants, mainly inviting participation where only the people that want to collaborate work with them. In addition to the already mentioned informal and semi-structured interviews, nine semi-structured interviews to elders from five different *ejidos* were implemented to better understand the current challenges faced by the management of the RBM. Different reasons from the interviewed inhabitants for collaboration with CONANP were identified that mainly have to do with interests, age and lack of infrastructure that limits the possibility for change to a rotational grazing system, and social conflict whether they were facing land tenure and organization problems that prohibited a successful implementation of a project proposed from the outside.

The main projects implemented by CONANP consist in faunal monitoring, soil restauration, and RBM vigilance. In addition, PRONATURA has mainly implemented restauration works in the area giving to the local inhabitants "payments for environmental services" that seek to promote local participation. The case of the ejido La Flor can be considered a conservation and development success story where ecotourism was established as an alternative way of life to ranching by promoting the establishment of a family business and giving a role to women in the socio-economic activity of the household. As we mentioned before there is a strong out-migration due to lack of access to education inside the RBM. An elder explained that in most of the small ranches inside the RBM, women are abandoning the ranch and young people do not want this type of life either. A complete socio-ecological evaluation of the ecotourism enterprise is needed but it can be said that the main perceived challenge was its limitation for growth and self-management. Members from other *ejidos* want to also implement ecotourism projects in their land but need the support of the different stakeholders participating in the RBM.

All the interviewed people are aware of the changing climate conditions and thus their need to lower their animal loads. It is worth mentioning that some changes are related with more fences in the RBM that limit the potential for cattle movement thereby limiting large herds of livestock moving extensively and freely throughout the range. While the interviewed ranchers want to maintain ranching, they are seeking for alternative sources of income either through research and conservation projects, the promotion of ecotourism, and/or the sustainable use and production of other natural resources like oregano, salt, mesquite, creosote bush, among others. All the elders perceived less rain since the early 1990s in the RBM. More detailed descriptions from two persons stated that the overall precipitation pattern seems to be constant, i.e., highly variable, with a couple of good rain years each 10 years. However, its distribution has changed notoriously since they are now very localized and abrupt (monsoon like). In addition, all the elders remembered several drought years where everything died, once even the "nopales" did. While there is certain disagreement, some people expressed that in the present the grasses seem to be shorter not growing as tall, and there are more scrubs of creosote bush in certain areas.

Conclusion

In the past, biosphere reserves promoted long-term ecological research and monitoring global change impacts; however, little attention has been given to the monitoring of social variables such as population, land tenure, or land use that directly affect other ecological processes and the conditions in and surrounding biosphere reserves. Today it is widely known that it is important to invest in maximizing the protection of the biological and cultural diversity of dryland SES and in the promotion of local knowledge that provides an ample spectrum of responses to global socio-environmental change (Chapin et al. 2009). In the RBM, while advances have been made to determine the local perspectives on the necessary natural resources, their economic and cultural value, the social systems in place to manage or control access to these resources, and the incentives to use the land in a sustainable manner, more research is needed. Furthermore, applied research is needed to continue to increase the incentives and opportunities for long-term monitoring of ecosystem goods and services and the effects of management of the land and its resources on the functioning of dryland SES. Through this case study, the team seeks to demonstrate how joint participatory and transdisciplinary research is crucial since each actor of a SES has its own perception of the land and its resources, its own value system, and diverse opinions about its future.

Acknowledgments The authors thank all dwellers of the Mapimi Biosphere Reserve, the technicians of CONANP, and collaborators of PRONATURA for generously sharing their knowledge and experience. We thank Héctor Sergio Cortina Villar from ECOSUR for fruitful discussions and José Alfredo Ramos Leal for valuable insights on geo-hydrological aspects of the reserve. EHS gratefully acknowledges the financial support from CONACYT (projects CB 2015-251388B, PDCPN-2017/5036). NMT thanks her Cátedra CONACYT ID Number 6133 as part of the project 615.

References

Agrawal A, Gibson CC (1999) Enchantment and disenchantment: the role of community in natural resource conservation. World Dev 27:629–649

Chapin III FS, Folke C, Kofinas GP (2009) A framework for understanding change. In: Chapin III FS, Kofinas GP, Folke C (eds.) Principles of Ecosystem Stewardship. Springer Verlag, New York, p. 3–28

Chambers R (1983) Rural Development: Putting the Last First. Longman, London

CONANP (2006) Programa de Conservación y Manejo de Reserva de la Biosfera de Mapimí, México. CONANP, México D. F

Coppock DL, Desta S, Tezera S, Lelo RK (2004) Pastoral risk management in Southern Ethiopia: observations from pilot development projects based on participatory community assessments. ENVS Fac. Publ, Davis, pp 1921–1931

Cornwall A, Jewkes R (1995) What is participatory research? Soc Sci Med 41:1667–1676

Davis DK (2016) The arid lands: history, power, knowledge. The MIT Press, Cambridge

Delhoume JP (1992) Le milieu physique. In: Delhoume JP, Maury ME (eds) Actas Del Seminario Mapimí. Estudio de Las Relaciones Agua-Suelo-Vegetación En Una Zona Árida Del Norte de México Orientado a La Utilización Racional de Estos Recursos Para La Ganadería Extensiva de Bovinos. Instituto de Ecología, A.C., Instituto Frances de Investigaciones Científicas para el Desarrollo en Cooperación (ORSTOM). Centro de Estudios Mexicanos y Centroamericanos, Coyoacán, p 396

di Castri F, Hadley M, Damlamian J (1981) MAB: the man and the biosphere program as an evolving system. Ambio 10(2–3):52–57

Durand L (2014) ¿Todos ganan? Neoliberalismo, naturaleza y conser- vación en México. Sociológica 82:183–223

Grunberger O, Reyes-Gomez V, Janeau JL (eds) (2004) Las playas del desierto chihuahuense

Guzmán M (2016) Las áreas naturales protegidas en el norte semiárido de México. Problemas de conceptuación y los desafíos de su gestión y manejo. In: Cañedo S, C Radding (eds.) Historia, medio ambiente y áreas naturales protegidas en el centro norte de México. Contribuciones para la ambientalización de la historiografía mexicana, siglos XVIII-XXI. El Colegio de San Luis, México, p 205–250

Halffter G (1978) Las Reservas de La Biosfera En El Estado de Durango: Una Nueva Polftica de Conservacidn y Estudio de Los Recursos Bidticos. In: Halffter G (ed) Reservas de La Biosfera en el Estado de Durango. Instituto de Ecología Publicación 4, Ciudad de México, pp 17–45

Halffter G (1981) The Mapimi Biosphere Reserve -Local Participation in Conservation and Development. Ambio 10:93–96

Hernández L (2001) Historia Ambiental de la ganadería en México. Instituto de Ecología, Xalapa

INEGI (2004) Cartografía Edafología Serie II Escala 1:25000

INEGI (2011) Capa Unión, Cartografía Uso del suelo y vegetación Escala 1:250 000 Serie V

Kaus A (1992) Common ground: ranchers and researchers in the Mapimí Biosphere Reserve. Dissertation, University of California, Riverside

Kaus A (1993) Environmental perceptions and social relations in the mapimi biosphere reserve. Cult Agric 45–46:29–34

López-Pardo JJ (2019) Análisis participativo de las obras de restauración: Estudio de caso ejido La Soledad. Master Thesis, Instituto Potosino de Investigación Científica y Tecnológica, San Luis

Mata-Páez RI (2019) Evaluación participativa del efecto de la intensidad de pastoreo en la funcionalidad ecohidrológica de los agostaderos de la Reserva de la Biósfera de Mapimí. Master Thesis, Instituto Potosino de Investigación Científica y Tecnológica, San Luis

Mata-Páez RI, Huber-Sannwald E, Martínez-Tagüeña N, Arredondo-Moreno J, Cázares-Reyes R, Reyes-Gómez VM, Urquidi-Macías JA (2018) La investigación participativa y su importancia para los estudios socio-ecológicos en zonas áridas: Estudio de caso del ejido La Soledad en la Reserva de la Biosfera de Mapimí, México. Memorias del IV Congreso Multidisciplinario de Ciencias Aplicadas En Latinoamérica, Mérida

Montaña C (1984) Ecological and socio-economic research in the Mapimí biosphere reserve. In: UNESCO, UNEP (eds) Conservation, Science and Society, contributions to the first international biosphere reserve congress, Minsk, Byelorussia/USSR, 26 September-2 October 1983. United Nations Educational, Scientific, and Cultural Organization, Paris, pp 520–533

Montaña C (1988a) Mayor vegetation and environment units. In: Montaña C (ed) Estudio integrado de los recursos vegetación, suelo y agua en la Reserva de la Biosfera de Mapimí. Instituto de Ecología, Coyoacán, pp 99–115

Montaña C (1988b) Estudio integrado de los recursos vegetación, suelo y agua en la Reserva de la Biosfera de Mapimí. Instituto de Ecología, Coyoacán

Moraine M, Melac P, Ryschawy J, Duru M, Therond O (2017) A participatory method for the design and integrated assessment of crop-livestock systems in farmers' groups. Ecol Indic 72:340–351

Nigh R (1989) El Desarrollismo Ecologista: Las Fantasias de La Conservacidn de La Naturaleza. Perfil de La Jornada 16:1–4

Rzedowski J (2006) Vegetación de México, México. CONABIO First Digital edition. https://www.biodiversidad.gob.mx/publicaciones/librosDig/pdf/VegetacionMx_Cont.pdf

Rzendowski J (1975) An ecological and phytogeographical analysis of the grasslands of Mexico. Taxon 24:67–80

Safriel U, Adeel Z, Niemeijer D, Puigdefabregas J, White R, Lal R, Winslow M, Ziedler J, Prince S, Archer E, King C (2005) Dryland systems. In: Hassan R, Scholes R, Ash N (eds) Ecosystems and human well-being: current state and trends, Millenium Ecosystem Assessment. Island Press, Washington, D.C, pp 623–662

Scopus (2019) Elsevier B.V. [WWW Document]. URL https://www.scopus.com/search/form.uri?display=basic

SMN (2009) Datos mensuales, estación Ceballos 00010005, Mapimí, Durango [WWW Document]. URL http://smn.cna.gob.mx/tools/RESOURCES/Mensuales/dgo/00010005.TXT

SMN (2010) Información Climatológica [WWW Document]. URL http://smn.cna.gob.mx/es/informacion-climatologica-ver-estado?estado=dgo

Stafford Smith DM, Cribb J (2009) Dry times. CSIRO Publishing, Clayton

Stringer LC, Dougill AJ, Fraser E, Hubacek K, Reed MS, Stringer LC, Dougill AJ, Fraser E, Hubacek K, Prell C (2006) Unpacking "participation" in the adaptive management of social – ecological systems: a critical review. Ecol Soc 11(2):39

Toledo V (2005) Repensar la conservación: ¿áreas naturales protegidas o estrategia bioregional? Gaceta Ecol 77:67–83

Walker B, Abel N, Anderies JM, Ryan P (2009) Resilience, adaptability, and transformability in the Goulburn-Broken Catchment, Australia. Ecol Soc 14(1):12

Web of Science (2019) Clarivate Analytics [WWW Document]. URL https://apps.webofknowledge.com/WOS_GeneralSearch_input.do?product=WOS&search_mode=GeneralSearch&SID=8ABQRfRJeCYX2lB3sFb&preferencesSaved=

Part III
Interculturality in Drylands

Chapter 11
"Women with Wings": An Experience of Participatory Monitoring in a Natural Protected Area

D. Pinedo, C. Leyva, M. Ballardo, M. A. Cordero, E. Estrada, A. Ocaña, D. S. Savín, Y. Savín, M. Silva, M. C. Tonche, and Y. Torres

Abstract The complexity of environmental issues focuses on the multiplicity of actors and interests that are integrated in the management of socio-ecological systems. This approach has been analyzed from different perspectives. In recent decades one of the approaches that have been incorporated into discourses, research, and public policies at the international and national levels is the gender perspective. The objective of this study is to evaluate the concept of gender, focusing on female leadership in the environmental management actions and competencies of the "Mujeres con Alas" ("Women with Wings") group. This community group was founded in 2015, in view of the need to integrate women in conservation activities within the protected area "Reserva de la Biosfera Bahía de Los Ángeles," located on the East coast of Baja California, Mexico. Currently, a group of 11 women carries out monitoring, bird tourism, and environmental education activities. For this study, an ethnography was made with semi-structured interviews and focal groups to document and explore the role of female leadership in environmental management. The group's strengths, obstacles, and areas of opportunity were identified throughout the bird conservation activities. With this analysis, it was shown how the "Women with Wings" have taken advantage of the given natural resources and they emerged as agents of change through the opportunities of environmental management, with collective actions in favor of both the development and well-being of their community and the natural resource conservation.

D. Pinedo
Facultad de Ciencias Marinas, Universidad Autónoma de Baja California, Ensenada, Mexico

C. Leyva (✉)
Facultad de Ciencias, Universidad Autónoma de Baja California (UABC), Ensenada, Mexico
e-mail: cleyva@uabc.edu.mx

M. Ballardo · M. A. Cordero · E. Estrada · A. Ocaña · D. S. Savín · Y. Savín · M. Silva
M. C. Tonche · Y. Torres
Grupo Mujeres con Alas, Domicilio conocido Bahía de los Ángeles, Ensenada, México

© Springer Nature Switzerland AG 2020
S. Lucatello et al. (eds.), *Stewardship of Future Drylands and Climate Change in the Global South*, Springer Climate,
https://doi.org/10.1007/978-3-030-22464-6_11

Keywords Gender · Female leadership · Participatory monitoring · Environmental management · Natural protected areas

Introduction

Gender and the Environment

Current socio-ecological problems present new challenges due principally to the multiplicity of actors and interests that interact in the management of natural resources (Maya and Ramos 2006). In the last decades, one of the approaches that have dominated the discourses, public policies, and research at the international and national levels to address these problems is the gender perspective. The gender approach emerged in the mid-1970s as a criticism of dominant development strategies. These criticisms were aimed at the need to recognize women as producers, mainly in agriculture and to abandon the idea that their role was exclusively "reproduction," that is, take care of the home, raising children, and attending husbands (Nieves 1998; Braidotti 2004). The focus of Women in Development, WID, was the first international approach in the 1960s that proposed the integration of women in the present development model, which was a western and predominantly patriarchal model (Soares et al. 2005).

Also, in the decade of the 1970s, the ecofeminist current emerged as a critique of the male dominated development model and with the premise that the relationship of women with nature is stronger than with men due to maternal and biological stimuli (Nieves 1998; Braidotti 2004). In the mid-1980s, both approaches were criticized even by feminist movements, as it was considered that women were only being incorporated into the Western model and that they had the right to create an alternative development model. This is how Gender and Development (GAD) emerged in the late 1980s, which represents a transition to not only integrate women in development but also to look for the potential that development initiatives have to transform inequalities in women, social and gender relations in order to empower women (Arellano 2003). Shortly afterwards, the Gender, Environment, and Sustainable Development movement emerged, in which the man–woman–environment relationship is analyzed. This approach indicates that men and women use and perceive resources in a differentiated way that is why their interests and priorities are also different (Arellano 2003). Rocheleau et al. (2004) proposed the concept of "feminist political ecology" where gender is a critical variable that grants access to resources and their control, by interacting with class, ethnicity, age, and culture, in order to understand the struggle waged by women and men to sustain ecologically viable ways of subsistence and the possibilities that society has to achieve sustainable development.

Thus, women participate as a key factor in community development processes and are recognized as intermediaries in the relationship between sustainable development and the environment, with population growth, migration, patterns of production and consumption, and social inequalities (Arellano 2003). In addition, throughout history women have played important social and environmental roles, and their political positions have had a great transforming power. For this reason, conservation of natural resources should also consider the division of gender work in order to achieve greater equity in the management of natural resources, which has a clear and differential relationship between men and women (Melero and Solís 2012). This could be possible as long as the participation of women in environmental management is conditioned to access, control, and decision-making as PNUD mentioned in 2011.[1]

Female Leadership

Traditionally, leadership has been associated with the male gender. This is observed in managerial positions or representatives of projects that are usually characterized with properties generally attributed to men as authority, control, and competitiveness. This characterization makes attributes considered to be more feminine, such as orientation or concern for others, not necessarily apt to be a leader; this has fomented biases against women in participation and access to leadership positions (Lupano and Castro 2011).

This work analyzes a successful example of female leadership in the group "Women with Wings" (Mujeres con Alas) to promote community and regional development in a Protected Natural Area. "Women with Wings" developed, through dialogue and collective action, supporting actions towards other social actors to understand the complexity of the socio-ecological systems and to build new ideas based on shared mental models and identities.

In addition to the construction of female leadership, the joint work of the CONANP[2] officials, academy members (UABC,[3] UV,[4] and UC Irvine), and "Women with Wings" represents an effort to implement the transdisciplinary collaboration for sustainability (Merçon et al. 2018) that responds to the need to get involved in bird conservation activities within the protected area.

For this study, ethnography was made with semi-structured interviews and focal groups to document and explore the role of female leadership in environmental management through the experience of the group "Women with Wings." The group's strengths, obstacles, and areas of opportunity were identified in the conservation activities.

---

[1] https://www.undp.org/content/undp/es/home/librarypage/hdr/human_developmentreport2011.html.

[2] Comisión Nacional de Areas Naturales Protegidas.

[3] Universidad Autónoma de Baja California.

[4] Universidad Veracruzana.

Case Study: Natural Protected Area (NPA) Establishment and Female Leadership

Approximately 550 km south of the city of Ensenada, Baja California there is a town called Bahía de Los Ángeles. It is a fishing village, with a population of around 600 inhabitants. Its ecological importance places it as the site with the highest marine productivity in the Gulf of California. This is the habitat for whales, dolphins, huge colonies of seabirds, whale sharks, sea turtles, as well as fish and invertebrates of ecological and commercial importance (Pronatura Noroeste 2014). In 2007, this area was declared a protected natural area (ANP for its acronym in Spanish) within the Biosphere Reserve of *Bahía de Los Ángeles, Canal de Ballenas y Salsipuedes* (Fig. 11.1).

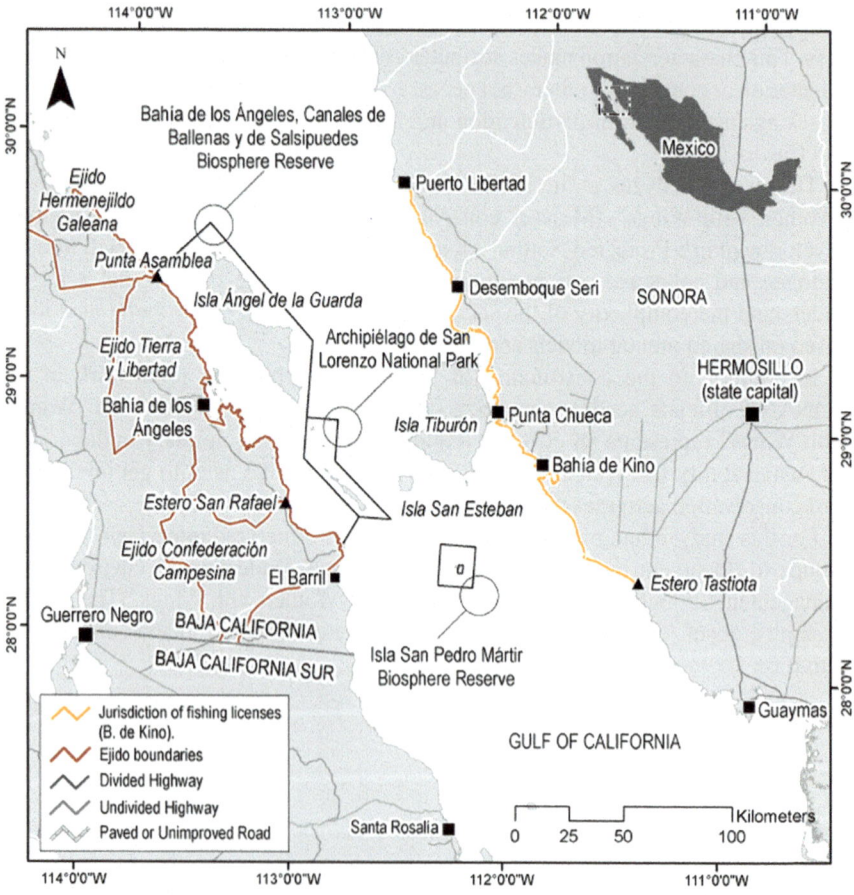

Fig. 11.1 Protected natural area map (SEMARNAT 2013). Source: Cinti et al. (2014)

Fig. 11.2 Logo and motto
of the "Women with
Wings" group (original
design based on the ideas
of the group)

The main economic activity is fishing, but for its unique natural and cultural features (flora, fauna, landscape, historical and archeological sites) within the region, this place has enormous potential in the development of tourist activities (Andere 2017).

The community group "Women with Wings" was created in 2015 with the support of the Conservation Program for Sustainable Development (PROCODES, for its acronym in Spanish). This program represents a national strategy to link biodiversity conservation with social and economic development through the participation of communities including women and men. "Women with Wings" is made up of 11 women who carry out bird monitoring, environmental education, and bird watching. The members of this group are women, mothers, housewives, wives, and active community members of their locality, who carry out collective and individual actions for the development and welfare of their family, their community, and the ecosystem, which makes them agents of social change (Fig. 11.2).

Skills to Build Their Own Story

The "Women with Wings" group was built through a process of learning techniques and skills to carry out bird monitoring activities. This process implied the coordinated participation from the group of women along with CONANP[5] officials,

---

[5] Comisión Nacional de Areas Naturales Protegidas.

academy members from UABC,[6] UV,[7] and UC Irvine. Learning to use a photographic camera to capture the birds during monitoring is one of the skills that the group has acquired. Its members have also improved in the use of this tool and currently include stranded birds and the landscape within their photographic records. Since these women recognize photography as a means of self-expression:

> Sí porque te ponían experiencias para que hicieras tu propia historia, algo de ti, algo que te guste...
> (They shared examples of personal experiences with us in order to create our own story, something about us, something that we were interested in)[8]

Its members have also acquired skills on bird identification with the support of published birding field guides:

> No sabíamos los nombres, pero de que las mirábamos, sí las mirábamos
> (We did not know the names (of the birds), but we were sure we saw them, yes, we did saw them)

Simultaneously to the identification of the birds, the group must fill in a field log to build a database, for which they learned to use the computer and to develop databases. Currently, this information is shared on eBird and Naturalista, online communities of learning and sharing information about birds and other elements of nature, where amateurs and professionals collaborate.

Capacities for Action

Capacities are related to the skills and resources to perform a task or fulfill an objective (Appendini and Nuijten 2002). "Women with Wings" meets every month to plan and schedule activities; these meetings are also linked to the development of other skills such as interpersonal communication and teamwork. Within the organization of activities, the group recognizes the different abilities of each member, in order to consider them during the formation of work teams:

> Ella siempre anotaba y yo siempre estoy con la foto y ahorita que agarra la cámara pues se le dificulta o a mi anotar porque pues estaba bien acostumbrada a la foto
> (She always took notes and I was always taking the pictures. Now that she has the camera, it is difficult for her to take pictures or for me to take notes, because I was so used to taking pictures)

Planning of duties was necessary due to the reduction of financial resources for fuel and travel expenses, either sharing paths or reducing the number of field trips per month to optimize fieldwork and money resources. In order to accomplish the objectives, set out on a monthly basis, they have developed a high capacity for discipline, which is reflected in a strong commitment by the women to attend to the group's monitoring, meetings, and activities despite the various tasks they perform

---

[6] Universidad Autónoma de Baja California.

[7] Universidad Veracruzana.

[8] All the paragraphs in italics are a transcription of the members of the "Women with Wings."

in their homes. The tasks demanded by the group have become part of their daily routine, even when it can be complicated at times to fulfill them:

> Porque salí embarazada cuando recién inicié, entonces todo el embarazo me la pasé... salía con mi panza, pero salía, pero ya los últimos meses fue cuando yo empecé a trabajar nada más aquí, a capturar toda la información que me traían... ya después cuando nació la niña que me la tenía que llevar también a campo...
>
> (I got pregnant just when I started to participate, but I participated the whole period in all activities with my pregnant belly; during the last months of my pregnancy I worked only here at the office and began to capture all the information they brought me. Once the girl was born I had to take her with me to fieldwork...).

Even with financial limitations, the group has continued with bird monitoring, constantly looking for alternatives to fulfill this responsibility:

> Pero ya nos habían advertido esto, que iba a tener un tiempo en que no iba a llegarnos nada y era como que tu prueba de fuego, si le sigues o te paras...
>
> (We had already been warned that there were going to be times when money would not come, so it was like a fire test: whether you continue or you stop...).

For the group, the ability to act (agency) is not understood as individual skill, but as a shared possibility (to be able to do) (i.e., Ema 2004). The capacity of agency is manifested in the actions that seek to contribute to the community, either through environmental education in schools or actions of care for the environment and the place where they live. The action of organizing a community bird monitoring group reflects this capacity:

> No pues, poder dejar algo, como enseñarles un poquito de educación ambiental, de enseñar algo a la comunidad...
>
> (I want to be able to leave something for the community, to teach them a little bit of environmental education, to teach something to the them).

Teamwork is recognized as one of the strengths of this group, which is linked to the built communication since its formation. Teamwork is associated with the support and empathy that are offered when working together:

> Y adaptarnos a los problemas que se nos presenten a cada familia, porque muchas pasaron, por decir las compañeras, una que se fue y la otra y ellas, fue el tiempo que le dio cáncer a su papá, se perdieron de cursos, se fueron a Ensenada, falleció, o sea fueron muchas cosas, nos tenemos que adaptar, ellas no podían, pero estamos nosotras que podemos representarlas en una junta o algo...
>
> (We have adapted to the problems we face within each family. One woman's father got cancer, went to Ensenada to get treatment for him but he died. She lost many training courses. Many things happened, but we have to adapt; for example, if some women cannot attend, then we represent them in a meeting or other activity)

These solidarity actions and support among the women are also associated with the sorority they have created between them:

> -Aguantarnos nuestro genio, o también hemos aprendido al tiempo de ella, de ella, o sea, ¿tú qué vas a hacer? y ¿tú que vas a hacer? si a veces hasta yo si veo que es más complicado lo de ella pues yo me adapto a lo mío
>
> -Sororidad también, diría alguien por ahí...
>
> -Es la palabrota... es nuestro lema
>
> -Luego estamos así ¡Sororidad, sororidad!
>
> (-We have to tolerate our way of being, we have also learned to do the activities when our times coincide by asking what the daily activities of each other are.

Fig. 11.3 "Women with Wings" performing bird monitoring. Coauthors of this chapter

> -Sorority, someone would call it like that,
> -which is a big word and also our motto,
> -Then we are saying: Sorority, sorority!).

Sorority is a gender political pact between women, and it means friendship between different women and peers, accomplices who set out to work, create, and convince, there is no hierarchy, but a recognition of the authority of each single woman (Lagarde 2009). This concept is inspired by solidarity practices between women, friends, neighbors, or relatives and it is recognized by the members of the group through the CONANP[9] officials, who intervene when it is necessary to remember the value of the voice of each woman (Fig. 11.3). Even without developing feminist political consciousness, they build relationships of mutual support and trust, not always conscious that this serves to sustain themselves within the patriarchal system (Royo et al. 2017).

Communication and teamwork skills are related to each other, since the development of communication between the women promotes teamwork. The women of the group recognize that the dialogue has allowed them to reach agreements and resolve misunderstandings:

> -Hablar las cosas
>   -El diálogo
>   -¿Qué nos han hecho? Que nos han juntado y sabes que hay que hablar aquí las cosas que no estamos contentas...
>   -¿Qué andar hablando las cosas allá? Las cosas se arreglan aquí...

---

[9] Comisión Nacional de Areas Naturales Protegidas.

(-To say things,
-Dialogue,
-What have they done to us? They have gotten us together, and we know that we have to talk things here and say what we don't like…
-Rather than talking about those things somewhere else, we fix issues here).

The communication is not only associated with the dialogue between the group, and it is also the ability to communicate and express themselves in front of other people and groups. Some members explain that being part of the group has allowed them to develop this capacity:

Porque a mí no me… no sabía explicar y pararme ahí enfrente que me hicieran preguntas y decía, no puedo

(Because I didn't know ... I did not know how to explain or stand there in front of other people who were asking me questions and I used to think I could not do it).

The environmental education activities carried out in the community schools allow the developing of communication skills, and an increment in confidence and the disclosure of the done work. The community/social feeling is associated with the ability to relate to others and the willingness to cooperate with them in order to achieve an effect on the community: "see with the eyes of another, hear with the ears of another and feeling with the heart of another" (Oberst 2002). Since its formation until now, the group has managed to organize its monitoring work; even with economic deficiencies they continue working since they have appropriated the project:

-O sea, como que ya lo agarramos… aunque no nos estén pagando
-Como comunitario
-Porque las amamos…
(-I mean, considering that we were working anyhow, even though they do not pay us,
-As a community member,
-Because we love birds...)

Best Practices

The environmental education within the "Women with Wings" group is a process that happens mainly in the family nucleus, when they participate in the bird watching activities they include their daughters, sons, and husbands. It is recognized that environmental education is necessary for the community and that their work within the family can multiply and create an impact at the community level:

Como que ya 10 mujeres, como quiera esas 10 mujeres tienen su familia, y esa familia tiene más familia, ya se corre la voz, ya hay más educación ambiental, ya cuidan más pues…

(We are 10 women already, and those 10 women have their family, and that family has more family, this helps spread the word; now there is more environmental education, and we protect the environment better...).

The influence of women in the family nucleus is strategic in the attitudes and habits that foster a paradigm shift for the care and use of natural resources (Galvis 2009).

Sharing with their families encourages interest and involvement in bird identification:

> Ahora los maridos llegan y: "a ver mija, tu guía versa, porque ahí anda un ave y verás vimos que tenía así y asa"
> (Now the husbands come and say: "let's see your bird guide, because there is a bird and looks like this and like that").

The group also recognizes the practices that damage the environment and they adopt new attitudes of appreciation towards birds and the ecosystems they live in:

> Simplemente de que tirábamos basura y ahorita ya no, ahorita es: "¡Hey! no la tires por la ventana, échala aquí en una bolsita y cuando haiga un bote de basura pues la tiramos"
> (We used to dump the trash everywhere and now no more, now we say: "Hey! Don't throw the trash out the window, here is a bag and when we see a trash can, we dump it there")

Being part of this group that carries out a conservation activity has influenced their habits. Women are the administrators of environmental resources in the domestic activities and they possess knowledge of local environmental processes, but the impact on their home and in the community is not recognized. However, women consider themselves more prepared to make changes in habits and they organize themselves to contribute to the protection of the environment (Arora 2011).

Foundations of a Social Identity

Identity is constructed from values, culture, and symbols that are shared among social actors, groups, and collectivities. It is formed by the accumulation of memories that a group or person carries with itself (Grayling 2002; Mendoza 2009). The collective memory of the group is formed by significant memories and anecdotes, from its origin to the present. The moment they decided to be part of a women group who performs bird monitoring is one of those memories:

> -Alguien le avisó a Saul Espinoza creo, le avisó a la Karla
> -A la Karla para que viniera la Andrea a ver de qué se trataba, que le gustaría que hiciéramos un grupo y fue así como que "no pues como que si" y ya que vimos lo fuimos planteando, pero más bien, la verdad uno se viene por lo económico, porque todavía no le agarras amor al arte... hasta que ya estas ahí así que así empezó... fue por la parte de que "no que hay un proyecto, que se baje un recurso" y pues órale, y pues que metiéramos papeles porque podíamos meter otras 10, otras 10, otras 10, y como se aprobara...
> (-I think someone told Saul Espinoza, and he told Karla.
> -Karla told Andrea to go and figure out what it was all about, finally she wanted us to form a group and that is how the group was created. At the beginning, honestly, we entered for economic reasons, because we still did not do it for the fun of it ...then, we learned we could submit paperwork to apply for funds and resources; that is how we started and then we could add more and more [women] 10 after 10 after 10 as approved).

With this memory, the group recognizes that the original motivation to start up this project was to receive financial support that would give them greater solvency and that in the process of formation they began to commit themselves and to enjoy the activities proposed.

Another memory they highlight is the Christmas count of 2017, as an activity of acceptance and involvement of their families in the monitoring. Each of the members shared with their families the activities they usually do in monitoring, teaching them how to use the tools, identify the birds and collect the data:

-El año pasado hicimos un conteo navideño donde los esposos se involucraban, era por familia el conteo navideño, entonces se dividieron las zonas, entonces estuvo bien padre, iba yo, mi esposo y mis dos hijos y no pues tú el GPS, tú los miralejos, yo lo apunto y pues andaban emocionadísimos, ella con su familia, ella con su familia, todo fue individual y en diferentes partes...
   -Y en diferentes sitios
   -Y terrestre y marino
   (-Last year we did a Christmas bird count, where the husbands got involved. Each family participated in Christmas count and zoning was assigned by family; it was incredible. My husband, my two children and I divided the work, one took the GPS, the other the binoculars, I registered the data. All the families were very excited
   -At different sites
   -Both in marine and terrestrial ecosystems).

This activity induced a change in the attitude of family members, because when all of them participated, they concluded that the activity was important for the community.

An important memory for the acceptance and recognition of the group by the community was the phenomenon of mortality of some birds in Bahía de Los Ángeles:

-Había personas que nos avisan, a mí en lo personal una prima de mi esposo me dijo "¿qué está pasando? hay mucho pato tirado" "¿ah sí? ¿en qué zona?", "no pues en esa" y ahí íbamos nosotras y ya me decía Saruhén, no pues llévate la pesola ¿no? una chiquita, y el vernier, y mide y llévate guantes y ya yo me llevaba al niño y yo y a mi esposo y hasta 100 zambullidores en bolsas, los que estaban más frescos mandarlos a estudiar, y ellos bien involucrados, ya ella haciendo otra zona o "no pues yo voy hoy",
   -Desde que pasó eso fue cuando la gente ya empezó como a avisar porque sí hubo mucho zambullidor muerto,
   -Era mucho,
   -Yo de ahí noté que como que ya cualquier cosa avisaban "hay un animal allá"... (-There were people who came and told us; a cousin of my husband said: "what's going on, there are a lot of ducks lying around." So, I asked her in what area and she pointed towards it. Then we went to check it out; we took the equipment to weigh and measure the ducks (a scale and the Vernier) and gloves. I went with my son and husband; we weighed and measured up to 100 grebes, those that were still fresh we sent for analysis. This is how the community got involved and we worked in different areas,
   -And people continued to indicate more sites with dead grebes,
   -There were many of them...
   -Ever since then, people began to inform us...).

The involvement of "Women with Wings" in documenting this phenomenon gave them the recognition of the community; since that moment they were identified as agents for environmental care.

The collective memory of the group has been formed through attitudes, and cognitive and effective practices. The group has its memory and discourse that allows it to appropriate the monitoring activities and to affirm its identity as a group, as well as cement their permanence in the community (Páez et al. 2007; Mendoza 2009).

Self-recognition is the reflective process by which a person acquires the notion of his/her self and his/her own qualities and characteristics. Each member of the group had his/her own process of self-recognition; however, the sense of belonging to a group has supported this process.

Since their formation, they have faced challenges at a personal, family, and community level. Now "Women with Wings" are able to identify qualities that they did not recognize before belonging to the group, such as the ability to achieve objectives and the freedom to carry out interesting activities. Being housewives does not determine your level of competence. In relation to the gender roles established in their families and in the locality:

-Pues me siento capaz, que a pesar de todo puedes hacer tus cosas, que, a pesar de los hijos, del esposo, de todas formas, que eres ama de casa y todo tienes, y sales puedes salir adelante, o sea todas tus cosas puedes cubrir…

-Pues me siento útil, primero que nada, porque pues es otra cosa que hacer aparte del hogar y de atender a tu esposo y a tus hijos. Pues haces otra cosa y pues que me gusta…

(-I feel self-confident that, in spite of everything, you have plenty of time to do your things, in spite of the children, the husband, being a housewife, having to do everything, you can get ahead, you can do all your activities ...

-I feel useful, first of all, because it is another thing to do apart from housework and attending your husband and children. You can do something else and I like that...).

Being part of this group that actively participates in conservation activities in the community has generated self-esteem and created certainty about the importance of the activities and their impact on the community and the family. Self-recognition is also a reflection of the results in the community, of the interest, and acceptance of the activities that the group performs:

Pues me siento importante, también, porque ya a la gente le interesa, le importa que sabes, de que así pues de un animal o algo ya te lo hacen saber, ya saben que a uno le importa las cosas que están pasando…

(I feel important because people here are starting to be interested, they care about what I know, if they have information about animals or what they know about them, they tell me or let me know about it since they know I care for…).

The monitoring activities, the coexistence with other women, the acceptance and recognition of the community give the women courage, decision power, and motivation to continue doing the activity that they like and that allow them to enjoy their freedom:

-No sé, pero ya somos Mujeres con Alas porque...

-Mujeres: Porque andamos volando

-Ya no nos tienen agarradas

-Nos sentimos libres

-Ahora si ya no nos van a parar, no podemos parar porque se va, va a decir tanto que invertí mi tiempo, mi espacio para que estas señoras paren cuando yo me vaya

(-Now we are "Women with Wings", because ...
-Women: Because we are flying.
-They do not stop us anymore.
-We feel free.
-They are not going to stop us anymore, we cannot stop because she's leaving; she's going to say: "so much of my time and place that I've invested for these ladies to stop when I leave...").

Gender Roles Within the Community

Gender roles are those social norms and behaviors that are considered socially appropriate for people of a given sex. Bahía de Los Ángeles is a fishing community, men go fishing and women take care of their children and their homes. It has been men who traditionally participate in conservation activities. Thus, gender roles are reflected in the opposition of their husbands to participate in activities outside the home:

Unas experiencias de mi compañera que es lo contrario pues ella si batalló para que su marido viera que si era verdad porque a lo mejor que decía "estás perdiendo el tiempo Alejandra no te vaya a dar equis pero ya cuando el vió era en serio..."

(One of the women struggled to have her husband see that we were involved in a serious activity, he said, "You're wasting time Alejandra", until he realized that it really was a serious activity...)

This opposition was not only within the family nucleus but also in the community. Because of social construction one associates women in the home, in the kitchen and with their children, not on the sea or at the beach, monitoring birds and least generating income from that activity:

-Era así de "en lugar de andar ahí, váyanse a limpiar su casa" comentarios así, entre la misma familia, eran comentarios como de "¿qué hacen ahí?" pero no saben que uno empieza a amar lo que hace, es como en cualquier cosa...
    -Ahora del diplomado decían "van a perder tantas horas ahí sentadas, para nada"
    -Las credenciales dicen que no nos las van a mandar...
    (-People said: "instead of going there, go clean your house", that sort of comments, Even within our own family they made comments like: "what are you doing there?" But they did not know that you start loving what you do, it's like with anything else you do...
    -When we took the workshop they said, "you are going to waste so many hours sitting there for nothing"
    -They say that they're not going to send us the certification cards)

Creating a group of women who carry out bird conservation activities shows dissatisfaction with the established gender roles; it represents the challenge of demonstrating through different actions the capacity they have as women, mothers, and housewives to carry out other activities that have an impact on the environment and in the community:

También pues demostrarle a la gente que no era nomas entrar a un grupo y ya y también que las mujeres podemos...

(We did it also to show people here that it was not just to enter a group and that was it, but also that we women can...)

The satisfaction is experienced when receiving the recognition from the community and families, feeling that the family is interested and involved in monitoring activities and environmental care:

Entonces el niño le decía a la niña porque son de la misma edad, que su mamá también era maestra, que era maestra de las aves, que no sé qué... o sea ya los niños empiezan a, empiezan a querer las aves, o sea empezamos de casa, ya ellos... en la escuela, a sus compañeros.

(Then a boy commented to a girl, they are the same age, that his mother also was a teacher, that she was a teacher of birds, that means, that the children begin to love the birds. That's because we teach them at home and then they do the same at school with their classmates)

These actions are rewards that allow them to stay motivated and to face any future challenges.

Final Considerations

Recently, "Women with Wings" were certified by the Ministry of Tourism as "Nature-oriented Tourism Guides with Specific Activity in Environmental Interpretation." Completing this certification despite economic and family challenges left a sense of well-being in the group; this is the culmination of a goal and the scope of a personal aspiration. "Women with Wings" have appropriated the monitoring work and also recognize that even without receiving an economic benefit, the love and commitment to birds, ecosystems and the community motivates the group to continue with conservation actions. The sense of belonging emphasizes its relationship with the well-being of people, it is defined as a feeling of belonging with a certain group or environment (Brea 2014). Its members have appropriated the community bird-monitoring project. Despite not having been an initiative of the community, it is now part of it, which has made this project successful. Each of the members of "Women with Wings," from their individual and collective actions, seeks the development and well-being of their family and community. As a group, it recognizes and accepts the differences of its trajectories, and seeks to organize themselves to create a positive impact on their community. Individually, from their role as mothers and wives, they engage in conservation tasks and modify habits so not to harm ecosystems. Also, they recognize the capacities they have, value and execute their power of decision to benefit themselves and all their family.

References

Andere RAE (2017) Determinación de los factores que intervienen en el crecimiento y desarrollo económico de Bahía de los Ángeles, Baja California, México. Universidad Autónoma de Baja California, Ensenada

Arellano MR (2003) Género, medio ambiente y desarrollo sustentable: un nuevo reto para los estudios de género. Rev Estud Género La Vent 2003:79–106

Arora JS (2011) Virtue and vulnerability: discourses on women, gender and climate change. Glob Environ Chang 21:744–751. https://doi.org/10.1016/j.gloenvcha.2011.01.005

Appendini K, Nuijten N (2002) El papel de las instituciones en contextos locales. Revista de la CEPAL 1(76):71–88

Braidotti R (2004) Mujeres, medio ambiente y desarrollo sustentable. Surgimiento del tema y diversas aproximaciones. In: Vázquez GV, Velázquez GM (eds) Miradas al futuro: Hacia la construcción de sociedades sustentable con equidad de género. Universidad Nacional Autónoma de México (UNAM), Centro Regional de Investigaciones Multidisciplinarias (CRIM), Programa Universitario de Estudios de Géneros (PUEG), Colegio de Postgraduados (CP) y el Centro Internacional de Investigaciones para el Desarrol, México, pp 23–59

Brea LM (2014) Factores determinantes del sentido de pertenencia de los estudiantes de Arquitectura de la Pontificia Universidad Católica Madre y Maestra, Campus Santo Tomás de Aquino. Universidad de Murcia, Murcia

Cinti A, Duberstein JN, Torreblanca E, Moreno-Báez M (2014) Overfishing drivers and opportunities for recovery in small-scale fisheries of the Midriff Islands Region, Gulf of California, Mexico: the roles of land and sea institutions in fisheries sustainability. Ecol Soc 19(1):15. https://doi.org/10.5751/ES-05570-190115

Ema LJE (2004) Del sujeto a la agencia (a través de lo político). Athenea Digit Rev Pensam e Investig Soc 5:1–24

Galvis BAH (2009) Influencia del núcleo familiar en la formación ambiental del niño-niña. Estudio de caso: institución preescolar liceo infantil casita encantada localidad Barrios Unidos, Bogotá, DC. Pontificia Universidad Javeriana, Bogotá

Grayling A (2002) El Sentido de las Cosas. Filosofía para la vida cotidiana, Colección. Editorial Crítica, Barcelona

Lagarde M (2009) La política feminista de la sororidad. Mujeres En Red. El Periódico Fem. pp 1–5

Lupano PML, Castro SA (2011) Teorías implícitas del liderazgo masculino y femenino según ámbito de desempeño. Ciencias Psicológicas 5:139–150

Maya D, Ramos P (2006) El rol del género en el manglar: heterogeneidad tecnológica e institucionales locales. Cuad Desarro Rural 3:53–81

Melero AN, Solís EC (2012) Género y medio ambiente. El desafío de educar hacia una dimensión humana del desarrollo sustentable. Rev Int Investig en Ciencias Soc 8:235–250

Mendoza GJ (2009) El transcurrir de la memoria colectiva: La identidad. Rev Casa del Tiempo 2:59–68

Merçon J, Ayala-Orozco B, Rosell J (eds) (2018) Experiencias de colaboración transdisciplinaria para la sustentabilidad. Serie Construyendo lo Común, número 1, CopIt-arXives y Red Temática de Socioecosistemas y Sustentabilidad, Conacyt, México

Nieves RM (1998) Desarrollo sustentable, ambiente y género: Antecedentes de su consideración en las reuniones internacionales. In: López Friné MI, Patricia H (eds) Género y Medio Ambiente. PRODEC, CIDHAL A.C., Cuernavaca, pp 235–239

Oberst Ú (2002) Salud mental y ética: El concepto de sentimiento de comunidad en la psicología de Alfred Adler. Persona 5:131–146

Páez RD, Marques J, Beristain CM (2007) Memoria social y colectiva y representaciones sociales de la historia. In: Psicología social. McGraw Hill, New York

Pronatura Noroeste AC (2014) La Reserva de la Biosfera de Bahía de los Ángeles ya tiene su programa de Conservación y Manejo, conócelo y respétalo. http://pronatura-noroeste.org/es/la-reserva-de-la-biosfera-de-bahia-de-los-angeles-ya-tiene-su-programa-de-conservacion-y-manejo-conocelo-y-respetalo/. Accessed 11 Jul 2018

Rocheleau D, Thomas Slayter B, Wangari E (2004) Género y ambiente: una perspectiva de la ecología política feminista. In: Vázquez García V, Velázquez Gutiérrez M (eds) Miradas al futuro: Hacia la construcción de sociedades sustentable con equidad de género. PUEG, CRIM, IDRC, CP, México, pp 342–404

Royo PR, Silvestre CM, González EL, Linares BE, Suarez EM (2017) Mujeres migrantes tejiendo democracia y sororidad desde el asociacionismo. Una aproximación cualitativa e interseccional. Investig Fem papeles Estud mujeres, Fem y género 8:223–243

Secretaria de Medio Ambiente y Recursos Naturales (SEMARNAT) (2013) Acuerdo por el que se da a conocer el resumen del Programa de Manejo del área natural protegida con la categoría de Reserva de la Biosfera la zona marina conocida como Bahía de los Ángeles, canales de Ballenas y de Salsipuedes. Diario Oficial de la Federación, 5 de noviembre de 2013

Soares D, Castoreña L, Ruiz E (2005) Mujeres y hombres que aran en el mar y en el desierto: Reserva de la Biosfera El Vizcaíno, BCS. Front Norte 17:67–102

Chapter 12
Sustainability Assessment in Indigenous Communities: A Tool for Future Participatory Decision Making

D. Galván-Martínez, I. Espejel, M. C. Arredondo-García,
C. Delgado-Ramírez, C. Vázquez-León, A. Hernández, and C. Gutiérrez

Abstract Developing an adequate set of indicators and a more consistent methodology to assess sustainability in a community is a task that requires understanding the level of viability of the systems involved and their contribution to sustainable development. The objective of this research was to design and apply an index to estimate the sustainability of rural indigenous communities characterized by intercultural processes. Three communities inhabiting the arid zones of Baja California, Mexico: Pa Ipai, Kumiai, and Cucapá are the study cases. To develop such an index, a review was made of the most common indicators used worldwide for measuring sustainability in rural communities. The method to reduce and focalize the list of indicators consisted in identifying indicators from the answers given to questionnaires applied to 166 individuals chosen at random in the three studied communities. The answers were analyzed applying the content analysis qualitative tool. The responses were assigned a weight ranging from 0 to 1 according to the remoteness (0) or proximity (1) to sustainability of response. The results depicted, in a spider graph, that the Pa Ipai community has the highest values towards sustainability, while Kumiai is on average and, the community of Cucapá with the lowest values and, therefore, far from achieving sustainability. For the particular case of the

D. Galván-Martínez · C. Gutiérrez
Programa de Doctorado en Medio Ambiente y Desarrollo, Universidad Autónoma de Baja California, Ensenada, Mexico

I. Espejel (✉)
Facultad de Ciencias, Universidad Autónoma de Baja California UABC,
Ensenada, Baja California, Mexico
e-mail: ileana.espejel@uabc.edu.mx

M. C. Arredondo-García
Facultad de Ciencias Marinas, Universidad Autónoma de Baja California, Ensenada, Mexico

C. Delgado-Ramírez
Escuela de Antropología e Historia del Norte de México, INAH, Chihuahua, Mexico

C. Vázquez-León · A. Hernández
El Colegio de la Frontera Norte, Tijuana, Mexico

© Springer Nature Switzerland AG 2020
S. Lucatello et al. (eds.), *Stewardship of Future Drylands and Climate Change in the Global South*, Springer Climate,
https://doi.org/10.1007/978-3-030-22464-6_12

indigenous communities' native to Baja California, the complexity of social and economic problems that arise can be attenuated if project proposals emanate from the same community, respecting the traditional social organization and family organization during their implementation. Our results provide elements to design and evaluate public policies on indigenous issues at the local level, so it is suggested that the indicators proposed in this research be contrasted with other rural indigenous communities of the drylands.

Keywords Sustainable development · Rural public policy · Interdisciplinary indicators · Mexican indigenous people · Baja California

Introduction

In terms of cultural identity, apart from the social and anthropological elements that give a group the keys to be unique are the links with the environment and the use of ecosystem services. This is a specific idea of what cultural identity is, provided that the concept and the idea of the environment is integrated into social and anthropological aspects. In addition to the above, the self-perception of indigenous communities leads to identify elements that describe their relationships with the environment and, therefore, identify themselves as part of nature, as an element that interacts, or on the other hand, a superior element owner of natural resources (Pretty et al. 2009). This construction of cultural identity tends to be inclusive (social and environmental aspects) in order to generate a clear and unarguable culturalism concept that is useful to represent the essence of a native community. The culturalism is an idea of the self-perception including those elements that are relevant to explain the history and values of a society or a community, and the relationships forged with the natural surroundings. While cultural identity is recognizable and understandable, intercultural will be possible to explain it as the relationship that arises from interactions between individuals and groups with particular cultural identities (Grimson 2001). It is important to observe the issues that arise from intercultural and interpret them through a comparative process to explain the relevance of the environment, how this is integrated into the economic cycle, how different communities explain their perception of the ecosystem, and how they perceive themselves in the environment. Therefore, the techniques used in an intercultural research become extremely important in order to obtain satisfactory communication.

To improve intercultural relations among researchers, developers, and the local people that will benefit from the project outcomes, a new way of starting project design has been proposed recently. Co-design and co-generation of knowledge is fully recommended (Schuttenberg and Guth 2015) and researchers from the human sciences become key persons in the research teams. In this chapter we selected a study case to show how traditional ethno-ecological studies start, how it changed while doing it, and how it must be developed if applied by the communities themselves. The research started by a question developed by a student (first author) as the typical questionnaire that natural scientists create when interested in ethno-ecological subjects. The answers to some selected questions were interpreted using a qualitative technique. Single or clusters of similar responses were used to select indicators of sustainability and then, these indicators were combined to "measure" the community sustainability. The final results were presented in the localities and the attendants mentioned to agree with the model and the indicators because they were understandable. A new form of presenting the results was performed and given as poster with the infographics to the communities. The spider graphs show the weak parts of their community sustainablity. Now, the communities willing to reach its sustainability have to organize and decide how they are going to get to the ideal model.

The communities are the people of the drylands that have developed life models according to the water scarcity (Morales 2003). They are indigenous groups of northern Mexico belong to the cultural zone of Arid America; characterized by being hunters, gatherers, and anglers (Gómez and Sánchez 2014). The word "Aridamérica" was described for the first time by the Kirchhoff (1954), due to the predominant desert climate, the feeding situations, and the water scarcity of these regions. According to Delgado (2008), Baja California is one of the states related to this cultural area of America, dominated by a hydrological, geographical, climatic, and cultural group of conditions. The prehispanic Indians of the peninsula come from the *Yumana* or *Yuma* family, which occupy the southwestern United States, their subsistence forms, and ability to adapt to the geographical environment of the arid zones characterize them (Rogers 1945). It has estimated that the population in the peninsula in pre-Hispanic times amounted to 50,000 inhabitants. Over time they were subdivided into *quilihuas* or *cahuillas* (*Kiliwa*), in the Sierra de San Pedro Mártir; the *cocapás* or *cucapás* (*Cucapá*), on the banks of the Colorado River; the *yumas* (*Paipai*), of Santa Catalina, in the mission of that name; the *diegueños* (*Kumiai*), in the Ensenada and Tecate region, looking for water and opportunities to stock up in inhospitable territories (Martínez 2011) (Fig. 12.1).

Cultural Indicators and Models

Although researchers have identified specific cultural indicators for indigenous communities (Del Val et al. 2008), we found no sustainability model integrating all aspects of sustainable development (i.e., ecological, social, and economic indicators) or including indicators of high relevance for indigenous communities, such as cultural or political indicators. Galván-Martínez et al. (2016) proposed the Model for the Estimation of Indigenous Community Sustainability (MEICS-MOSUC by its Spanish Acronym) as an integral tool for measuring communities' achievements towards sustainability. In a previous stage, the authors reviewed currently applied theoretical approaches to sustainability and sustainable development in rural and indigenous communities, the indicators of sustainability available in Mexico, and the accessibility to local level information. The review allowed for choosing three

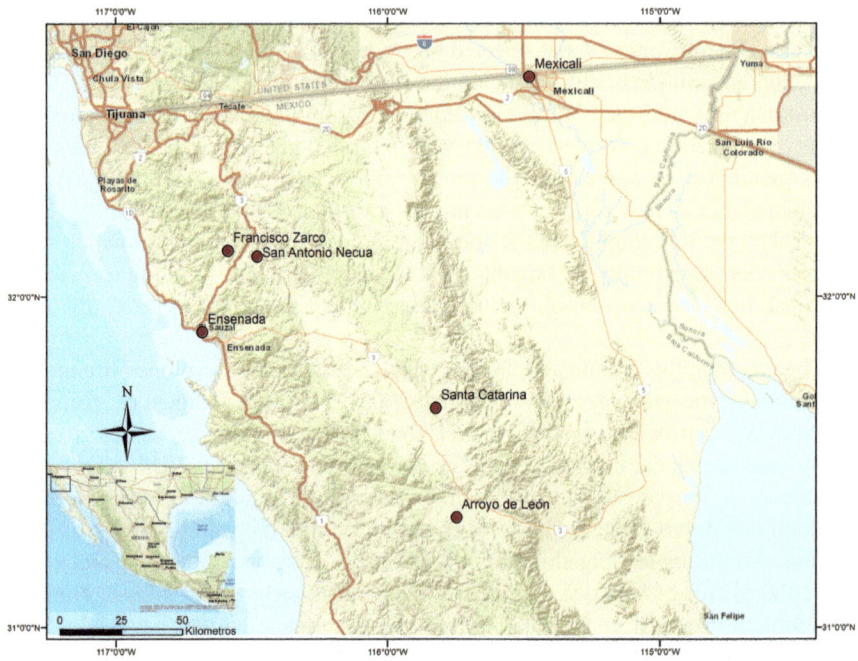

Fig. 12.1 Location of studied Indigenous communities of Baja California, Mexico

theoretical approaches that, beyond their divergences, are all focused in quality of life, welfare, and challenges faced by people living in rural areas: The Community Model of Sustainable Development (CMSD) (Toledo 1996; Barkin 1998); the Buen Vivir (BV) (Gudynas and Acosta 2011), and the Sustainable Livelihoods Model (SLM) (Chambers and Conway 1992). After developing a conceptual synthesis of the three chosen approaches, their differences and similarities were analyzed for appraising the importance of each component of the selected models to be integrated in a definition of indigenous community sustainability. Afterwards, retaking the information pyramid (Hammond 1995; SCOPE 1995), the MEICS index of community sustainability was defined integrating five sub-indexes and 20 ecological, social, economic, political, and cultural indicators.

The Future Use of Community Sustainability Indicators and Models

Developing an adequate set of indicators and a more consistent methodology to assess sustainability in a community, a country, or the world is a difficult task that requires understanding the level of viability of the systems involved and their contribution to sustainable development (Liu et al. 2017). Independently of these difficulties, joining the Sustainable Development Goals 2030 (UNDP 2019), Mexico committed itself to develop actions aimed at generating indicators to measure the progress made towards sustainable development and to evaluate policies and strategies aimed at that objective (UNDP 2015). Since the former Program of Action for Development Sustainable Agenda 21, Mexico developed the National System of Environmental Indicators that includes a set of basic and key indicators that provide a brief description of the current situation of the country's environment and natural resources, the pressures that affect them, and the institutional responses for its conservation, recovery, and sustainable use (SEMARNAT 2015). The system has been updated according to the Agenda 2030. The Mexican government committed to implement the Agenda 2030 for Sustainable Development and move from the Millennium Development Goals (MDGs) to the Sustainable Development Goals. Therefore, the country is committed to include the different social actors including the indigenous peoples in the implementation of the Agenda, specifically in Objective 1 (without poverty), Objective 10 (reduced inequalities), and Objective 3 (cities and sustainable communities) (Sustainable Development Knowledge Platform, 2017[1]). Currently Mexico has developed some disaggregated indicators for state, municipal, and main cities (Information System for Sustainable Development Goals[2]); however, the rural communities remain without estimates of sustainability particularly in drylands.

Description of the Case Study

The objective of this study was to measure MEICS index values of three rural indigenous communities inhabiting in the arid zones in the state of Baja California, Mexico. The results provide local level information needed for indigenous communities and external decision-makers—jointly—to analyze, reorient, and evaluate the actions necessary to make advances towards sustainability, and to provide elements for designing and evaluating public policies aimed at local level indigenous issues.

Study Area

The model was developed integrating three of the four indigenous communities of Baja California: The Cucapá occupy the settlement of El Mayor Cucapá. The Cucapá territory includes part of the alluvial plain of the Colorado River to the east, the rocky massif of the Sierra el Mayor, and the Laguna Salada Basin. The Kumiai people are settled in San Antonio Necua in the municipality of Ensenada. The Pa Ipai community inhabits the settlement of Santa Catarina that is located southeast of

---

[1] https://sustainabledevelopment.un.org/.
[2] Agenda 2030 http://agenda2030.mx/#/home.

Ensenada (Fig. 12.1). The other indigenous community named Kumiai was not visited for security reasons, but some people living in Ensenada were interviewed and their answers were used for the discussion of the results.

Methods

The theoretical indigenous community sustainability assessment model MEICS previously proposed by Galván-Martínez et al. (2016) was subjected to practical evaluation among the three rural indigenous communities, implementing the pyramid of information proposed by SCOPE (1995) and Hammond (1995) in two stages (Fig. 12.2 left). Stage 1 corresponds to previous design of the MEICS by a top-bottom process (published in Galván-Martínez et al. 2016), and stage 2 was the approach adopted in this paper, followed by a bottom-top process (Fig. 12.2 right). Stage 2 consisted in identifying indicators from the answers given to questionnaires applied to individuals chosen at random in the three studied communities, and afterwards assigning numerical values to them in ranges defined by the results of the questionnaires.

The questionnaires were designed based on the theoretical aspects considered by the three above-mentioned approaches to sustainability assessment (CMSD, BV, and SLM). Analyzing differences and similarities relevance criteria were defined for each aspect in order to allow the evaluation of the importance of each element

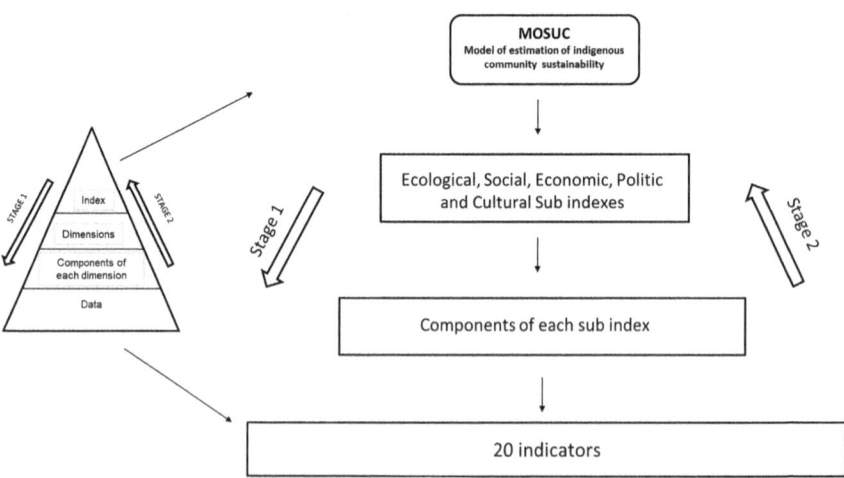

Fig. 12.2 Information Pyramid, modified from SCOPE (1995) and Hammond (1995) (left), and the Model of Estimation of Community Sustainability (MEICS) designed by Galván-Martínez et al. (2016) (right)

for being integrated into the model. Therefore, 20 aspects were selected and transformed into indicators, and converted into a question. Such questions were later polished aided by expert advice until the final questionnaire was integrated with 18 questions providing data of the 20 selected indicators (Table 12.1).

A total of 166 questionnaires were applied, 51 in the Kumiai community of San Antonio Necua (29 women and 22 men, between 18 and 70 years), 56 in the Cucapá community of El Mayor Cucapá (34 women and 22 men, between 18 and 83 years), and 59 in the Pa Ipai community of Santa Catarina (38 women and 21 men, aged between 18 and 73 years). The answers were analyzed applying the content analysis qualitative tool, and the responses were assigned a weight ranging from 0 to 1 according to the remoteness (0) or proximity (1) to sustainability of response. For example, in the case of Indicator 9: Hybridization of medical knowledge, weight was assigned based on content of answers as explained in Table 12.2.

Subsequently, the values for each indicator were averaged, and then added, according to the number of indicators in each sub-index (ecological sub-index, 5 indicators; social sub-index, 5 indicators; economic sub-index, 4 indicators; political sub-index 2 indicators, and cultural sub-index 4 indicators were added), the result is the value of each sub-index. Since indigenous culture is the point at issue of the proposed definition of community sustainability (Galván-Martínez et al. 2016) therefore the obtained values of the ecological, social, economic, and political sub-indexes were weighted multiplying them by the cultural sub-index value,[3] and the results of the products were added to obtain the values of the MEICS index by using the algorithm:

$$MEICS = \Sigma \left[Ecol, Pol, Soc, Econ \left(Cult \right) \right]$$

where MEICS is the sustainability index of the analyzed community, Ecol is the ecological sub-index value, Pol is the political sub-index value, Soc is the social sub-index value, Econ is the economical sub-index value, and Cult is the cultural sub-index value.

This methodological approach differs from that of cultural capital that makes reference to the factors that provide human societies with the means and adaptations to deal with the natural environment and actively modify it (Berkes and Folke 1994). As mentioned before, the methodology itself was not participatory since the actors did not select or argue with the researchers all the decision made during the research, but the final outcomes liked because understandable and can be used as a starting point of a participatory process for the construction of the community sustainability.

---

[3] An analogous methodology to that applied here can be found in Seingier et al. (2011).

Table 12.1 The list of sustainability indicators obtained from the native communities

Theory approach	Element of importance for sustainability	Sub-index	Indicator	Value Cucapá	Value Kumiai	Value Pa Ipai
SLM	Maintains or encourages the goods on which livelihoods depend	Ecologic	1. Type of use	0.57	0.68	0.71
			2. Management practices	1	1	1
			3. Conservation	0.4	0.61	0.64
CMSD	Adequate or no destructive use of natural resources		4. Population's perception	0.69	0.71	0.77
BV	New relation with nature, not only as reservoir or repository		5. Adaptation strategies	0.49	0.51	0.29
SLM	X	Political	6. Respect territory	0.75	0.42	0.78
CMSD	Capacity to create their own productive organization		7. Socio-ecological conflicts	0	0.5	0.5
BV	Respect to the indigenous leadership					
SLM	May deal and recover from stress and changes	Social	8. Social participation	0.48	0.65	0.68
	Provides for future generations					
CMDS	Increases quality of life		9. Hybridization in medical knowledge	0.61	0.73	0.77
	Social control (food, health, education, housing, information, and recreation)		10. Hybridization in alimentary knowledge	0.55	0.71	0.75
	Local participation					
BV	Improvement of quality of life		11. Hybridization in educational knowledge	0.81	0.95	0.91
	Food sovereignty		12. Future	0.51	0.66	0.86
	Encouragement of participation					
	Intercultural education					
SLM	Economically sustainable when it is effective	Economic	13. Livelihoods strategies	0.71	0.67	0.69
CMDS	Regulation of the community's economic interchanges of the community		14. Sowing/harvesting	0.29	0.64	0.44
	Food sovereignty		15. Breeding/hunting	0.2	0.82	0.51
	Diversity of productive activities					
BV	Built of a just, democratic, solidary economic system			0.75	0.38	0.51

			Economic sovereignty Community's economy balanced	16. Economic linkage			2
SLM	X			17. Being indigenous			
CMDS	Cultural control (decision making of the community to preserve their cultural values) Preserve cultural heritage	Cultural		18. What do you like of being indigenous?	2	2	
BV	Assertion of indigenous knowledge			19. Preservation of intangible heritage		2	
	Respect to indigenous world view			20. Preservation of tangible heritage			

Table 12.2 Example of assigned weight of answers

Content of answer	Value of answer
No access to medicine	0
Prefer allopathic medicine	0.5
Prefer traditional medicine	0.5
Use both	1

Results

The results of the model performance applied to the indigenous communities of Baja California, Mexico are shown in Fig. 12.3 and Table 12.3 and explained in the following text.

Ecological Sub-Index

In general, the studied indigenous communities are well adapted to their ecological environment and have extensive knowledge about the natural resources in their surroundings. The former assumption coincides with answers given by the representatives of the three studied groups. The Cucapá, Kumiai, and Pa Ipai have high ecological sub-index values since they use natural resources (Santos 2013) following traditional management practices and seeking to care for them by means of conservation strategies. Although interviewees perceive changes occurred in environmental resources, they mentioned that communities have implemented adaptive strategies. Most of the interviewed Cucapá in El Mayor Cucapá perceive a remarkable decrease of terrestrial animal populations that has probably resulted from the decrease in water flow of the Hardy River and wastewater discharges in the Laguna Salada Basin. Some also mentioned a diminished abundance of fish and migratory birds like ducks, herons, pelicans, and seagulls. As a result of the degradation of their habitat, the Cucapá have modified their traditional economic activities or adopted new ways of life to reduce their dependence on natural resources, such as seeking employment outside the community instead of continuing fishing as they did traditionally (Santos 2013). Some of the interviewed Kumiai have also observed a decrease in plant populations adjacent to their community, which worries them due to the intensity of the process. While some interviewed Pa Ipai report that the natural resources of their community have changed, some considered their livelihood has not been severely impacted. However, the plundering and increased aridity of Yuman land has led to abandonment of subsistence activities directly linked to primary production. That is the case of harvesting from the wild of the *palmilla* or Mojave yucca (*Yucca schidigera*) by the Pa Ipai of Santa Catarina and the Kiliwa of Arroyo de León.

MEICS

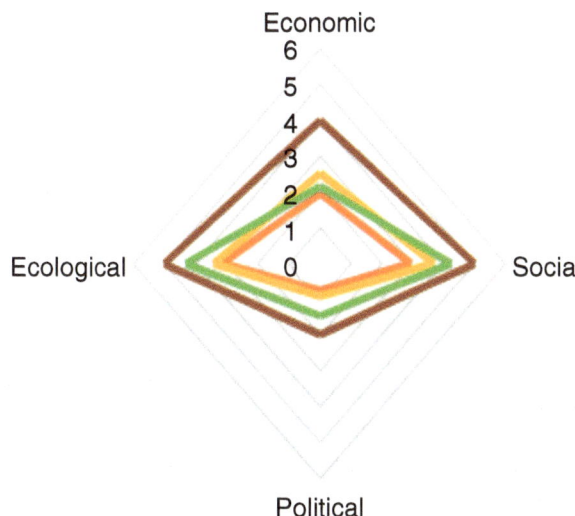

Fig. 12.3 Sustainability index expressed in a spider graph

Table 12.3 Sustainability values obtained

	Economic	Social	Political	Ecological	Cultural	Total
Cucapá	1.95	2.96	0.75	3.15	4	35.24
Kumiai	2.51	3.7	0.92	3.41	4	42.16
Pa Ipai	2.2	4.15	1.48	4.21	4	48.16
Ideal	4	5	2	5	4	64

See Fig. 12.3

According to some interviews, they perceive modifications in their surroundings, but a Cucapá and a Kumiai mentioned that they have implemented adaptation strategies such as going further away to gather scarce resources or replacing missing resources with others. Some interviewees mentioned that in certain cases, the government has applied reforestation programs and projects aimed at resources of cultural or economic importance to indigenous communities. In addition, some interviewed Cucapá mentioned that they try to adapt when the closed fishing seasons occur by finding alternative jobs. On the contrary, some interviewed Pa Ipai said that the changes in their community have not affected them strongly and that there was no need to develop any adaptation strategy. Nevertheless, Garduño (2015) quotes that primary productive activities in the Pa Ipai community, such as harvesting of *palmilla*, subsistence beekeeping, and other traditional practices have experienced a downtrend. This author also mentions that while primary economic activities take place within the Yuman settlements, waged jobs must be found outside them, resulting in 71.6% of the Yuman economically active population being migrant (Garduño 2016).

Political Sub-Index

The three communities analyzed here face socio-ecological conflicts; in order to explain such conflicts a political sub-index is used relating the prohibitions affecting the right to use their natural resources. The Paipai community of Santa Catarina recorded the highest value of political sub-index, since all interviewees mentioned that the territory restricted to their community is respected by their neighbors. In addition, most of the interviewees consider that socio-ecological conflicts are in the process of being resolved by the authorities. Contrary to what happened with the Pa Ipai Jamao community where they were dispossessed of land.

The Kumiai had a lower value because most of the interviewed people consider their neighbors do not respect their territory. According to one interviewee, the Kumai face territorial conflicts with the owners of a winery, which pretends to adjudicate land located within indigenous community territory. The Cucapá had the lowest value since the interviewed people refer three types of socio-ecological conflicts: water sources, intrusion into their territory, and fishing restrictions. The water sources conflict perceived by some interviewed Cucapá is about the decrease in water level of the Hardy River by the US Hoover dam and the pollution of the Salada Lagoon by wastewater discharge. Some interviewed Cucapá mentioned that this conflict involves legal restrictions denying their recognition as native settlers—and, therefore, having differentiated rights—that prevent them from fishing in the Colorado River.

Social Sub-Index

The Pa Ipai had the highest value due to the communities' optimistic vision of the future and to equitable participation in decision making; an indispensable element for the achievement of development in rural areas. Analyzing the interviews, we can say that the Kumiai have also achieved a combination of modern and traditional knowledge in education, food, and medicine.

Kumiai community recorded a low value due to their lack of participation in decision-making issues arising from agrarian law; in addition, most interviewees have a less optimistic perception of the future. A positive aspect is that the indigenous language is taught in primary school, and residents are satisfied with teaching. The Kumiai emphasize the fundamental role of language teacher, whom they perceive as a pillar in safeguarding their cultural values.

In contrast, the Cucapá native community recorded the lowest value attributable to factors such as low participation in decision making, the pessimistic perception of the future, and the prevalence of modern uses in medical aspects, education, and nutrition. Villareal (2014) described that the current diet of Cucapá includes many processed foods, causing high levels of glucose in the blood and several cases of type II diabetes mellitus (Camarena 2009).

Economic Sub-Index

It is important to mention that economy of the three groups is mainly based on primary sector activities. The community of Kumiai registered the highest economic sub-index, and the underlying reason is the supplements of its economy, like hunting or raising animals and planting or harvesting plants. This is probably due to the influence of the predominantly agricultural economy in the region and, to a lesser extent, to the development of animal husbandry (intensive and extensive). The interviewees of San Antonio Necua considered that their economy is attached to the Valle de Guadalupe because they are hired as farmers or laborers in neighboring ranches.

Furthermore, the Pa Ipai community presented a lower value due to the limited activity of cattle breeding, which is discouraged by the cold and dry climate (Wilken 1998). However, the Pa Ipai interviewees mentioned that they complement their economy collecting pine nuts (*Pinus pinea*, piñon), some of them have orchards that provide seasonal vegetables and supplement their income with federal government programs.

The Cucapá have the lowest value since they do not complement their economy with agricultural activities of self-subsistence. Nowadays they try to continue fishing, but the fluctuations in the volume of the river and the pollution of residual waters of the Laguna Salada basin (Villareal 2014), plus the fishing seasons imposed by the national authorities, make it difficult to continue their fishing activity. Therefore, some of the Cucapá interviewees mentioned that they look for employment in surrounding ejidos or even at the city of Mexicali in order to obtain monetary income. They also mentioned that they supplement their income with federal government programs aimed at combating poverty, reducing socioeconomic inequalities, and expanding opportunities for indigenous communities.

Cultural Sub-Index

This is the index where intercultural issues arise. There is a mixture of old traditions and modern activities that coincide. Important is that the three communities agree that their traditions and oral expressions are what they like most about their indigenous culture. However, each community faces particular challenges scenarios. Regarding language, intercultural issues would be the proportion of Spanish and native language spoken. Among the Cucapá most interviewees are proud of their traditions and oral expressions, which contrasts with the small number of speakers who speak fluently, a circumstance that has led some authorities say they are no longer "fully indigenous" (Muehlmann 2012).

In the same way, the Kumiai most interviewees mentioned that what they like most about their culture are oral traditions and expressions (language and stories). However, Wilken (2012) found that 71% of fluent speakers were older adults (60 years of age or older), revealing a strong association between age and language use. This implies making urgent efforts of documentation and revitalization of the Kumiai language. In that sense, Leyva (2014) emphasizes that "will be the households, stakeholders in perpetuating the transmission of knowledge, and youth attracted to preserve their culture." The Pa Ipai referred that the traditions and oral expressions are the aspects that they liked most of their culture, some respondents mentioned enthusiastically that a National Institute of Indigenous Language program is developing in their community for promoting the practice and teaching of the Pa Ipai language. Being fully bilingual would be a key issue for intercultural assessments.

The Cucapá protect their culture by transmitting to the new generations, through fostering the participation of children and youth people in cultural festivals. It is important to emphasize the role of the Cucapá women, not only in the care and transmission of their culture, but also in defense of their territory, projects execution, and leadership in fishing organizations (Villareal 2014). Intercultural behaviors mix traditional and modern activities. Most fishers combine technologies with modern businesses. As well, in the daily cooking these intercultural influences arise as shown by co-author Gutiérrez (2017) in the Kumiai women. Figure 12.3 shows that the Pa Ipai are closer to reaching sustainability, although they can increase the obtained value by developing strategies to protect their environment in order to increase the value of the ecological sub-index, such as implementation of a Unit of Management for the Conservation of Wildlife. Such conservation strategy has been successful in indigenous communities like the Maya in Yucatan (Retana-Guiascón et al. 2011) because it allows them to use the resources in a sustainable way and implementing strategies to preserve them for the future generations.

Recommendations and Future Perspectives

The MEICS model developed and applied among three indigenous rural locations in Baja California, Mexico proved to be a useful tool for the evaluation of sustainability, since the values of the MEICS index obtained are relevant information on four indicators of sustainability: economic, ecological, social, and political aspects. Our results indicated that, according to the interviewees, the Pa Ipai community has the highest values towards sustainability, while Kumiai is on average and, finally, the community of Cucapá with the lowest values and, therefore, far from achieving sustainability. It should be mentioned that perceptions, knowledge, and strategies of local social actors should be preponderant in the diagnoses (Soares and Garcia 2014). Therefore, for the particular case of the indigenous communities, native to Baja California, the complex of social and economic problems that arise can be attenuated if project proposals emanate from the same community, respecting the traditional social organization and family organization during their implementation (Wilken 1998).

In the economic sub-index, the Pa Ipai could request the advice of civil organizations that can teach farming methods and animal husbandry for self-subsistence (only during the spring-summer season and on a small scale using resistant species). In relation to the social sub-index, it is suggested that the community request advice from academics or civil organizations to plan the growth of their community in a systematic way and with the least impact on their natural resources. The Kumiai are half way of reaching sustainability, to become closer to that goal, they could improve their economic sub-index value by giving impetus to the community's ecotourism center, since this form of tourism sensitizes visitors and privileges sustainability, conservation, and environment appreciation (both natural and cultural) (Toledo and Ortiz 2014). Ecotourism is also a viable alternative to increase indigenous income in a culturally sustainable way in San Antonio Necua (Bringas and González 2004), where it would improve not only the value of the economic sub-index, but also of the social sub-index by means of contributing to greater cohesion and community participation, and to heighten the value of their cultural traditions and customs. To fulfill the ideal value of the political sub-index by the Kumiai community, it could be useful if they link their political demands to international networks. There is successful networking among indigenous peoples in the world that provide legal advice and help with political negotiations (Hall and Fenelon 2004).

The community of Cucapa achieved improvements in sustainability. To enhance their sustainability in the social dimension, the Cucapá could seek to rely on institutional programs that allow them to design strategies to improve their nutrition and could help them to achieve a better general health. A result of intercultural influences is malnutrition as demonstrated in other countries with indigenous people (Kuhnlein and Receveur 1996; Smith and Smith 1999). Intercultural food and cooking is an aim to develop within the indigenous communities of Baja California. In addition, for the economic dimension, it is suggested that the Cucapá seek advice to develop small-scale self-subsistence economy strategies, activities that elsewhere have proven to be a good alternative for rural economies (Barkin 2001). The results of this research provide elements to design and evaluate public policies on indigenous issues at the local level, focused on regional intercultural development, specially where binational indigenous communities occur. We suggested that the indicators proposed in this research be incorporated into local programs developed in rural indigenous communities to monitor and evaluate their way to sustainability. Similarly, considering MEICS as a management tool would allow partnerships with decision-makers to jointly and creatively find long-term solutions to the problems of environmental management where intercultural issues prevail. Also, intercultural topics influence the quality of life of the indigenous communities, encourage the participation of the people regarding their mixture of modern culture and old traditions, and must be the basis to design projects and programs appropriate for each community.

Acknowledgments The authors would like to express their gratitude to the indigenous communities of El Mayor Cucapá, San Antonio Necua, Santa Catarina for facilitating the information during the fieldwork. Also, to the Red of Patrimonio Biocultural from Conacyt for financing fieldwork on the communities. Also, to Alejandro García Gastelum who supported the research at the starting point. And to Sergio Zárate for the English review.

References

Barkin D (2001) Superando el paradigma neoliberal: desarrollo popular sustentable. In: Giarraca N (ed) ¿Una Nueva Ruralidad en América Latina? CLACSO, Consejo Latinoamericano de Ciencias Sociales, Buenos Aires, pp 81–99

Barkin D (1998) Riqueza, pobreza y desarrollo sustentable. Editorial Jus y Centro de Ecología y Desarrollo, México

Berkes F, Folke C (1994) Investing in cultural capital for sustainable use of natural capital. In: Jansson AM, Hammer M, Folke C, Costanza R (eds) Investing in natural capital: the ecological economics approach to sustainability. Island Press, Washington

Bringas RNL, González AIIJ (2004) El turismo alternativo: una opción para el desarrollo local en dos comunidades indígenas de Baja California. Econ Soc y Territ 15:551–590

Camarena OL (2009) La salud desde lo social: Caso pueblo indígena cucapá. 206. http://docencia. colef.mx/system/file/ponencias/mesa. Accessed 10 Apr 2019

Chambers R, Conway G (1992) Sustainable rural livelihoods: practical concepts for the 21st century. Institute of Development Studies, Brigton

Del Val J, Rodríguez N, Rubio M, Sánchez C, Zolla C, Cunningham M (2008) Los pueblos indígenas y los indicadores de bienestar y desarrollo. Pacto del Pedregal. http://www.inegi.org. mx/inegi/contenidos/espanol/eventos/IXeieg/doctos/30desep/sesion4b/informe.pdf. Accessed 22 Aug 2008

Delgado G (2008) Historia de México: legado histórico y pasado reciente, 2nd edn. Prentice Hall/Pearson, México

Galván-Martínez D, Fermán-Almada JL, Espejel I (2016) ¿Sustentabilidad comunitaria indígena? Un modelo integral. Soc y Ambient:4–22

Garduño E (2015) Yumanos Pueblos Indígenas de México en el siglo XXI. Comisión Nacional para el Desarrollo de los Pueblos Indígenas, México

Garduño E (2016) La frontera norte de México: Campo de desplazamiento, interacción y disputa. Front Norte 28:131–151

Gómez AA, Sánchez VS (2014) La Frontera Cultural Meso-Aridoamericana: Construcción De Imaginarios Nacionalistas En La Historia Mexicana. Rev Xihmai 9:85–104

Grimson A (2001) Interculturalidad y comunicación. Grupo Editorial Norma, Bogotá

Gudynas E, Acosta A (2011) La renovación de la crítica al desarrollo y el buen vivir como alternativa. Utopía y Prax Latinoam 16:71–83

Gutiérrez SC (2017) La cocina tradicional del grupo Kumiai de Ensenada Baja California: su preservación en un contexto de globalización. Dissertation, Universidad Autónoma de Nayarit

Hall TD, Fenelon JV (2004) The futures of indigenous peoples: 9-11 and the trajectory of indigenous survival and resistance. J World-Systems Res 10:153–197. https://doi.org/10.5195/jwsr.2004.307

Hammond AL (1995) Environmental indicators: a systematic approach to measuring and reporting on environmental policy performance in the context of sustainable development. World Resources Institute, Washington

Kirchhoff P (1954) Gatherers and farmers in the greater southwest: a problem in classification. Am Anthropol 56:529–550

Kuhnlein HV, Receveur O (1996) Dietary change and traditional food systems of indigenous peoples. Annu Rev Nutr 16(1):417–442

Leyva GA (2014) Documentando una lengua: el caso kumiai. Rev Digit Univ 15:2–10

Liu G, Brown TM, Casazza M (2017) Enhancing the sustainability narrative through a deeper understanding of sustainable development indicators. Sustain 9(6):1078

Martínez PL (2011) Historia de Baja California. Instituto Sudcaliforniano de Cultura. Archivo Histórico Pablo L. Martinez, La Paz

Morales MP (2003) Cultura y territorialidad: aportes etnológicos para la gestión ambiental comunitaria: estudio de caso, comunidad Kumiai de San José de la Zorra (México). Abya-Yala, Ecuador

Muehlmann S (2012) Von Humboldt's parrot and the countdown of last speakers in the Colorado Delta. Lang Commun 32:160–168. https://doi.org/10.1016/j.langcom.2011.05.001

Pretty J, Adams B, Berkes F, de Athayde SF, Dudley N, Hunn E, Maffi L, Milton K, Rapport D, Robbins P, Sterling E, Stolton S, Tsing A, Vintinner E, Pilgrim S (2009) The intersections of biological diversity and cultural diversity. Conserv Soc 7:100–112

Retana-Guiascón OG, Aguilar-Nah MS, Niño-Gómez G (2011) Uso de la vida silvestre y alternativas de manejo integral: El caso de la comunidad maya de Pich, Campeche, México. Trop Subtrop Agroecosyst 14:885–890

Rogers MJ (1945) An outline of Yuman prehistory. Southwest J Anthropol 1:167–198

Santos MM (2013) Érase una vez un valle. In: Leyva C, Espejel I (eds) El Valle de Guadalupe. Conjugando tiempos. Universidad Autónoma de Baja California, Mexicali, p 129

Schuttenberg HZ, Guth HK (2015) Seeking our shared wisdom: a framework for understanding knowledge coproduction and coproductive capacities. Ecol Soc 20:226–236

Scientific Committee on Problems of the Environment (SCOPE) (1995) http://www.scopenvironment.org. Accessed 26 Apr 2019

Secretaria de Medio Ambiente y Recursos Naturales (SEMARNAT) (2015) Indicadores verdes. http://apps1.semarnat.gob.mx/dgeia/indicadores_verdes/. Accessed 23 Jul 2015

Seingier G, Espejel I, Fermán-Almada JL, González OD, Montaño-Moctezuma G, Azuz-Adeath I, Aramburo-Vizcarra G (2011) Designing an integrated coastal orientation index: a cross-comparison of Mexican municipalities. Ecol Indic 11:633–642

Smith P, Smith M (1999) Diets in transition: hunter-gatherer to station diet and station diet to the self-select store diet. Hum Ecol 27:115–133

Soares D, Garcia A (2014) Percepciones campesinas indígenas acerca del cambio climático en la cuenca de Jovel, Chiapas – México. Cuad Antropol Soc 39:63–89

Toledo V (1996) Principios etnoecológicos para el desarrollo sustentable de comunidades campesinas e indígenas. Etnoecológica 6:7–41

Toledo V, Ortiz B (2014) México, regiones que caminan hacia la sustentabilidad. Universidad Iberoamericana Puebla, Puebla

United Nations Development Programme (UNDP) (2019) http://www.undp.org/content/undp/en/home/sustainable-development-goals.html. Accessed 26 Apr 2019

United Nations Development Programme (UNDP) (2015) http://www.mx.undp.org/content/mexico/es/home/post-2015/sdg-overview.html. Accessed 26 Apr 2019

Villareal RJ (2014) Efectos de la degradación del río Hardy en el Mayor Cucapá, 1950–2014. Colegio de la Frontera Norte, Tijuana

Wilken MA (2012) An ethnobotany of Baja California's Kumeyaay indians. Dissertation, San Diego State University

Wilken RM (1998) Desarrollo sustentable de las comunidades indígenas de Baja California. Instituto CUNA y FANCA, México

Chapter 13
International Recognition of the Biocultural Protection in Dryland Regions: The World Heritage Property in the Tehuacán-Cuicatlán Biosphere Reserve

L. Vera and S. García

Abstract To know and deal with the current challenges of the socio-ecological environment, it has been necessary the use of mechanisms to protect, preserve and make visible the bioculturality of communities. These challenges are multidimensional and intersectoral, and relate to phenomena such as the increase in the intensity of natural phenomena, changes in the ecosystems, and the rapid loss of biodiversity involving loss of ecosystem services, public health and economic spillover exacerbated by the population increase, and the pressure involved in the current development model. An example of a dryland region with socio-ecological pressures such as those mentioned above is the case of the Tehuacan-Cuicatlán Biosphere Reserve, with a World Heritage Mixed Property of UNESCO. Traces of ancient communities have been found in this area that managed to adapt and evolve within drylands. Since 1999 the area has been considered of great importance because of its different schemes and instruments for conservation at the international, national, and local level. Above all, it is an important example of how international conventions urge to protect and preserve the identity of the people, their ecosystem, and their ways of life in dryland regions.

Keywords Adaptation · Bioculturality · Dryland regions · World Heritage

L. Vera (✉)
Instituto Mora, CONACYT, Mexico City, Mexico

S. García
Independent Consultant, Mexico City, Mexico

© Springer Nature Switzerland AG 2020
S. Lucatello et al. (eds.), *Stewardship of Future Drylands and Climate Change in the Global South*, Springer Climate,
https://doi.org/10.1007/978-3-030-22464-6_13

Introduction

The diversity of ecosystems and cultures in Mexico is very wide and unique. Therefore, it has been relevant to locate areas and recognize their importance, especially those that represents a breakthrough in evolutionary processes from settlements in some inhospitable region. Promoting the adoption of protection schemes with recognition at different scales such as international, regional, national, and local demonstrates the political and social commitment that help to have an universal language and international recognition of the linkage between social and environmental processes.

The protections of the biodiversity and ecosystems are as important, as the protection of indigenous cultures in the world. In Mexico, there is a relation between ethnic groups and dryland regions from which we could find potential knowledge to manage and use efficiently the ecosystem. It also exists as a complex connection between dimensions of biological, agricultural, and linguistic diversity (Toledo et al. 2008). However, even when indigenous peoples have contributed to the maintenance of ecosystems, biodiversity, and the development of agrobiodiversity, many of them are marginalized, since the public policies addressed to them have not been sufficient (Ramos Vázquez 2011).

In the Convention concerning the Protection of the World Cultural and Natural Heritage (UNESCO 1972) the term Mixed Property reflects the interaction between human and nature, referring to those sites that illustrate the evolution of human society and its settlements over time. This evolution is conditioned by the limitations of its natural environment and the consecutive and successive social, economic, and cultural forces which intertwine to tell the same story of development and evolution that persists over time (WHC 2018). Also, these forces can be recognized as an element of interculturality.

In this regard on the unbreakable ties that human and nature have created throughout history, in this article we urge to show a case of a specific dryland socio-ecological system, in which a community needed to involve social projects simultaneously examines, monitors, and manages the bio-geo-physical natural components and the social/cultural components, erasing disciplinary barriers between the ecological and social issues. All these are components of a biocultural memory and this has been able to be transferred and adapted in different cultures and times, so these techniques can be considered today an element of interculturality.

The experience within the Tehuacán-Cuicatlán Biosphere Reserve traces of a long interaction between human and nature can be found. Currently there are records of 4000 years nomadic peoples, who managed to move to a sedentary life in an inhospitable environment, which despite the environment and thanks to the ingenuity of the community gave rise to an area with attributes of outstanding universal value, which in 2018 have been recognized by UNESCO as a World Heritage Mixed Property.

The unique biocultural attributes of the Reserve, shows the evolution of ancient civilizations in an inhospitable environment because of its arid conditions due to long periods of drought, and because of its great and endemic biodiversity that still lasts on current times. Some of these biocultural evidences are reflected on the diverse indigenous languages, pottery, the extraction of salt using ancestral techniques, the infrastructure for water management in dryland zones (see Hernández Garcíadiego 2005) that contributed to domesticate plants, among others (CONANP-INAH 2016), and it is a key site in the origin and development of agriculture in the region of Mesoamerica. History helps us see the signs of an adaptation process in this dryland through the management of resources that were carried out by human being to survive. Now a days, this resources still provide to settled communities within the Reserve and its periphery, so it can be taken up as a successful adaptative inhospitable environment.

The recognition for the area and its attributes has attracted multiple actors at the local, national, and international level which have promoted their registration to different protection schemes which, without a doubt, are an important reference in the planning of activities to preserve the cultures, conservation, and management of the site, as well as an ideal scene to review and consider the sustainable development precepts and the objectives set in the International Development Agenda (Agenda 2030). We will refer to three different zones in the text: The Tehuacán-Cuicatlán Valley which comprises a little bit larger area of 10,000 km^2 that are equivalent to 1,000,000 hectares which includes the floral province with the same name; Tehuacán-Cuicatlán Biosphere Reserve which has an approximate area of 490,186 hectares; and the World Heritage Property Tehuacán-Cuicatlán Valley: original habitat of Mesoamerica that has a land of 344,931.68 hectares. All these zones are interrelated.

The Tehuacán-Cuicatlán Biosphere Reserve: A Dryland Region of Outstanding Universal Value

The Tehuacán-Cuicatlán Valley is located in the arid–semiarid zone more to the South of North America, in the intersection of Neotropic and the Neartic, in the center of the cultural region of Mesoamerica, in the local province named Tehuacán-Cuicatlán (CONANP-INAH 2016) (Fig. 13.1). In this zone, arid conditions are determined by its geology, making an evident effect of a wall with the mountains that obstruct the passage of humidity from the Gulf of Mexico which is increased to be located in the tropic where the rate of evaporation is higher. These special features make The Valley, the most southern area of Mexico with an arid ecosystem (INAH 2019). In 1998, an area of The Valley with an approximate area of 490,186 hectares was decreed as a Natural Protected Area (NPA) with the category of management and named as Biosphere Reserve (BR) Tehuacán-Cuicatlán (CONANP-SEMARNAT 2013).

Due to its geographic and climatic conditions, the BR contains biomes with high levels of endemism and a large of endangered species, a rarity of plant species and

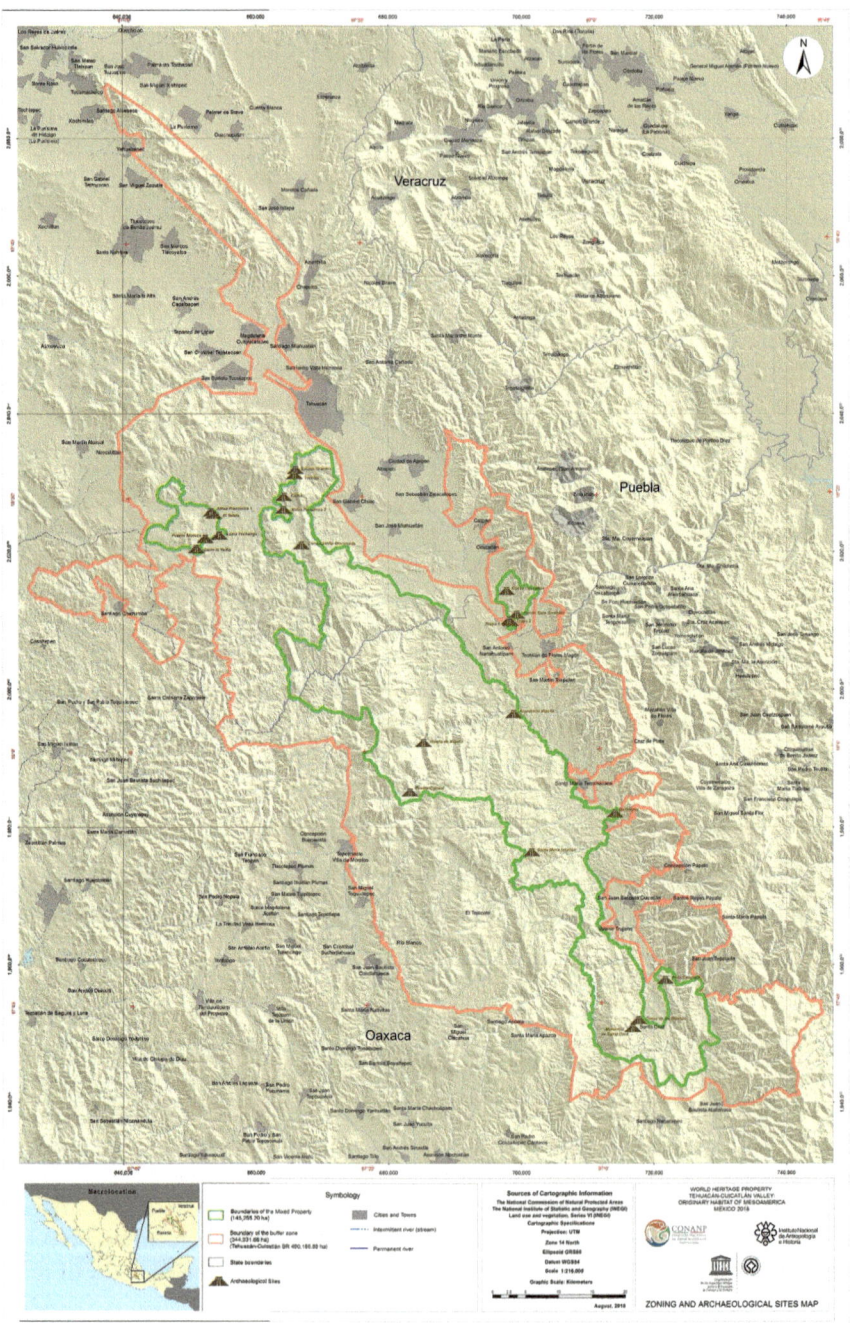

Fig. 13.1 Polygonal of the Tehuacán-Cuicatlán biosphere reserve and of the world heritage mixed property (CONANP-INAH 2016)

flora communities. The extraordinary biodiversity includes more than 3000 vascular plants, in which 10% are endemic to The Valley. Its fauna diversity exceeds any other arid area of the planet and, likewise, it is an exceptional center of agrobiodiversity.

Since the decree of the Federal NPA, the reserve has managed to preserve its biodiversity, the community of the ecological and evolutionary processes that take place their, as well as the cultural and historical heritage associated with them. This holistic management was possible with the implementation of politics, measures and strategies, and restoring specific actions through scientific, cultural and management research projects, which have allowed the communities that live there, to continue their development (CONANP-SEMARNAT 2013). The protection of the richness of the area has resulted in the protection of different national and international instruments to guarantee its preservation for future generations as part of their heritage and people's identity. At the national level, its protection has been given since the establishment by the NPA Decree in 1998, the publication of its Management Program in 2013, as well as the Federal Law on Monuments and Archeological, Artistic and Historic Areas, the General Law of Ecological Equilibrium and Environmental Protection, Land Management and Programs, among others.

At the International Level the BR was registered in 2012 in the World Network of Biosphere Reserves of the Man and the Biosphere Program (MAB) of UNESCO and considered as a management model for sustainable development. In 2018 a portion of the Biosphere Reserve was recognized by its inscription to the World Heritage List of the Convention concerning the Protection of the World Cultural and Natural Heritage of UNESCO, recognizing this portion as a World Heritage Mixed Property. The protection of its exceptional biocultural attributes at world level is enriched by the importance of being immersed in an arid environment and local communities.

The preservation obligations that have been contracted from the UNESCO designations are a great challenge for the Mexican government, being an area with a high population number where its growth and development threatens the conservation of its cultural and natural attributes. However, the inheritance of its bioculturality is at risk, since the new generations are losing interest in the preservation of their local cultural activities, which hastens its deterioration or possible loss considering also the increase of extreme natural phenomena due to the derived effects of climate change.

As researchers had noted, the history of occupation and use of the area is dated back to the first hunters 4000 years ago, who found the geology of the place, an area of protection before predators, being the orography a strategic place to hide and look for prey that helped to the adaptation and development. Since then, it has occurred a large and varied human occupation with the Mixtec, Chocholtec, Cuicatec, Ixcatec, Mazatec, and Zapotec cultures, whose common base are the languages of the Otomangue family (INAH 2019). It has been documented that in the transition from nomadic to sedentary, one of the archaeological surprises was that the corn was not the first cultivated plants. One of the main sources of food was the maguey. Proof of this are the remains of chewed maguey found in several of the excavations carried out in The Valley. This food was first cooked which provided a high sugar content (MacNeish 1967).

The towns that inhabited the region started to build works for water retention due to aridity conditions, which led to develop technologies to settlements and the early domestication of crops as a management system of water through canals, wells, aqueducts, and dams, being the oldest system in the continent that allowed the emergence of agricultural settlements (CONANP-INAH 2016). Furthermore, the area is one of the main diversification centers for the *cacti* family that are endangered all around the world; it is the house of the densest columnar cacti forests in the world, forming a unique landscape that it also includes agaves, yuccas, and oaks (CONANP-INAH 2016). It has been documented that in the transition from nomadic to sedentary, one of the archaeological surprises was that the corn was not the first cultivated plants. One of the main sources of food was the maguey. Proof of this are the remains of chewed maguey found in several of the excavations carried out in The Valley. This food was first cooked which provided a high sugar content (MacNeish 1967).

To mention some examples of the use of biodiversity, succulents have been recorded in the area as the largest all around the world, whose fruits have been used as food and the intensive management of some species has generated a targeted selection to its domestication. The *magueyes* are another plant widely used since the beginning as a source of fiber and food. Recently, ecological, floral, and linguistic evidence was gathered that pointed The Valley as a domestication site for pepper and it is possibly where it was named as chilli (INAH 2019). All this evidence has been located throughout the BR, thanks to the intervention of diverse disciplines that have been given the duty to gather information and evidences, in particular by studies from archaeologists, historians, anthropologists, biologists, engineers, etc., at least since 1940. The knowledge of the domestication techniques, use and exploitation of the biodiversity, and the development of technologies for water management, pottery, salt production, among others allows to build the past that reflects the changes that have taken place in the area, both anthropogenic and environmental, without excluding the climate change.

Mixed Property Tehuacán-Cuicatlán Valley: Original Habitat of Mesoamerica

After a research process involving social and natural disciplines, in July 2018 the World Heritage Committee recognized Tehuacán-Cuicatlán Valley as a Mixed Property on the World Heritage List (WHC 2018). Among the criteria considered for the designation of outstanding universal value (CONANP-INAH 2016), include:

- *Criterion (iv): be an outstanding example of a type of building, architectural, or technological ensemble of landscape which illustrates (a) significant stage(s) in human history.* The technological ensemble of water management of the Tehuacán-Cuicatlán Valley, along with other archaeological evidences such as the remains found in caves, plant domestication sites, salt ponds, and pottery marks a stage of the utmost importance for the Mesoamerican region: the appearance and development of one of the oldest civilizations in the world. Located throughout The Valley, these technologies bear unique evidence of the constant adaptation of humans to the environment and reflect their innovative capacity to face the adverse environmental conditions in the area. Also, the vast production of salt and pottery allows us to understand the social organization of the area since pre-Hispanic times, as well as the modernization and commercial development of The Valley.

- *Criterion (x): contains the most important and significant natural habitats for in situ conservation of biological diversity, including those containing threatened species of Outstanding Universal Value from the point of view of science or conservation.* The Tehuacán-Cuicatlán Valley: original habitat of Mesoamerica is the arid or semiarid zone with the greatest biological diversity in North America. It is a world biodiversity hotspot that contains biomes characterized by high levels of endemic and endangered species, rare flora, and plant communities. Its faunistic diversity surpasses that of any other drylands of the planet and, moreover, it is an outstanding agrobiodiversity center. It is also a world center of diversification for numerous groups of plants. The Property includes 22 archaeological sites that exemplify the adaptation process of human being to an inhospitable place (CONANP-INAH 2016).
- Caves, which are the reflection of the first settlements from nomadic to sedentary, with rock paintings of zoomorphic and anthropomorphic figures that illustrate what those civilizations would see in their environment. There are also records of their eating habits and subsistence patterns. Some of these materials correspond to animal bones, kitchen tools such as metates, different types of seeds and fruits. Perhaps the most surprising evidences is the adaptation process, domestication of plants such as avocado, corn, pepper, and amaranth, hunting, the weaving of baskets, and the use of ceramics.
- Terraces, which were built to have flat areas where to make their settlements and squares.
- Archaeological sites, they show the development of civilizations in settlements that were used civically and ceremonially.
- Water management. Considering it is an arid area, the main and surprising development of water management strategies are remarkable. In the Dam's area, it was found an structure which was built to control water. It is considered one of the largest dams in Mesoamerica, approximately equivalent to the volume of the Moon Pyramid in Teotihuacán with 24 m high, 106 m wide and 400 m long, and which represents the expansion of water irrigation in The Valley. Aqueducts and canals that supported the water distribution comprise a wide network.
- Salines, the salt exploitation was another legacy which allowed people to adapt in The Valley. This activity was one of the most recognized by the old settlers, developing a technique that it continues to be used nowadays.

The Work of UNESCO and the International Development Agenda (Agenda 2030)

Mexico has been very active on international environmental forums, especially in the environmental conservation ones. The most important forum is the Earth Summit in 1972, the first international statement that has recognized environmental goods as world heritage, like all goods, such as air, water and forests, among others; these goods are subject to common law, but also impose a universal duty of protection and care, as Zortea and Lucatello (2016) establish. Nowadays, at the international level, there are numerous Treaties and Programs that Mexico has ratified, among which some are related to two worldwide movements: the preservation of cultural sites and the ones that demand the preservation of nature.

The relevance of the following section is the emphasis of the international narrative of development and conservation, and its inherent link with the current 2030 Agenda, which highlights social problems that cross-address many environmental issues, both in interrelated processes to achieve sustainable development. (To learn more about environmental mainstreaming in projects derived from international conventions and international cooperation, see Zortea and Lucatello 2016.) In this regard, UNESCO has promoted a series of Treaties and Programs to support these issues of global interest, among those that stand out for this chapter: the Convention concerning the Protection of the World Cultural and Natural Heritage (also referred to as the World Heritage Convention), since the category of Mixed Properties the inescapable link between Culture and Nature.

The cultural and natural heritage is part of the incalculable and irreplaceable Property, not only of each nation but also of all humanity. That is why the loss of one of the most precious assets, as a result of its degradation or disappearance, is an impoverishment of the heritage of all the peoples in the world. It can be considered that one of the elements of this heritage due to its remarkable qualities has an Outstanding Universal Vale (OUV) and therefore deserves special protection against the growing dangers that threaten them (CONANP-INAH 2016).

In order to achieve the protection of the cultural and natural heritage, the General Conference of UNESCO approved on November 16th, 1972, the World Heritage Convention that until 2019 has been ratified by 193 States Parties (Countries). It is one of the legal instruments for the rescue, preservation, and safeguard of the heritage of humanity (WHC 2018) and aims to identify, protect, preserve, revalue, and transmit to future generations the cultural and natural heritage of outstanding universal value (OUV) at a worldwide level which is established under ten criteria (six cultural and four natural), the conditions of integrity and/or authenticity, protection and management that the site has and guarantees the safeguarding.

In Mexico, the Supreme Law is at the same level as the Political Constitution of the United Mexican States, so its compliance is an obligation of the Government of Mexico. UNESCO has considered necessary to focus its work on the social bases, that is, on the culture of men and women. The General Director of UNESCO, Irina Bokova, pointed the need to strengthen its labor to unleash full potential of human ingenuity as a source of resilience in a tie of change, as well as a source of creativity and growth. In this sense, it is necessary to make from culture a vector of sustainable development to consider education to achieve a change of customs, values, and patterns to support and sustain this process.

UNESCO and its conventions were given the task of reforming and adopting the SD principles to apply them in their respective processes, promoting innovative, holistic, and effective program implementation structure and modalities, as well as achieving greater interdisciplinarity. To get this, a medium term strategy was made for the period of 2014–2021, in whose UNESCO mission is reconfigured as a specialized agency of the United Nations that works through education, science, culture, communication, and information, to contribute to the consolidation of peace, the eradication of poverty, and sustainable development and intercultural dialogue. The strategy developed considers nine strategic objectives which are linked to the 2030 Agenda and the SDGs, and it will also be fundamental to establish the basis of a sustainable culture, particularly with the conservation schemes of the world heritage, such as strategies 6 and 8 that refer to the support of inclusive social development through the promotion of intercultural dialogue to bring cultures closer, the promotion of ethical principles, and the promotion of cultural expressions.

UNESCO believes that only "through a straight cultural component a human centered development approach can be achieved that yields results that are sustainable, inclusive and equitable" (UNESCO 2014, p. 25); In particular, Strategic Objective 7 on protection, promoting, and transmitting Heritage, it emphasizes the link between heritage and the challenges humanity faces today, such as climate change and natural disasters, the loss of biological diversity, drinking water, conflicts, unequal access to food, education and health, migration, urbanization, social marginalization, and economic inequalities.

In particular, on the World Heritage Convention in 2015, the General Assembly of the States of the States Parties approved the policy document for the incorporation for the perspective of sustainable development in the process of the World Heritage Convention (21st meeting of the General Assembly of the Convention on Heritage). They also recognized the growing interest from the investment sector for the conservation of World Heritage Properties and strongly encourages all banks, investment funds, the insurance industry and other relevant private and public sector companies to integrate into their sustainability policies, provisions for ensuring that they are not financing projects that may negatively impact World Heritage Properties and that the companies are investing in subscribe to the "No-go commitment," and invites them to lodge these policies with the UNESCO World Heritage Center (p. 13).

Conclusions

At the global level, it has been recognized and advanced in terms of environmental conservation, and Mexico has not been unaware of this evolution, responding to global changes and the demands that the environment deserves for its protection. In the last decades it has increased and enforced the institutional, legal, and management framework, as well as the commitment of the organized society and academy, which has brought a greater knowledge of our territory and with it a legitimate responsibility for its conservation.

In this way, the international context has always been a promoter and guide to national policy agendas. Clear examples of this have been the United Nations Conference for Environment and Development where the CBD was generated, being this the first agreement on a global level that establishes as a central point, conservation and sustainable development for the survival of the humanity and the environment. Likewise, the international recognition that is granted to the NPA is another tool for their protection, such as the designation of a Ramsar Site or a World Heritage Properties. All these instruments have been well accepted by many of the UN countries and have been integrated into their national conservation schemes, and Mexico. It is one of the countries that has taken advantage of these conventions to promote the protection of their ecosystems, biodiversity, and culture.

The designations of the NPA as World Heritage Properties have brought with them not only a list of commitments to be fulfilled, but also international support for the conservation of these sites and their Outstanding Universal Value. Some of the advantages of being included in the World Heritage List are:

- Being part of a worldwide group that attracts greater conservation interest and greater obligation of governments, which can greatly help the country to raise awareness in favor of the preservation of the NPA, as well as raise the interest of specialists, with new research or monitoring initiatives that support the management of the area.
- The NPA that belongs to this List is benefited by receiving more dissemination, which increases their potential capture of national and international economic resources.
- A very particular type of tourism is encouraged that is focused on visiting only World Heritage Properties and that is very sensitive to the value that these Properties have for humanity.
- Access to the World Heritage Fund, which is annually enabled to identify, preserve, and promote World Heritage Properties in its territory, as well as to provide emergency assistance to repair damage caused by natural disasters or caused by human societies, as well as for staff training.

Nevertheless, in Mexico there is still a long road to take advantage of the benefits and other actions that must be undertaken, such as:

- Strengthen the dissemination and teaching on the World Heritage Convention and its related instruments, in order to facilitate the understanding and application of the same to the actors involved and interested in the reserve.
- Encourage decision makers to inform themselves and act coherently and in compliance with the obligations acquired by signing the Convention concerning the Protection of the World Cultural and Natural Heritage, and thereby reduce the threats that may arise.

- Reinforce the role of local communities, particularly with children and young people, fostering the bond and commitment to World Heritage for the benefit they derive from it.
- Expand the training of the different actors, in order to guarantee the preservation and management of the site.
- Increase the budget assigned to conservation work, perhaps by giving incentives to those areas that have this type of international designation. And also seek greater international financing support for them.

References and Documents Consulted

Bezaury-Creel J, Graf-Montero S, Barcklay-Briseño K, de la Maza-Hernández R, Machado-Macías JS, Rodríguez-Martínez del Sobral E, Rojas-González de Castilla S, Ruíz-Barranco H (2015) Los Paisajes Bioculturales: un instrumento para el desarrollo rural y la conservación del patrimonio natural y cultural de México. CONANP, Ciudad de México, p 40

CESOP (2006) Centro de Estudios Sociales y de Opinión Pública. Medio ambiente. http://archivos.diputados.gob.mx/Centros_Estudio/Cesop/Eje_tematico/pagina_nueva_2.htm. Accessed Nov 2018

CONANP-INAH (2016) Nominated file. Tehuacán-Cuicatlán Valley: originary habitat of Mesoamerica. https://whc.unesco.org/en/list/1534. Accessed Nov 2018

CONANP-SEMARNAT (2013) Programa de Manejo Reserva de la Biosfera Tehuacán-Cuicatlán, 1st edn. Secretaría de Medio Ambiente y Recursos Naturales, Ciudad de México

FMCN (2009) Estimación y Actualización de la Tasa de Transformación del Hábitat de las Áreas Naturales Protegidas SINAP I y SINAP II del FANP

Hernández Garcíadiego R (2005) El secreto tecnológico del Ssitema Hidroagroecológico más antigui de Mesoamérica. El Complejo Purrón". Alternativas A.C., México. http://www.alternativas.org.mx/El%20Complejo%20de%20Purron.pdf. Accessed Mar 2019

INAH (2019) Arqueología Mexicana/Revista Bimestral XXVI(155)

IUCN (2018) Outstanding universal value, a compendium on standards for inscription of natural properties on the world heritage list. http://whc.unesco.org/archive/2008/whc08-32com-9e.pdf. Accessed 20 Jan 2019

MacNeish RS (1967) A summary of the subsistence. In: Byers DS (ed) The prehistory of the Tehuacán Valley. Environment and subsistence, vol I. University of Texas Press, Austin, pp 290–331

Ramos Vázquez A (2011) Biodiversidad, conservación y marginación indígena en México. Globalización. http://rcci.net/globalizacion/2011/fg1159.html. Accessed Mar 2019

SEMARNAP (1998) Decreto por el que se declara área natural protegida, con el carácter de reserva de la biosfera, la región denominada Tehuacán-Cuicatlán, ubicada en los estados de Oaxaca y Puebla. Secretaría de Medio Ambiente, Recursos Naturales y Pesca. Diario Oficial de la Federación, Ciudad de México

SRE (2018) Secretaria de Relaciones Exteriores. Tratados celebrados por México. https://aplicaciones.sre.gob.mx/tratados/introduccion.php. Accessed Nov 2018

Toledo V, Boege E, Barrera-Bassols N (2008) The biocultural heritage of Mexico: an overview. Terralingua. https://terralingua.org/the-biocultural-heritage-of-mexico-an-overview/. Accessed Feb 2019

UNESCO (1972) Convention concerning the Protection of the World Cultural and Natural Heritage. http://portal.unesco.org/en/ev.php-URL_ID=13055&URL_DO=DO_TOPIC&URL_SECTION=201.html Accessed Oct 2018

UNESCO (2014) Medium-term strategy 2014–2021. https://unesdoc.unesco.org/ark:/48223/pf0000227860. Accessed 15 Jan 2019

UNESCO (2015) Report of the director-general. https://unesdoc.unesco.org/ark:/48223/pf0000234528

UN-ODS. https://www.un.org/sustainabledevelopment/

Villaseñor JL, Dávila P, Chiang F (1990) Fitogeografía del Valle de Tehuacán-Cuicatlán. Bol Soc Bot Méx 50:135–149

WHC (2017) Operational guidelines for the implementation of the world heritage convention. UNESCO World Heritage Center, Paris

WHC (2018) World Heritage Center. The world heritage convention. http://whc.unesco.org/. Accessed Nov 2018

Zortea M, Lucatello S (2016) El mainstreaming Ambiental en los proyectos de cooperación internacional y Desarrollo. Instituto Mora: Universidad Iberoamericana, Ciudad de México

Chapter 14
The Agadir Platform: A Transatlantic Cooperation to Achieve Sustainable Drylands

A. Rizzo, A. Sifeddine, B. Ferraz, E. Huber-Sannwald, D. L. Coppock, E. M. Abraham, and L. Bouchaou

Abstract For the purpose of achieving sustainable development in the context of a changing climate, the development and implementation of tripartite cooperation tools, into a transatlantic cooperation framework, is the crux of a project to bring about a transdisciplinary platform focused on research, technology, and innovation in drylands. It finds its roots in the Agadir Declaration of May 2016. The objective

A. Rizzo (✉)
UMR ESPACE-DEV (IRD, Université de Montpellier, Université de la Réunion,
Université de Guyane, Université des Antilles), Montpellier, France
e-mail: alessandro.rizzo@ird.fr

A. Sifeddine
UMR LOCEAN (IRD, CNRS, MNHN, Sorbonne Université), Departamento de Geoquimica-
UFF-Brazil, UNAM-IRD-Mexico, Ciudad de México, Mexico

B. Ferraz
Centro de Gestão e Estudos Estratégicos (CGEE), Brasilia, Brazil

E. Huber-Sannwald
División de Ciencias Ambientales, Instituto Potosino de Investigación
Científica y Tecnológica, San Luis Potosi, San Luis Potosí, Mexico

D. L. Coppock
Department of Environment and Society, Utah State University, Logan, UT, USA

E. M. Abraham
Instituto Argentino de Investigaciones de las Zonas Aridas (IADIZA-CONICET-Universidad
Nacional de Cuyo), Mendoza, Argentina

L. Bouchaou
Laboratoire de Géologie Appliquée et Géo-Environnement (LAGAGE) Faculté des Sciences,
Université Ibn Zohr, Agadir, Morocco

© Springer Nature Switzerland AG 2020   227
S. Lucatello et al. (eds.), *Stewardship of Future Drylands and Climate
Change in the Global South*, Springer Climate,
https://doi.org/10.1007/978-3-030-22464-6_14

of the platform is to set up a "hub or rear base" at the University of Ibn Zohr in Agadir to develop transdisciplinary research and training mechanisms on climate change and its impacts on the functioning of ecosystems and their goods and services in arid and semiarid regions. Currently, the main challenge to achieve sustainable development resides in ensuring that decision-making processes are supported by science. How to translate scientific knowledge on complex long-term issues at the national, cross-regional, and transatlantic scale into better informed public policy remains an open question for multi-sectoral partnerships. The main thread underlying this chapter relates to the establishment of interface models between science and policy: what challenges will the Agadir Platform assume to bridge various forms of interdisciplinary science and policy expertise to inform decision-makers on long-term wicked problems related to drylands socio-ecological systems?

Keywords Governance · South–South · Tripartite cooperation · Open data · Public policy · Sustainability

Introduction

The adoption of a new climate agreement by the 196 Parties to the United Nations Framework Convention on Climate Change (UNFCCC) in Paris in December 2015 has been regarded worldwide as a historic achievement. Assuming the Paris Agreement is implemented consistently, it will mark a watershed in climate governance, and lead to profound transformations in the global economy and society.

With the publication of the Fifth Assessment Report of the Intergovernmental Panel on Climate Change (IPCC) in 2014 (Burkett et al. 2014), it became evident that a business-as-usual approach was no longer viable and that a complete overhaul of environmental policy was required. Since then, policy elaboration has grown to encompass all areas of human development, and accordingly, climate change adaptation has been brought to the fore by the Paris Agreement, stressing the need for international action and cooperation, particularly with regard to vulnerable developing countries.

To date, global environmental governance has largely been decentralized, and its mechanisms have assumed a variety of forms beyond multilateral agreements. The issue is therefore to determine how governance distributed in the hands of a wide range of actors (state and non-state) can be made effective locally (Chan et al. 2016). The functioning of complex systems, such as drylands, can be explained through the concept of "emergence," that is, how a system intertwined with its environment reacts to changes in the latter, in a way it would not do so in isolation. Emergence, when referring to function in an environment is related to the concept of collective behaviors. Any system and its environment can thus be seen as parts of larger systems encompassing their interactions and feedback mechanisms. In the context of collective behaviors, "emergence" refers to the way collective properties arise from the properties of its constituent parts, for instance, global trends from local structures and relationships (Bar-Yam and Kantor 2018).

"Emergence," in policy studies, refers to the behavior resulting from local interactions based on locally defined rules, with a particular focus on the extent to which local behavior occurs in spite of central government rules and policies. The importance of emergence when dealing with global challenges cannot be overstated. Decision-making related to sustainability is knowledge-intensive, and fostering collaboration between academic and policy-oriented work while developing assessment methods that accommodate diversity of function, goal, and output, is essential to tracking environmental long-term performance beyond the Paris Agreement, the sustainable development goals (SDGs), other UN conventions and conferences (UNCCD, UNCDB, and other initiatives as well), and international commitments (OECD, Global science Forum 2011). Regarding the future of governance, it has become plain that success can only be brought about by extending the common perspective beyond the UN, to the full panoply of existing and potential approaches.

In a recent review of global rangeland systems, Coppock et al. (2017) found that improving governance for natural resource management was the top future research and outreach priority as considered by experts. Can current governance conditions guarantee sustainability, especially in social–ecological systems like drylands, which are most vulnerable to environmental risks? Otherwise, how can said conditions be made conducive to sustainable development? Bearing this in mind, which capabilities should be reinforced globally, nationally, and locally to promote good governance solutions? Within this perspective—and in full awareness that the lack of reliable and relevant data in developing regions is one of the main factors stifling the development of projects which would otherwise be eligible for different climate funds[1]—how can knowledge co-generation and sharing and international cooperation facilitate access to science, technology, and innovation data and information to overcome this distortion?

The initial section of this chapter discusses the major features of a transatlantic approach to address sustainable dynamics in drylands through a work of bolstering the science–policy interface within the framework of multilateral commitments. The second section presents the model implemented and the major challenges the Agadir Platform has to face to locally and globally to be effective.

The Context: Morocco, Among Mediterranean and African Countries Particularly Exposed to the Effects of Global Changes

Even though the Kingdom of Morocco has a status of low carbon emitter, its geographical position confines it to a great natural vulnerability to climate change (desertification, floods, water scarcity...). This vulnerability is exacerbated by the increase of arid and semi-arid

---

[1] Twitter: @LucGnacadja (11/12/2018).

zones, hence worsening the desertification phenomenon and impacting water resources. With agriculture still largely dominated by rainfed crops and therefore dependent on rainfall, vulnerability to the impacts of climate change is real. With the increasing needs of the population and industries, these vulnerabilities will go increasing while climate change will exacerbate the ecological footprint of activities.[2]

Morocco has recently come up with a range of high-level documents and policies aimed at enabling and encouraging climate resilient development.[3] Because it is particularly exposed to the risks associated with climate change, Morocco's national vision, as stated in the INDC—Intended Nationally Determined Contribution, aims to mitigate its effects by endeavoring to "preserve its territory and civilization in the most appropriate manner, effectively responding to the vulnerabilities of its territory and implementing an adaptation policy that builds resilience for all of its population and its economic actors to face these vulnerabilities" (INDC 2015). In line with the concept of sustainable development, it reinforces the point that there needs not be a trade-off between environmental sustainability and economic development.

This is the policy framework that inspired the development of a platform for international cooperation on sustainability in drylands to be located in Agadir, Morocco, principally led by *Ibn Zohr* University. Bearing in mind the relevance of North–South, South–South, and triangular regional and international cooperation, this platform, focusing on the study of the arid and semiarid regions, is a welcome complement to the emerging framework of international cooperation. Further, Morocco anticipates a strategic stake in its Souss-Massa[4] region and the city of Agadir will stimulate policy-driven research in the Sahelian countries and open the avenue towards collaboration with emerging countries in Latin America—like

---

[2] Hakima El Haité, Former Minister Delegate to the Minister of Energy, Mines, Water and Environment, in charge of Environment.

[3] The new Constitution (2011) promotes sustainable economic development, in the spirit of environmental protection. Moreover, the National Charter for Environment and Sustainable Development (Charte Nationale de l'Environnement et du Développement Durable 2009) enacted through the National Strategy for Sustainable Development (NSSD) for 2017–2030 also focuses on climate resilient development strategies. In 2014, Morocco developed its National Climate Change Policy (MCCP) as a coordination tool for measures and initiatives on climate change, providing an operational framework for the development of medium- and long-term strategies. In 2015, Morocco submitted its Intended Nationally Determined Contribution (INDC) presented to the United Nations Framework Convention on Climate Change (UNFCCC) in perspective of the COP21 held in Paris (2015). The main objective identified by the Moroccan authorities has been to make the "territory and civilization more resilient to climate change while ensuring a climate transition to a low-carbon economy" (INDC 2015).

[4] In the Souss-Massa region, located in southwestern Morocco, three factors determine its semiarid Mediterranean climate, namely relief, ocean coast, and the Sahara. Thus, the north of the region, dominated by Atlas, is characterized by a semiarid to humid climate progressing towards the plain. The latter, which occupies the sunken relief of the Atlas and the basins of Souss and Massa, has an arid climate despite a wide opening on the Atlantic. Finally, the southern and southeastern part of the region that makes up the north side is covered by the Sahara Desert climate. The precipitations are highly variable in space and time with a rainfall average of 200 mm/year (Bouragba et al. 2011; Seif-Ennasr et al. 2017).

Mexico and Brazil—equally affected by climate change, land degradation, and desertification. This platform is a significant part of the royal guidelines seeking to strengthen Morocco's scientific cooperation towards sustainability for drylands in Africa and Latin America.

International Cooperation Framework: A Transatlantic Approach to Achieve Sustainable Drylands—The Case of the Agadir Platform

It is a common opinion that international cooperation in general and the tripartite cooperation in particular can be promoted and monitored as a means of implementation of the SDGs[5] and the Addis Ababa Action Agenda[6] on aid effectiveness. Through tripartite cooperation, partners involved share and cogenerate knowledge, learn together, facilitate capacity development, collaborate and jointly create solutions to development challenges. Combining efforts based on complementary knowledge can be key to achieving good results and to taking the 2030 Sustainable Development Agenda forward. All countries can potentially be providers, facilitators, and beneficiaries of knowledge sharing in tripartite cooperation, so it transcends divides between different types of cooperation, geographic boundaries, knowledge systems, cultures, and mental models, among others (UN-ESC 2008).

With this perspective, co-generation of useful knowledge for the sustainability of Northern Africa and Latin America drylands lies at the crux of the Agadir Platform which was founded in December 2016. Its main objectives are to coordinate national participatory research networks, such as RISZA (*Red Internacional para la Sustentabilidad de las Zonas Áridas*—International Network for the Sustainability of Drylands)[7] in Mexico, collaborative, transdisciplinary research efforts and training opportunities, and to build a cross-regional coordinated research framework centered around the SDGs for the sustainable development of local socio-ecological systems. Hence, the Agadir Platform represents a novel science–policy–society interface for dryland development in the transatlantic context. Knowledge sharing

---

[5] https://www.un.org/sustainabledevelopment/sustainable-development-goals/ (UN, 2015).

[6] https://www.un.org/esa/ffd/wp-content/uploads/2015/08/AAAA_Outcome.pdf.

[7] The International Network for the Sustainability of Drylands is comprised of various members from diverse sectors like academic and government institutions, NGOs, civil associations, students, and the civil society including indigenous communities. These members are organized to analyze collaborative actions that strengthen agenda for the dryland regions and respond to the UN sustainable development goals. For these actions, the network requires the implementation of multilateral and inter-institutional collaborations, which are inclusive considering all the stakeholders including the decision-makers. It also provides the possibility to establish strong collaborations with members of Latin America, North Africa and Europe (source: www.risza.com.mx).

will be central as a means to ease the introduction of models of mitigation and adaptation for a sustainable future. Such a platform should greatly improve the way regional (global) transatlantic challenges are identified, defined, and addressed. The combined input of stakeholders across the political and societal spectra both nationally and at the scale of the Agadir Platform will offer an invaluable opportunity to ensure early recognition of risks, tipping points, climate feedback, among others that may require a tripartite (global) joint response portfolio.

One of the main difficulties lies in discerning, in which problems require a local, national, regional, or transatlantic/global response. Systematic, proactive global scanning has been crucial in bringing new emerging problems—and new responses to long-standing problems—to the attention of decision-makers. Scientifically speaking, the platform seeks to address the following caveats to sustainable development trajectories of representative socio-ecological systems: (1) the analysis of past, present, and future climate volatility at the local, regional, and transatlantic scale in order to reduce uncertainties for future climate projections and scenarios; (2) the calibration of biological markers and ecological indicators, in particular, environmental and climatic contexts and conditions; (3) the understanding of the functioning and linkage of local hydrological cycles in relation to dryland use and cover changes with regional climate variability; (4) assessment programs of the integrity of natural ecosystem and agricultural systems and the condition of their ecosystem goods and services and to develop adaptive management practices; (5) the investigation of coastal and marine resources and their interfaces; (6) the discovery of conceptual models for renewable energy projects carefully considering the biophysical and socioeconomic contexts; and finally (7) understanding livelihood resilience and development in diverse cultural, socio-political settings of human settlements, as a basis for innovative adaptation strategies in arid and semiarid regions.

Box 14.1 Climate–Ocean–Land Interconnections Characterizing the Transatlantic Region

Oceans absorb, store, and distribute energy by acting upon the temperature and circulation of the atmosphere. Their ability to store heat is much more efficient (absorbing 93% of excess energy) than those of landmasses (3%) and the atmosphere (1%). For those reasons, they are considered the moderators of climate and of its changes, and any alteration in ocean circulation engendered by external forces (volcanic, solar radiation, greenhouse gas, etc.) can escalate into changes in global atmospheric currents. These alterations are therefore at the root of the heterogeneous dynamics and spatial–temporal distribution of precipitations, of which the vicissitudes can be the source of severe droughts and in turn cause desertification.

Long-term studies on the tropical and subtropical climate variability are justified by the societal requests to understand the distribution and intensity of seasonal rainfall mode or the duration of dry periods and drought with signifi-

(continued)

cant impacts on the economy of the developing and poor countries. There is also interest in the future behavior of the climate variability in a world submitted to the effects of global warming. For several years, the Pacific has been considered as the most important tropical and extratropical climate regulator at different time scales and mainly during the El Nino Southern Oscillation (ENSO) phenomenon. Effectively, this phenomenon produces extreme events (inundations, droughts) at the scale of the Pacific Basin but also at the global level through complex teleconnections systems (Ropelewski and Halpert 1989). For these reasons and during many years, the interest and efforts of the scientific community have been focused on the study of the Pacific Ocean and its variability to the detriment of the Atlantic Ocean. From the 2000s, scientific studies conducted by international teams have shown that the Atlantic Ocean also modulated the climate (Knight et al. 2005). But in reality, the climate in the tropics is the result of the integration of these main mechanisms (including Monsoon systems) whose duration and specific amplitude can balance, dilute, or favor the expression of their actions on tropical systems.[8] Recent studies on the role of decadal and multidecadal variability in the Atlantic Ocean have shown that these modes of multidecadal variability (AMO-Atlantic Multidecadal Oscillation, and NAO-North Atlantic Oscillation) may lead to drought in Mexico, Brazil, and Morocco but with very heterogeneous and antiphase expressions (Méndez and Magaña 2010). For example, a positive mode of AMO causes droughts in Amazon region (Marengo et al. 2011), Northern Mexico (M), and inundations in Southern Morocco. The heterogeneity of spatial and temporal expressions of these transatlantic events, the associated impacts, and the underlying mechanisms remain poorly understood. Therefore, it is of seminal importance to establish the long-term (past, present, and future) climate variability envelope within the framework of the Agadir Platform encouraging transatlantic cooperation and research activities on both sides of the Atlantic in Latin America and in North and West Africa, and on the role of the Atlantic and its impacts on dryland continental and coastal marine ecosystems. Within the framework of a transatlantic collaboration effort among Brazilian, French, Mexican, and Moroccan scientists, spearheaded by RISZA and the CHARISMA[9] project, preliminary results of the analysis of instrumental data during the last century confirm those trends (increase in temperature and decrease in precipitations) in Mexico, South America, and Morocco.

---

[8] It concerns also the Atlantic and Pacific variabilities from the interannual to millennium time scales.

[9] **CHA**ngement et va**RI**abilité**S** cli**MA**tiques passés et actuelles au Maroc: forçages, réponses, imp**act**s et rétroactions—Bases pour la proposition de solutions d'adaptations (*Past and current climate change and variability in Morocco: forcing, responses, impacts, and feedback—Basis for proposing adaptation solutions*).

In arid and semiarid regions, natural climatic variations and anthropogenic disturbances lead to the deterioration of the ecosystem, and therefore have a strong impact on water and soil-restoration cycles. These disruptions have considerable effects on local societies with limited access to food that meets quality standards and trigger disease and epidemic outbreaks, thereby undermining food security and human health of a large population, respectively. Arid and semiarid lands make up 40% of the planet's land surface and have in recent decades experienced a densification in human activity, as well as prolonged periods of drought and desertification (Coppock et al. 2017). They are regulated by the complex interactions of biophysical, social, political, and economic factors emerging as a result of alteration, degradation, and loss of ecosystem services. The most recent climate model simulations, based on contrasted representative concentration pathways (RCPs)—RCP 8.5 and RCP 4.5—show that global climate change could cause dryland biomes to expand by 11–23% by the end of the twenty-first century (Huang et al. 2015). If this occurs, dryland biomes could cover more than half of the global land surface. Climate change will lead to extended droughts, regional warming, and, combined with a growing human population, an increased risk of land degradation and desertification in the drylands (Bastin et al. 2017). Despite the problems afflicting these regions both environmentally and socially, they are considered productive systems because of the richness and diversity of their natural resources, and because they contribute more than 40% of the world's food output.[10] This confluence of risks clearly mandates the full attention of experts in understanding the mechanisms that govern climate change and its impacts on both the environment and society at the local, regional and global scales.

Once the challenges are defined and the need for action identified, the next stage will involve the implementation of proposed solutions. Governments of countries involved in this transatlantic framework are persuaded to pledge their engagement, but cannot act alone. Challenges into the transatlantic region often are interdependent and their dynamics complex, yet many research and global-assessment programs are managed separately, a symptom of a lack of coordination in the policy sphere. Governments, civil society, and the private sector need to consider how to integrate the many disparate global challenge frameworks in order to coordinate research efforts, thereby maximizing coherence and minimizing repetition. Good governance, empowerment, and accountability will be essential to these international collaborative frameworks, and governance schemes should be flexible and support innovation and risk-taking in research (Table 14.1).

---

[10] CGIAR research program on Dryland system: http://drylandsystems.cgiar.org.

Table 14.1 Main biophysics and socioeconomic characteristics of drylands in the counties involved into the transatlantic cooperation framework (the Agadir Platform)

	Argentina	Brazil	Mexico	Morocco
Drylands area (km²/%) Hyper-arid Arid Semiarid Dry-subhumid	The continental portion of Argentina has an area of 2,758,829 km². According to the water regime, the sub-humid and semiarid lands cover 27.50% of the territory, and the arid region—the one with the greatest extension—represents 51.50% of the surface. The relationship between precipitation and potential evapotranspiration defines areas with a predominance of deficit water regimes over 75% of the territory	The geographical area of the Brazilian semiarid region extends to eight states of the Northeast and the North of Minas Gerais[a], totaling a territorial extension of 980,133,079 km. However, considering the territorial dimension of large regions, the Northeast presents 56.46% of its territory in the semiarid portion, the Southeast with 11.09%, and the country reaches 11.53%. The semiarid region was delimited on the basis of the 800 mm isoieta, the Thornthwaite aridity index of 1941 (municipalities with an index of up to 0.50), and the drought hazard (over 60%)	Mexican drylands cover over 1,107,770 km² (55.5%) of its continental surface mostly in the north central part of the country; they include the three driest climate types, with semiarid regions being the most abundant, followed by arid and hyper-arid zones (Baja California)	In Morocco, drylands cover more than 80% of the territory (approximately 640,750 km²). Most of these lands are classified into three subtypes of aridities: hyper-arid, arid, semiarid. They correspond, respectively, for areas in isohyets less than 100, 100–400, and 400–600 mm/year. Only 7% of the country receive more than 600 mm/year

(continued)

Table 14.1 (continued)

	Argentina	Brazil	Mexico	Morocco
Vegetation types	Natural vegetation or semi-natural composed of shrub and herbaceous steppes, dry forests, shrubs, and grasslands. Agriculture based on irrigation. With deep transformations by anthropic action: crops, grazing, and urbanization	Caatinga[b] is the predominant ecosystem in the semiarid region, whose flora is composed of trees and shrubs characterized by rusticity, tolerance, and adaptation to the climatic conditions of the region. The floristic composition is not uniform and may vary according to the volume of precipitation, the quality of the soils, the hydrographic network, and the anthropic action	Natural semiarid grasslands cover 6.1% (118,320 km^2) and highly diverse desert scrub communities (14 different matorral, Mezquital, and gypsum community types) with over 6000 species cover 30% of drylands. Rainfed and irrigated agriculture make up the rest	Natural vegetation or semi-natural composed of steppes, shrubs, and grasslands. Agriculture based on rainfall
Land use types/ production systems	Pastoral activities, irrigated agriculture (vitiviniculture, olive growing, fruit trees), mining, oil, recreation, tourism, and conservation	In the recent process of transformation through which the semiarid economy passes, there are several contradictions and a strong competition between modern agriculture, capitalized and protected from drought by irrigation systems and family farming, dependent on environmental conditions, without adequate capital and political incentives	Livestock grazing, rainfed and irrigated agriculture (corn, beans, cereals, alfalfa, sorghum, apple trees, nut trees), mining, fracking	Mainly for pastoral activities. They are predominantly dominated by: bare soil, herbaceous vegetation, deciduous shrub vegetation, and savanna mosaic/ shrub or forest. Bare soil represents 57% of the surface drylands areas

(continued)

Table 14.1 (continued)

	Argentina	Brazil	Mexico	Morocco
Human population Size Growth Rural/ Urban	30% of total Argentine population. Most provincial states with drylands have average per capita incomes below the national average, and the percentages of households with unmet basic needs double the national average. 83% of the Argentine population is urban	The results of the demographic census revealed that the resident population in the Brazilian semiarid region reached the mark of 22,598,318 inhabitants representing 11.85% of the Brazilian population or 42.57% of the Northeastern Brazils population. Survey of the total population residing in the semiarid region shows that 61.97% of its inhabitants live in urban areas and 38.03% in rural areas^c	The rural population represents 23% of the Mexican population (131 million); there exist poor estimates of the rural population in Mexican drylands due to highly fluctuating numbers in seasonal and permanent migration. Total human population growth lies by 1.24%, population density in drylands lies between 1–60 habitants/km², most of the area has 0–20 habitants/km²	47% of total Moroccan population, with a majority in rural areas
Land tenure	The inhabitants of the drylands face very serious problems of land tenure, litigation of titles, occupations, smallholdings, and latifundia	The reality of the land structure corresponds mostly to properties of two, three, or ten hectares, for the majority of the population. While properties of two, three, or ten thousand hectares of only one owner are in more favorable areas for production	Most dryland is communal land (ejidos), recent amendments in Agrarian reform permit privatization; private land; small proportion federal land (natural protected areas and biosphere reserves)	Mostly private followed by collective lands

(continued)

Table 14.1 (continued)

	Argentina	Brazil	Mexico	Morocco
Livelihood types	The scarce productive alternatives, the low value of production and marketing difficulties, produce scarcity and migration	In the rural areas the families continue with the traditional subsistence crops, always attentive to the rains and planting in several stages to reduce the risks of loss of the crops. Permanent crops—fruit growing—are less significant in the region than temporary crops, being developed mainly in the irrigated perimeters	Ranchers, pastoralists, agriculturalists, fishermen, artisans, subsistence farming, wage labor in agricultural and mining sector	These areas contribute to the livelihoods of thousands of low-income rural people and protect the country from desertification
Indigenous groups	Populations of native peoples such as the wichis, huarpes, araucanos, and mapuches inhabit the drylands with community systems of land use	Semiarid region is home to dozens of indigenous ethnic groups, with more than 70,000 people. At present, there is an ethnic diversity as Pankararu, Truká, Kiriri, and Xucuru peoples, among others	High ethnic diversity Cucapá, Comcaac, Kiliwa, Kikapu, Kumiai, Mayo, Pápago, Pima, Tarahumara, Tepehuano, Wikarika, Yaqui	
Principle threats to socio-ecological resilience	Desertification/land degradation	Sustainable practices are not always present which result in environmental degradation allied to water scarcity, human pressure, land degradation, and desertification processes	Desertification/land degradation, drug trafficking, mining, migration, exhaustion of water resources, invasion of mega development projects	Water scarcity, human pressure, and global changes

(continued)

Table 14.1 (continued)

	Argentina	Brazil	Mexico	Morocco
Main economic values	Drylands contribute 49% of the agricultural production of the country	Services currently account for approximately 54% of the region's total economy, followed by industry (29%) and agriculture (17%). On the other hand, the implementation of projects on water accumulation and irrigation provided the growth of irrigated fruit production. In addition, some states were able to internalize the process of industrialization (textile industry, clothing, footwear and leather, mineral, graniteira, food and beverage, plaster and construction)	Livestock production, agriculture, alternative energy, production of traditional alcoholic beverages (mezcal, pulque, sotol), tourism	Agriculture, pastoralism, and tourism

[a]Regarding the territorial extension of the states, data show that 92.97% of the territory of Rio Grande do Norte is in semiarid, Pernambuco 87.60%, Ceará 86.74%, Paraíba 86.20%, Bahia 69.31%, Piauí 59.41%, Sergipe 50.67%, Alagoas 45.28%, and Minas Gerais 17.49%

[b]Caatinga is considered one of the Brazilian biomes most altered by human activities. Despite its biological importance and the threats to its integrity, about 5% of its area is protected in federal conservation units, which allows to classify the Caatinga as one of the Brazilian ecosystems less protected and more threatened

[c]There has not been a significant change in the share of the semiarid population in the Northeastern states during the last few years, but a slight decrease in the participation of about 1% of them, except in Pernambuco State, where this share increased by 0.7%. Recent data seek to consider the migration of part of the population in economically active age of the semiarid to more dynamic urban centers. The proportion of people aged between 15 and 59 grows confirming the historical movement of migration

International Research Cooperation Model: The Conceptual Framework of the Agadir Platform

The framework of the 2030 Agenda for Sustainable Development—adopted by world leaders in September 2015, and officially coming into force on the 1st of January 2016—widely acknowledges the interconnectedness between climate change and global warming and sustainable development and human well-being. Thus, in order to fulfill its goals and targets, the measures employed must tackle the limited adaptation capabilities of certain geographic areas, especially in developing regions, and strive to mitigate vulnerabilities with a view of contributing to sustainable development.[11] For every nation, but particularly for developing and least developed countries, climate change is congruent with many environmental and socioeconomic issues, including but not limited to: biodiversity loss, desertification, deforestation, decline in resource availability, energy distribution, income generation, security, and health. This tight interweaving of issues poses the risk of negative interactions between climate change and sustainable development, but also suggests that improving one could in turn reduce vulnerability of the other.

Across the 17 goals of the 2030 Agenda, as many as 42 of the targets as well as the entire 17th goal are dedicated to the challenges of implementation. However, these targets are largely unconcerned with the interconnectedness of these goals, and leave the door open, at best to unrealized synergies, but also to potentially perverse outcomes. Therefore, greater attention ought to be paid to these interdependencies, notably across sectors, across societal actors, and among low-, medium-, and high-income countries (Stafford-Smith et al. 2017).

Indeed, the outcome of solutions that depend on the efforts of countries and organizations to deal with the economic, social, and political ramifications of climate change cannot be measured with cause-and-effect linearity (Stiglitz et al. 2006). The issue at hand is therefore to determine how the relationship of national sustainable development planning vis-à-vis its environmental consequences can be used to promote policy coherence and enforce coordination among fragmented sectors. Owing to its complexity and unpredictability, climate change has increasingly been deemed a wicked problem (Head 2008), that is, a problem occurring at the confluence of human and environmental concerns, to which new solutions only beget new problems (Sun and Yang 2016). When elaborating major proposals for change, keeping this image in mind ensures one is cognizant of the particular dynamics engendered by contending inputs, variable outcomes, and their degree of interaction over time.

Research has recently revealed the necessity of moving away from linear solutions—solutions that involve setting targets for the reduction of greenhouse gas (GHG) emissions for each country, industry, and organization—and of expecting people to change their practices for the sake of common good (Sun and Yang 2016).

---

[11] Article 7 of the Paris Agreement.

Viewing climate change and global sustainability as wicked problems enables us to interpret their challenges and solutions in more dynamic and complex terms, including social messes and industrial and organizational fragmentation (Conklin 2016). As climate change and its consequences now appear inevitable, it seems sensible to undertake adaptive measures as a way to cushion against negative effects and explore potential opportunities. Adaptation, as an organizational principle, enables us to understand the relationship between the system and its composite parts. Indeed, adaptation capabilities encompass myriad transdisciplinary aspects of the environment, and this multi-sectoral approach is a welcomed starting point for a reflection on the systemic, reciprocal ties between environment and society.

The path to adaptation, however, can take on a variety of forms reflecting the diversity in kind and in degree of risks and vulnerabilities, which complicates yet substantiates the need for monitoring. The current dearth of a standardized method for tracking adaptation, especially in arid socio-ecological systems, means it is foremost to develop tools with which it can be systematically measured over time and across regions. Nevertheless, to reduce exposure and susceptibility to risk, adaptation will likely involve policy, legislative, administrative, institutional, and financial responses, and therefore, even appropriate measurement techniques will not guarantee the implementation of the measures. Further, the effects of adaptation may not come to light for many years or even decades, and will fluctuate according to uncertain or unknowable future climatic and socioeconomic conditions. Understanding the conditions of adaptation is in line with target 17.8 of the Sustainable Development Agenda, which stresses the importance of operationalizing "technology and innovation capacity-building mechanism for least developed countries...."

Earlier measurement systems have seldom taken the aforementioned challenges and limits into account. If a unified approach to quantifying adaptation as a continuous process (adaptation pathways, Wise et al. 2014) is to be developed, it must emerge alongside with tools and strategies designed to evaluate the process of adaptation; ensure that this monitoring subsequently translates into iterative actions (Sabel and Victor 2016); establish a comparative framework across scales and regions; identify where the greatest deficiencies lie so as to guarantee that resources are allocated accordingly; and keep public governance systems updated on the progress and remaining frailties in the adaptation process. Designing such tools is all the more urgent in light of the Paris Agreement's stipulation that "each party shall, as appropriate, engage in adaptation planning processes and the implementation of actions, including the development or enhancement of relevant plans, policies and/or contributions"[12] and that "the provision of scaled-up resources should aim to achieve a balance between adaptation and mitigation, ...considering the need for public and grant-based resources for adaptation."[13] This statement is in line with the

---

[12] Article 7 of the Paris Agreement.
[13] Article 9 of the Paris Agreement.

need to respect "each country's policy space and leadership to establish and implement policies for poverty eradication and sustainable development," as stated in target 17.15 of the 2030 Agenda, which emphasizes the need for international commitments to bolster capacity-building in developing countries and in turn facilitate national endeavors to enhance sustainable development[14] and policy coherence.

Above all, complying with the recent global environmental and development commitments is a matter of suitable governance (Adger et al. 2009). However, governance poses specific challenges, such as, for instance, how to deal with institutional fragmentation, since global challenges involve almost all domains of policy and governance levels; or how to account for persisting uncertainties regarding the nature and scale of risks and proposed solutions; and how to design short-term policies based on long-term projections. The main question remains: how to develop strategies that bolster sustainable development in the context of climate change, and increase the ability of society to cope with and initiate social transformations away from unsustainable trajectories towards new social–ecological pathways strengthening ecosystem stewardship, and its associated values (Olsson 2003; Chapin et al. 2009).

However, change is brought about by slow-moving variables (Chapin et al. 1996; Carpenter and Turner 2000) that nevertheless bear a strong influence on internal dynamics, and therefore, tracking and understanding these variables may be the most crucial step. Indeed, it will illuminate the rates, directions, and magnitudes of the key processes underlying the long-term evolution of societies and ecosystems (Stafford Smith and Huigen 2009).

Although large quantities of data are currently being collected, there remain significant gaps and deficiencies in our ability to monitor drylands (Verstraete et al. 2009) principally in reason of the high variability of data on human and social dimensions caused by the tight coupling of human and environmental processes in arid and semiarid regions. Existing systems of observation in drylands have typically been designed for ad hoc purposes, without much concern for integration or interoperability with other structures; in some cases, they have initially developed or adapted for different environments. As a result, many key variables in drylands are not always properly monitored.

Much data that could be critical on multiple fronts of policy-making currently lies fallow or fragmented. Many governments, but particularly the poorest and most marginalized, still lack access to adequate demographic and territorial data. The current situation will evidently fall short of expectations on sustainability, both globally and locally. Short of an overhaul in the way data is treated and collected, the international community, acting through bilateral or multilateral ODA (Official Development Assistance) will not be able to support the most vulnerable people. Thus the importance of capacity-building as a way of improving the availability of high-quality, reliable, punctually reported data in the developing world by 2020.[15]

---

[14] Target 17.9 of the Sustainable Development Agenda (2030 Agenda).

[15] Target 17.18 of the Sustainable Development Agenda (2030 Agenda).

Therefore, it is necessary to identify relevant data types to facilitate the development and tracking of metrics. As a general rule, sources must (1) provide information in global, national, and local contexts; (2) be systematically collected and follow standardized guidelines and protocols of data collection and processing such that a comparative analysis of high-quality data will reflect real trends at different spatial and temporal scales; (3) provide sufficient details of metadata; (4) include all pertinent knowledge systems; and (5) be reported consistently, and collected at regular intervals; and (6) be made accessible to all stakeholders or groups of interest. As this will require the combination of many different types of datasets and information, it reinforces the need for collective participation at the design, implementation, monitoring, and evaluation stages, to define the questions, needs, and interests to be addressed; select data and analytic tools, and in particular work with local communities to bolster their involvement during the whole process including data collection, and in the subsequent interpretation and application of findings (Maaroof 2015).

By nature, complex systems display non-linear features. This is particularly true for such systems as drylands, for which it would be unwise to assume that human behavior will be predictable and follow simple and linear patterns. A number of gaps in the knowledge base can be identified, most notably: (1) a limited understanding of barriers and challenges; and (2) the want of metrics to evaluate the state of the system, and to assess its progress and effectiveness. Consequently, while most data is technically public, navigating the means of access to it can prove a challenge in itself, and mining it for relevant insights often requires technical expertise and training that organizations and governments with limited resources cannot systematically afford.

Efficient use of data can only be effected through the collaboration of many actors, including scientists and those with on-the-ground experience, leveraging their strengths to plumb the technical possibilities and the context in which this knowledge can be implemented. The various stakeholders, those who own data and those who depend on it, should ideally coalesce into a data ecosystem that will stoke the development and implementation of various policies. Among the main stakeholders we find: (1) governments and public institutions; (2) international and regional organizations; (3) philanthropy; (4) charities; (5) the private sector and industry; (6) academics and scientists; and (7) civil society as a whole. The challenge will be to bring together these different actors, as stakeholders tend to act within rather than among systems and procedures, and it will be crucial that platforms are developed and managed effectively so that the availability of data benefits integrated approaches to sustainability.

Data-driven development strategies, as tools for formulating policy, can greatly facilitate the implementation of SDGs. However, many emerging and developing countries are still struggling to collect and manage much smaller datasets and statistics, where data is still largely unintegrated, fragmented, or of poor quality, and statistics are often top-down without feedback to communities. New forms of interinstitutional relationships must arise before data, resources, human talent, and

Box 14.2 International Research Platform Model: The Agadir Platform—Source: The Authors

In October 2018, during RISZA's international event "Observando juntos al desierto desde multiples miradas" in the Mapimi biosphere reserve of Mexico, a collective brainstorming session took place with the goal to define a common model for an international scientific platform, namely the Agadir Platform. Figure 14.1 illustrates the final diagram with its various concepts, tasks, and items, and their relationships to the implementation and operation of the platform. In this framework, three main caveats to implementation were identified, respectively, relating to: (1) governance, (2) support structures, and (3) data availability.

1. Governance models in such partnerships can take on various shapes. Although it is doubtful that one single framework will be suited to the full gamut of activities, the brainstorming session was an attempt to better understand how research on global challenges can be optimally organized. One of the conclusions was that such research infrastructure projects would tremendously benefit from a more diverse interest-base, involving also government representatives, NGOs, civil society, indigenous groups, especially marginalized communities. This, in turn, should foster flexibility in governance structures, though the influence of research and innovation in the private sector ought not to be neglected. The priority when identifying a governance model for the Agadir Platform was to find a structure in which scientific expertise, social innovation, industrial strengths and interests, and political objectives would converge to strengthen collective engagement. The main purpose of this exercise was to provide solutions to the peculiar challenges of arid and semiarid socio-ecological systems in order to define inclusive sustainable development trajectories.

2. As part of the project, four pillars (criteria) to its development and implementation were delineated. Firstly, political support. Here, the opening contention was that national governments must continue to invest in research and partnerships so as to avail themselves of new sources of innovation and growth, thereby maintaining vital connections in the global research landscape. Political support enhances a country's ability to assimilate scientific developments as per its national interest. International activities and collaboration should play a key role in national scientific and innovation strategies, so that countries are poised to benefit from the intellectual and financial leverage. Secondly, varied incentive models, such as innovative research agendas and strategic planning, advocacy and outreach, and collaborative membership, will add to the platform's social capital. The challenge for governments, scientists, civil society, and the other entities involved will be to make the most of global science: how to

(continued)

ensure that the fruits borne of this science are fully taken advantage of to solve today's global problems and prepare for to the opportunities and challenges of the future? Thirdly, it has become abundantly clear that a system of innovation (SI) must lie at the foundation of the Platform's strategy and agenda. Such a system must be holistic in the sense that it encompasses a broad range of determinants of innovation and allows for the inclusion of organizational, social, and political, as well as economic factors. It must be truly interdisciplinary and absorb the perspectives of various standpoints and professional outlooks. Finally, the need for human capital and funding. Every country has a stake in solving global problems, but national and local capacities for the implementation and application of science are highly variable. Many efforts are funded directly by governments, but philanthropists, the private sector, and other actors, such as innovation hubs, can play a vital role in supporting risk-taking research and promoting behavioral change. Cross-sectoral coordination will be most likely maintained when all participants are involved, from planning to evaluation, and when budget allocations are explicitly allocated for this purpose.

3. Tackling global problems requires an understanding of their local manifestations. In areas where traditional scientific infrastructure is weak, local knowledge or non-peer-reviewed research may be a welcome addition to otherwise scarce data, particularly in order to develop cost-effective, participatory, and sustainable adaptation strategies. Thus, bottom-up, micro-level data, generated by citizens and locally relevant sources, will complement the largely top-down, macro-level approach to development data. Synergistic interactions at the confluence of small data and traditional social indicators within mature data ecosystems will unlock the full potential of data for development. Here, the key challenge to implementation for the platform lies in developing a permanent observation system that crosses micro- and macrodata on drylands within the focused regions.

decision-making capability can be fully leveraged. Spaces will be needed to serve as forums for technical, cultural, and institutional development. These spaces should emphasize collaboration as a way to build upon the power of individuals, organizations, businesses, and institutions, to investigate challenges and come up with solutions, galvanizing a global culture of learning and action.

Several research performing organizations, acting locally or globally, have launched new initiatives, RISZA is one of them, and governance mechanisms aimed at multilateral scientific cooperation to address global challenges, which may offer some important insights into how best move forward. While basic scientific research will continue to be moved by the drive and curiosity of individual scientists and the

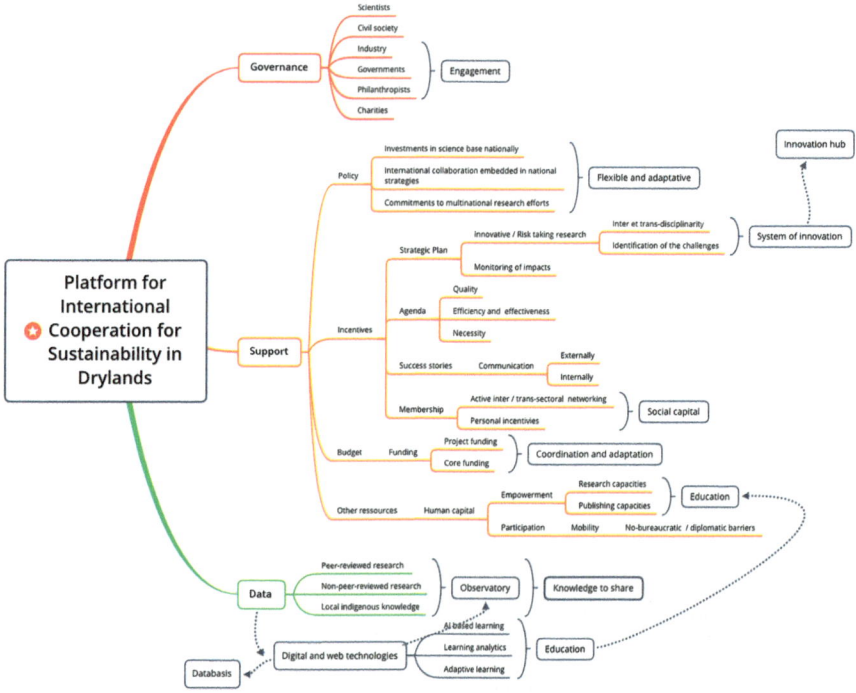

Fig. 14.1 International research platform model. Graphic realized with XMind: Zen

goals of funding agencies, it has also been accepted that research agendas can benefit from being advised by a more diverse range of interests, with major involvement of civil society and indigenous communities. This would foster engagement, but would entail governance structures flexible enough to support it. Thus, ensuring that industrial strengths, scientific capabilities, and policy objectives are all on a par is a priority for future governance structures, especially in developing world.

Good governance, transparency, and accountability are pivotal to successful collaborative frameworks. Global challenges are often interdependent and dynamics between them are complex. Research programs that focus on these challenges are often managed separately, reflecting an endemic fragmentation as well as a lack of coordination in the policy sphere. Governments, civil society, academia and research performing organizations, the private sector, philanthropic and charitable organizations, all need to consider how to integrate the many disparate global frameworks to better coordinate research and education efforts, and reward innovative and creative endeavors in the sustainability sphere. Furthermore, the collaboration of academics and other ex-academia stakeholders, as a form of transdisciplinary research, will be an essential part of a solution-oriented, sustainable research framework (Irwin 2018).

Governance scheme such as these will better manage transparency issues that surround data access, and may radically change the way scientists work together. Within such a governance framework, it will be paramount to ensure continuous assessment of the design and framing of research questions through to the production, diffusion, exploitation, and assessment of new knowledge. This will help galvanize and maintain the involvement of all stakeholders. Greater overall engagement in knowledge (research–education–innovation) generation and sharing is a real opportunity for a more widespread assessment of socio-environmental challenges, as illustrated by the "citizen science" approaches (Liu and Kobernus 2016). Many models of partnerships to generate useful and policy-relevant knowledge bring together scientists, governments, corporations, philanthropists, charities, and civil societies, to address regional to global challenges. This kind of transdisciplinary research is increasingly becoming a data-driven enterprise—due to meteoric improvements in digital systems' capabilities in acquiring, storing, analyzing, and distributing information while consistently reducing costs—and a multi-stakeholder governance model will help facilitate data and resource sharing.

A growing number of government agencies, funding agencies, and international organizations are supporting the demand for greater data sharing, particularly in sustainability research concerned with global challenges. However, despite growing support for open data sharing in many parts of the world, there remain challenges to implementation especially for researchers in developing and least developed countries in Africa and Latin America. One of the main purposes of data exchange is to expand the number and diversity of scientists dealing with societal issues, but this is only significant if researchers can use and interpret data from local indigenous communities. Thus, researchers using open data should strive to conduct relevant ethical studies for local communities and report the results back to them in order to promote the development, transfer, dissemination, and diffusion of environmentally sound technologies.[16]

Innovative mechanisms, new incentives, and policies designed to enhance convergence and advance transdisciplinary sustainability research must arise to relax the current hindrances to the efficiency and performance of scholars and academic institutions. Such initiatives should be bottom-up, led by researchers with investments from their own governments, particularly in developing areas. International commitments are required to strengthen Southern countries' investments in research capacity and to bring forth collaborative research platforms that will serve as a consistent front against global challenges. Finally, effective communication tools and outreach strategies that suit the needs of different stakeholder groups will be key to the empowerment of sustainability research and its applications.

Bearing this in mind, global challenge programs should consider incorporating a capacity-building element to help minimize disparities and improve scientific literacy

---

[16] Target 17.7 of the Sustainable Development Agenda (2030 Agenda).

across regions. Addressing global problems requires an understanding of their local and regional manifestations (Cherlet et al. 2018), as well as how they are represented. In regions, where traditional scientific infrastructure is weak, this may involve drawing on local indigenous knowledge or non-peer-reviewed research, especially in the development of adaptation strategies which are cost-effective, participatory, and sustainable. This is the main education and training perspective included within the Agadir Platform.

Sources of funding for participatory transdisciplinary research must be identified and/or generated. Long-term projects that require large facilities, expensive technologies, and upfront investment cannot come to the fore short of national governments pledging long-term financial commitments or creating appropriate incentive frameworks. Where programs rely on multilateral consensus, every financial contribution is vital to maintaining accountability and inclusivity. Accordingly, the platform in Agadir has been equipped for the analysis of the physical, chemical, mineralogical, and biological components of various arid and semiarid zones. It will complement other research centers in similar regions through its participation in a network of analogous centers focused on development challenges. At present, funding is mainly and directly provided by *Ibn Zohr* University and the Souss-Massa region's administrative entities and national networks like the Mexican RISZA. Further, the platform is aligned with the royal guidelines that aim to consolidate and expand Morocco's strategic cooperation with other regions of Africa, Latin America, and Southern Europe.

Conclusions

Because the world is undergoing changes from political, economic, societal, and environmental standpoints, it is imperative to support the development, implementation, and consolidation of transdisciplinary projects bringing together research, training, education, and innovation, especially those that focus on drylands in Africa, Latin America, and Southern Europe and elsewhere.

Based on the Agadir platform, the decision-making process aims to improve sustainable development policies and solicit the support of strong, innovative and solution-based scientific agendas. In that context, science, technology, and innovation policies are about identifying and articulating new missions for development goals—especially those who dialogue with inventive implementation arrangements.

Sharing experiences, knowledge, and world views to better understand the creeping forces affecting social–ecological systems in arid and semiarid regions is a key element in the identification of sustainable solutions that will ease the transition towards a better world for future generations in line with the 2030 Sustainable Development Agenda, and is the core objective of the Agadir Platform.

Considering the above exposed, conclusion remarks on Agadir initiative lead at least to: (1) encourage interdisciplinary research to increase the knowledge base needed for sustainable development pathways; (2) foster increased international collaboration and exchange of knowledge and scientific capacity on the global level by intensifying science based platforms; (3) establish platforms for free and open data sharing; (4) reinforce the science–policy–action process to ensure that the best scientific knowledge is available to decision-makers and end users; and (5) understand the implementation of a sustainable development policies agenda as a continuous learning process that needs close and regular scientifically based revision and long-run finance perspective.

Acknowledgements The authors greately appreciate stimulating discussions and dialogues with members of the RISZA network and the Agadir Platform during many workshops and group meetings in Mexico and Morocco, in particular special thanks to Dr. Omar Halli and Dr. Omar Akhayat, Director and Vice-Director, respectively of the Ibn Zhor University in Agadir. Elisabeth Huber-Sannwald gratefully acknowledges financial support from CONACYT (projects 280605, 293793, PDCPN-2017/5036). Special thanks also to the CHARISMA project for the constructive inputs and recommendations. The authors also thank IRD researchers and engineers who played an important role into the construction and the implementation of this project.

References

Adger NW, Dessai S, Goulden M, Hulme M, Lorenzoni I, Nelson DR, Naess LO, Wolf J, Wreford A (2009) Are there social limits to adaptation to climate change? Clim Chang 93:335–354

Bar-Yam Y, Kantor D (2018) A mathematical theory of interpersonal interactions and group behavior

Bastin JF, Berrahmouni N, Grainger A, Maniatis D, Mollicone D, Moore R, Patriarca C, Picard N, Sparrow B, Abraham EM, Aloui K, Atesoglu A, Attore F, Bassüllü C, Bey A, Garzuglia M, García-Montero L, Groot N, Guerin G, Laestadius L, Lowe AJ, Mamane B, Marchi G, Patterson P, Rezende M, Ricci S, Salcedo I, Sanchez-Paus Diaz A, Stolle F, Surappaeva V, Castro R (2017) The extent of forest in the dryland biomes. Science 6338:635–638

Bouragba L, Mudry J, Bouchaou L, Hsissou Y, Krimissa M, Tagma T, Michelot JL (2011) Isotopes and groundwater management strategies under semi-arid area: case of Souss upstream basin. Appl Radiat Isot 69(7):1084–1093

Burkett VR, Suarez AG, Bindi M, Conde C, Mukerji R, Prather MJ, St. Clair AL, Yohe GW (2014) Point of departure. In: Field CB, Barros VR, Dokken DJ, Mach KJ, Mastrandrea MD, Bilir TE, Chatterjee M, Ebi KL, Estrada YO, Genova RC, Girma B, Kissel ES, Levy AN, MacCracken S, Mastrandrea PR, White LL (eds) Climate change 2014: impacts, adaptation, and vulnerability. Part A: global and sectoral aspects. Contribution of working group II to the fifth assessment report of the intergovernmental panel on climate change. Cambridge University Press, Cambridge, pp 169–194

Carpenter SR, Turner MG (2000) Hares and tortoises: interactions of fast and slow variables in ecosystems. Ecosystems 3:495–497

Chan S, Brandi C, Bauer S (2016) Aligning transnational climate action with international climate governance: the road from Paris. RECIEL 25(2). ISSN2050-0386

Chapin FS, Torn MS, Tateno M (1996) Principles of Ecosystem Sustainability. The American Naturalist 148(6):1016–1037

Chapin FS III, Folke C, Kofinas GP (2009) A framework for understanding change. In: Chapin FS III et al (eds) Principles of ecosystem stewardship. Springer Science, New York

Cherlet M, Hutchinson C, Reynolds J, Hill J, Sommer S, von Maltitz G (eds) (2018) World atlas of desertification. Publication Office of the European Union, Luxembourg

Coppock DL, Fernández-Giménez M, Hiernaux P, Huber-Sannwald E, Schloeder C, Valdivia C, Arrendondo JT, Jacobs M, Turin C, Turner M (2017) Rangelands in developing nations: conceptual advances and societal implications. In: Briske D (ed) Rangeland systems: processes, management, and challenges. Springer Earth System Sciences. Springer, Berlin, pp 569–641

Head BW (2008) Wicked problems in public policy. Publ Pol 3(2):101–118. Curtin University of Technology. ISSN 1833-2110

Huang J, Yu H, Guan X, Wang G, Guo R (2015) Accelerated dryland expansion under climate change. Nat Clim Chang 6(2):166–171

Irwin, T (2018) The Emerging Transition Design Approach. 10.21606/dma.2017.210.

Knight J-R, Allan R-J, Folland C-K, Vellinga M, Mann M (2005) A signature of persistent natural thermohaline circulation cycles in observed climate. Geophys Res Lett 32. https://doi.org/10.1029/2005GL024233

Liu H-Y, Kobernus M (2016) Chapter 7: citizen science and its role in sustainable development: status, trends, issues, and opportunities. In: Analyzing the role of citizen science in modern research (advances in knowledge acquisition, transfer, and management). https://doi.org/10.4018/978-1-5225-0962-2.ch007

Maaroof A (2015) Big data and the 2030 sustainable development agenda. Report for UN-ESCAP. https://www.unescap.org/sites/default/files/1_Big%20Data%202030%20Agenda_stock-taking%20report_25.01.16.pdf

Marengo J, Tomasella J, Alves L-M, Soares W-R, Rodriguez D-A (2011) The drought of 2010 in the context of historical droughts in the Amazon region Jose A. Geophys Res Lett 38:2011. https://doi.org/10.1029/2011GL047436

Méndez M, Magaña V (2010) Regional aspects of prolonged meteorological droughts over Mexico and Central America. J Clim 23(5):1175–1188 https://doi.org/10.1175/2009JCLI3080.1

Morocco Intended Nationally Determined Contribution (Indc) Under The UNFCCC, (2015). Available at: https://thecvf.org/wp-content/uploads/2018/08/Morocco-INDC-submitted-to-UNFCCC-5-june-2015.pdf

Olsson P (2003) Building capacity for resilience in social-ecological systems. Stockholm University, Stockholm

Organization for Economic Cooperation and Development (OECD), Global Science Forum (2011) Report on opportunities, challenges and good practices in International Research Cooperation between developed and developing countries, Paris, France

Ropelewski CF, Halpert MS (1989) Precipitation patterns associated with the high index phase of the southern oscillation. J Clim 2:268–284

Sabel CF, Victor DG (2016) Making the Paris process more effective: a new approach to policy coordination on global climate change. Policy analysis brief. The Stanley Foundation, Muscatine

Seif-Ennasr M, Hirich A, Zine El Abidine EM, Choukr-Allah R, Zaaboul R, Nrhira A, Malki M, Bouchaou L, Beraaouz E (2017) Assessment of global change impacts on groundwater resources in Souss-Massa basin. Water resources in arid areas: the way forward. Springer, Cham, pp 115–140. https://doi.org/10.1007/978-3-319-51856-5_8

Stafford-Smith M, Huigen J (2009) From desert syndrome to desert system: developing a science for desert living

Stafford-Smith DM, Griggs D, Gaffney O, Ullah F, Reyers B, Kanie N, Stigson B, Shrivastava P, Leach M, O'Connell D (2017) Integration: the key to implementing the sustainable development goals. Sustain Sci 12:911–919

Stiglitz J, Ocampo JA, Spiegel S, French-Davis R, Nayyar D (2006) Stability with growth: macroeconomics, liberalization and development. Oxford University Press, Oxford. https://doi.org/10.1093/0199288143.001.0001

Sun J, Yang K (2016) The wicked problem of climate change: a new approach based on social mess and fragmentation. MDPI. Sustainability 8:1312. https://doi.org/10.3390/su8121312

United Nations (2015) Transforming our world: the 2030 agenda for sustainable development. A/RES/70/1

United Nations Economic and Social Council (UN-ESC) (2008) Background study for the development cooperation forum. Trends in South-South and triangular development cooperation. New York, USA

United Nations Framework Convention on Climate Change (UNFCCC) (2015) Decision 1.CP/21, Adoption of the Paris Agreement. http://unfccc.int/paris_agreement/items/9485.php

Verstraete, M, Stafford Smith M, Scholes R (2009) Designing an integrated global monitoring system for drylands. Proceedings, 33rd International Symposium on Remote Sensing of Environment, ISRSE 2009. 898–901

Wise RM, Fazey I, Stafford-Smith DM, Park SE, Eakin HC, Van Gardenen AERM, Campbell B (2014) Reconceptualizing adaptation to climate change as part of pathways of change and response. Glob Environ Chang 28:325–336

Chapter 15
The Atlas Workshops of Agdz, Morocco: A Model Region for a Scientific–Artistic Dialogue

U. Nehren, S. Lichtenberg, H. Mertin, N. Dennig, A. El Alaoui, H. Zgou, S. Alfonso de Nehren, and C. Raedig

Abstract The town of Agdz in southeast Morocco was chosen as an applied case for the cooperation project "The Atlas Workshops of Agdz: A model region for a scientific-artistic dialogue." The aim of the project was to use scientific and artistic approaches to identify transformation processes in the socio-ecological system and the associated impacts on the cultural heritage of the *Imazighen* (Berbers). Transdisciplinary and intercultural teaching and learning methods provided international students with a holistic view of system contexts and changes and enabled them to analyze and reflect on them. Based on the project results, this chapter discusses transformation processes in Agdz in the course of global change, which pose new challenges for the region. We examine aspects of oasis agriculture, water use, traditional earthen architecture, and the *Ahwash* ritual, an intangible cultural heritage of the *Imazighen*. Moreover, we discuss the extent to which transdisciplinarity and interculturality can contribute to a better understanding of the system and foster

U. Nehren (✉) · S. Alfonso de Nehren · C. Raedig
Institute for Technology and Resources Management in the Tropics and Subtropics, TH Köln, Köln, Germany
e-mail: udo.nehren@th-koeln.de

S. Lichtenberg
Department of Physical Geography, University of Passau, Passau, Germany

H. Mertin
Freelance musician, sound performer and music ethnologist, Cologne, Germany

N. Dennig
Dindum Kulturkommunikation e.V., Pulheim-Manstedten, Germany

A. El Alaoui
Natural Resources and Environment Research Team (NR & E), Department of Chemistry, Faculty of Science and Technology Errachidia, Moulay Ismail University, Meknes, Morocco

H. Zgou
Polydisciplinary Faculty of Ouarzazate, Ibn Zohr University, Agadir, Morocco

© Springer Nature Switzerland AG 2020
S. Lucatello et al. (eds.), *Stewardship of Future Drylands and Climate Change in the Global South*, Springer Climate,
https://doi.org/10.1007/978-3-030-22464-6_15

253

mediation processes, and what role music and dance play in these. This takes place against the background of the overarching objective of sustainable rural development and socio-ecological resilience.

Keywords Southeast Morocco · Sustainable development · Socio-ecological system · Interculturality · Transdisciplinarity · Inquiry-based learning · North–South dialogue

Introduction

The globalized world is characterized by an increasing speed of physical and information flows and complex interactions and networks, which leads to rapid changes in social–ecological systems that have emerged over long periods of time. This leads to uncertainty among both the affected population and politicians and planners, since there is little time available for monitoring and guiding complex dynamic processes and adapting to new political, social, and environmental conditions. Prominent examples of these globally observable transformation processes, commonly referred to as "global change," are population growth and increasing urbanization (United Nations 2014), climate change (IPCC 2014), loss of habitats and biodiversity (Cardinale et al. 2012), soil degradation (UNCCD 2017), and chemical pollution of water, soil, and air (Rockström et al. 2009).

According to the United Nations (2018), 55% of the world's population already lives in cities today—compared with only 30% in 1950—and this figure will rise to 68% by 2050. By contrast, the world's rural population will peak at 3.4 billion in a few years' time and then fall to 3.1 billion by 2050. However, rural regions will develop differently depending on geographical, socio-economic, cultural, and infrastructural factors. While rural areas with good access to markets and resource-rich areas will attract new settlers, remote areas in the global arid regions without rich mineral resources will hardly attract immigrants. Their low carrying capacity makes them the regions with the lowest human well-being (MA 2005). These regions often suffer from rural exodus and an aging society. Traditional knowledge passed down from generation to generation is increasingly at risk as young people move to big cities in search of work and better education, and those who stay in rural areas often seek modernity in their living environments and deliberately break with traditions or simply forget them. This threatens the material and immaterial cultural heritage based on traditional knowledge.

In the project "The Atlas Workshops of Agdz, Morocco: A model region for a scientific-artistic dialogue" we address sustainable rural development and cultural heritage in the context of higher education. The project is part of the program "University Dialogue with the Islamic World" that aims at strengthening cooperation between higher education institutions in Germany and Islamic countries. Against this background the four participating universities have jointly defined a project scope with principles and objectives (Table 15.1).

Table 15.1 Partners, principles, and objectives

Participating Universities	
Germany	Institute for Technology and Resources Management in the Tropics and Subtropics (ITT), TH Köln—University for Applied Sciences, Cologne
	Centre for Contemporary Dance, University for Music and Dance, Cologne
Morocco	Faculté des Sciences et Techniques (FST), University Moulay Ismail, Meknès
	Faculté Polydisciplinaire de Ouarzazate (FPO), University Ibn Zohr, Agadir
Principles	
Transdisciplinarity	Complex and dynamic transformation processes can only be adequately analyzed on the basis of a transdisciplinary approach that integrates concepts and methods from various disciplines with the involvement of local actors. Students and young scientists from different disciplines and cultural backgrounds should therefore work together on issues of sustainable rural development, resilience, and cultural heritage in order to understand the underlying causes of socio-cultural and environmental change and develop strategies to address related challenges
Inquiry-based learning in real-world laboratories	Real-world laboratories facilitate learning processes and foster cultural exchange. We chose the town of Agdz in southeastern Morocco, where the German NGO "Dindum"—Association for Cultural Communication has been working for about 20 years and has built up trusting relationships with the local community. Agdz has a rich heritage of the *Amazigh* (Berber) culture and at the same time is exposed to strong social changes
Art and interculturality	Art, especially music and dance, is an integral part of Moroccan culture. In the Amazigh culture of southeast Morocco, performances such as the *Ahwash* ritual (lit.: the dance of the community) carry deep-rooted cultural knowledge in an astonishing continuity. Nevertheless, they are also subject to change. Consequently, music and dance is at the same time a relevant element and research subject, as well as a medium of intercultural communication. We therefore address music and dance in three different ways: (1) as an artistic expression, (2) as an enabling medium for intercultural communication, and (3) as an object of research
Objectives	
Foster transdisciplinary thinking and intercultural communication	Students and young researchers should be trained in transdisciplinary thinking and intercultural communication using modern teaching methods in an inspiring "real-world" environment
Enhance artistic dialogue	Dancers and musicians from Germany should enter into an artistic dialogue with traditional *Ahwash* artists and this way exchange artistic expressions and create forms of nonverbal artistic and intercultural communication

(continued)

Table 15.1 (continued)

Strengthen science and art interfaces	Science and art should be linked in such a way that both sides learn from each other; scientists should examine the extent to which music and dance should be included in socio-ecological research, while artists should reflect their work from a cultural and socio-ecological perspective
Raising awareness of cultural heritage	Key actors from Agdz should be involved to raise awareness for the value and relevance of their culture and the need to preserve it; this should trigger processes to revitalize the tangible and intangible cultural heritage

A Brief Introduction to Amazigh History

The *Imazighen* [singular: *Amazigh*] or Berbers are the indigenous people of a region called *Tamazgha*, which encompasses Morocco, Algeria, Tunisia, Libya, Northern Mali, the Canary Islands, and large parts of Mauritania, Niger, and Egypt (Almasude 2014). The *Amazigh* languages belong to the Afro-Asiatic language family and are spoken by about 40 million people (Lafkioui 2018), of which about 11 million live in Morocco (Ennaji 1997). They can be grouped in several sub-groups or varieties, among them the "*Tamazight*," which is spoken in the Middle Atlas and east of High Atlas (Ennaji 1997).

The origin of the *Imazighen* has been controversially discussed which, among other things, is probably due to their very large settlement area and their ethnic diversity. From an archeological perspective, prehistoric *Amazigh* colonization in the Maghreb region can be traced back to at least 10,000 years ago (known as Neolithic Capsian culture; Ilahiane 2006). In historical times, first references to people known as Berbers are found in Egyptian, Greek, and Roman records (Shoup 2011). For many centuries, *Amazigh* groups settled along the Mediterranean coast from Egypt to the Atlantic Ocean, where the semi-nomadic tribes have experienced several conflicts with invaders and colonists such as Phoenicians, Greeks, Romans, Vandals, and Byzantines, before the Arabs conquered the Maghreb. The various pre-Arabic invasions left some imprint on the *Amazigh* culture as well as it vice versa influenced other cultures.

The Islamization of the Maghreb by the Arabs started in 647 when Rome/Byzantium lost Libya and Tunisia and was concluded in 709 when all of North Africa was under control of the Arab caliphate. Unlike previous conquests, the Islamization and Arabization of the region had fundamental and lasting effects on the *Amazigh* society, and in large parts replaced former social norms and political idioms (Maddy-Weitzman 2011). However, the ruling Arabs treated converts as "second-class Muslims," taxed them heavily, and even enslaved them, which finally led to the Great Berber Revolt of 740 (Elfasi and Hrbek 1988). After the revolt, several tribal kingdoms and later larger Berber dynasties (from 1060 to 1549) emerged in the Maghreb and in Al-Andalus.

In Morocco, from 1549 onward, the Berber dynasties were replaced by the descendant Arab dynasties that claimed descent from the prophet Muhammad. The first was the Saadi dynasty (1549–1659) followed by the Alaouite dynasty (since 1666). The rulers of these dynasties resided in the four royal cities of Fez, Marrakech, Meknes, and Rabat. After the Middle Ages, large parts of North Africa including Algeria and Tunisia got under the control of the Ottoman Empire. Morocco, however, became an independent power under the Saadi dynasty, and was the only Arab state to successfully assert itself against the Ottomans. Already in the early nineteenth century Morocco was increasingly exposed to European claims to power, which finally led to the colonization of northern and southern Morocco by the Spaniards, resulting in three wars (1893, 1909, and 1921). France also increasingly exercised political influence, which led to the division of the country in 1912, with a small part in the north becoming a Spanish protectorate, much of the country, however, a French protectorate (Maddy-Weitzman 2011).

After the World War II, negotiations began to end the protection treaty, and in 1956 Morocco was finally declared independent as the "Kingdom of Morocco." With the independence, the question arose as to what role the *Amazigh* culture could play in the new state, and how it could become part of a homogeneous national identity—among others against the background of "tribal-oriented speakers of unwritten Berber dialects" (Maddy-Weitzman 2001). In the second half of the twentieth century the *Amazigh* contribution to Moroccan history, culture, and collective identity has been increasingly acknowledged. Jay (2016), for instance, points out many positive aspects of the flourishing *Amazigh* culture in arts and media in the last decades; among others the creation of the *Institut Royal de la Culture Amazighe* in 2003 and the development of an "*Amazigh* theatre" that renews and disseminates oral heritage such as myths and narratives using local dialects. Furthermore, a strong *Amazigh* movement also carries the danger of ethnical conflicts and challenges the existing political and socio-cultural order (Maddy-Weitzman 2006). Despite the opportunities and risks posed by a growing *Amazigh* collective self-consciousness and identity, Maddy-Weitzman (2001) judged this development in Morocco and Algeria as irreversible and predicted that this process will have a significant impact on North African affairs.

Agdz and Its Environmental, Socio-Cultural, and Political Challenges

Agdz is located in the Zagora Province in the region of Draa-Tafilalet (Fig. 15.1). "Agdz" means "resting place" and refers to its historical relevance along the caravan route from Marrakesh to Timbuktu. The town played a major role for the exchange of goods across the Sahara and is considered an important meeting point of cultures. In the last two decades, Agdz has experienced a considerable increase in population from 5870 in 1994 to 10,681 in 2014 (Haut-Commissariat au Plan 2019). This development is reflected in the infrastructure development with new roads and housing estates in the urban area. The region is characterized by harsh climatic and

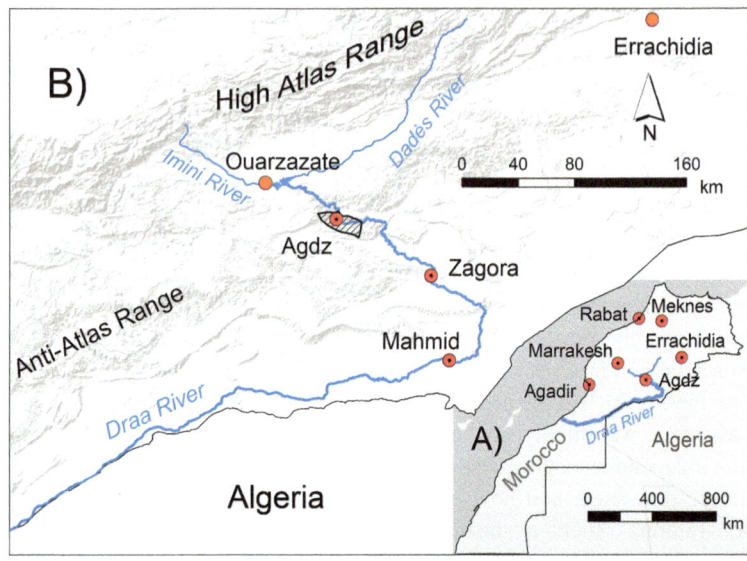

Fig. 15.1 Study area map (*A*) Location of Agdz and Draa River in Morocco; (*B*) Draa valley with Agdz community (hatched) and the two partner universities in Ouarzazate and Errachidia (in orange). Most of the year the Draa River falls dry after Mahmid and does not reach the coast. Sources: Esri, HERE, Garmin, FAO, NOAA, USGS, ©OpenStreetMap contributors, and the GIS User Community

environmental conditions. The climate in the Middle Draa valley is arid to hyper-arid with an annual maximum precipitation of about 200 mm in the High Atlas Mountains in the north and less than 30 mm in the south of the basin (Johannsen et al. 2016). Temperatures range from −1 to −7 °C in winter to over 40 °C in summer (Graf 2010).

The Draa River originates in the High Atlas Mountains at the confluence of the Dades and Ouarzazate Rivers and runs over a 200 km long belt of six palm tree oases to finally drain into the salt lake Lak Iriki. Since 1972, the river water flow is managed through the Al Mansour Eddahbi reservoir (Graf 2010). The communities in the Middle Draa valley depend on agriculture to sustain their livelihoods, which makes them highly dependent on the scarce water resources (Box 15.1). According to Heidecke (2009), the total water demand for agriculture in the entire Middle Draa valley sums up to 300 million m³ per year, whereas only about 3% of the total water withdrawal is used for human consumption, but about 97% for agriculture. Due to variability in rainfall, the inhabitants build more wells to extract groundwater, trying to stabilize the water availability. Overuse of groundwater resources in combination with persistent drought events, and possible future impacts of climate change are a great challenge for the Middle Draa Valley (Karmaoui et al. 2014, 2016; Johannsen et al. 2016).

Box 15.1 Oasis Agriculture

The Draa River sustains lush oases, where synergic human–nature interactions have formed and maintained agroecosystems. In Agdz, the oasis shows a common scheme with predominately small plots that consist of three strata: (a) palm trees, (b) small fruit trees and shrubs, and (c) herbaceous vegetation. The date palms (*Phoenix spec.*) reach heights of 15–30 m (Karmaoui et al. 2014). In the shrub stratum, fruit trees like pomegranate (*Punica granatum*), olive tree (*Olea europaea*), fig (*Ficus carica*), apples (*Malus spec.*), and henna (*Lawsonia inermis*) are planted, while the herbaceous stratum is dominated by the forage alfalfa (*Medicago sativa*), corn (*Zea mays*), and vegetables (Fig. 15.2). Over centuries dryland inhabitants have coped with the challenge of limited water availability by developing complex water harvesting and distribution techniques and community-based management practices (Adeel et al. 2008). The water is collected and channeled in the mountains using a tunnel system, which in South Morocco is called *Khettara* (in other regions, it is known as *Qanat* or *Foggara*). Using a gently sloping underground channel the water is transported from an aquifer or well to the settlements and cultivated areas that are often located several kilometers away from the water source. In the oasis, traditionally open irrigation canals known as *Saguias* (or *Sagias*) lead to the individual fields. The water distribution inside the oasis is managed by the community with every family receiving the same amount of time to irrigate their crops. Loam walls, palm trees, and shrubs create a cooler microclimate and provide shade for the herbs. This way, water loss through evapotranspiration is reduced. Today, the oasis system is subject to land use intensification and land consolidation, which have significant impacts on water availability and water management.

Besides agriculture, tourism has become another important economic factor. Tourists are attracted by the rich heritage of the *Amazigh* culture, in particular the impressive earthen architecture (Box 15.2), the deep-rooted dances and cultic ritual of the *Ahwash* (Box 15.3), and the handicraft traditions such as weaving and pottery. However, increasing numbers of tourists put additional pressure on the scarce water resources, even though the proportion of water use for tourism in the Middle Draa Valley is only 0.3% of the total consumption (Karmaoui et al. 2016). In the worst-case, tourism can lead to cultural tension and destruction of cultural heritage as shown by Keenan (2002) for various locations in the Sahara region. Furthermore, there are also numerous examples around the globe where appreciation of cultural elements by tourists has led to a revitalization of the heritage (Vorlaufer 1999).

In recent years, the region of Agdz has also experienced economic growth in the local and regional transport and trade sector. Moreover, money transfers by men working partly in the cities of the north or in Europe also play a significant role for the local economy (Heidecke and Heckelei 2010). The social structure of the

Box 15.2 Earthen Architecture

The entire Draa valley is lined by more than 300 *ksur* (singular *ksar)* and numerous *kasbahs* and *dârs,* which form one of the most impressive landmark heritages of earthen architecture (Fig. 15.3a), dating back to the fifteenth century and built to protect the settled *Amazigh* population from the attacks of the nomadic *Amazigh* tribes (Baglioni et al. 2015). The *ksar* is a kind of fortress with high walls and defense towers. Inside its walls it consists of an inner settlement structure with several *dârs* (houses), which creates a fresh and dark tunnel network and reduces the exposure of the houses to heat and sand storms. Thus, it leads to favorable thermal comfort inside the buildings. The *kasbahs* are fortified houses which often cover large areas, including several buildings, public and private areas. They are owned by powerful and wealthy *Amazigh* families (Mileto et al. 2012). All buildings were constructed with local materials using the masonry techniques with rammed earth (*alleuh*) for the lower floors, earthen bricks (*toub*) for the upper floors, and date palm wood for the horizontal structures. The use of these local materials in the traditional way is in harmony with nature and adapted to the climate conditions. Unfortunately, today many historic buildings have already fallen into disrepair, symbolizing the loss of tangible and intangible heritage (Fig. 15.3b).

Box 15.3 Ahwash

Ahwash (also spelled *Ahwach* or *Ahouach*) is a ritual collective performance of *Amazigh* communities in southern Morocco (see Fig. 15.4). It combines poetic chants, lyrics, rhythm, and dance and is performed by a large number of male and female participants (from 20 to more than 100) as a celebration of the community. However, in some region's groups comprise only males. The ritual is passed from generation to generation, but there are no written documents that prove the exact geographical origin and time of its emergence. Some scientists find evidence that the *Ahwash* contains pre-Islamic components (e.g., Schuyler 1979) and may have originated in the region of Telouet in the High Atlas between Marrakesh and Ouarzazate and spread from there to the surrounding villages (Boudraa and Krause 2009). In the isolated villages the *Ahwash* may then have further developed relatively independently, so that the formations and elements of performance differ from village to village. However, there are several common elements that distinguish the different varieties from other performative rituals in South Morocco. Therefore, the communities speak of a common *Ahwash* which decisively contributes to the local *Amazigh* identity. There are characteristic instruments that are used during the performance, such as the *dindum* (big stand drum), the *tbel* (smaller drum), and the *tara* (a kind of tambourine). *Ahwash* goes far beyond a folkloric music and dance performance by

comprising spiritual elements that are sometimes even perceived as super-
natural and fulfilling important functions in the community. However, there is
also an increasing trend towards the commercialization of *Ahwash*, both as a
tourist performance and as a form of popular music played by local radio
stations.

Fig. 15.2 Oasis of Agdz (Photo Credit: U. Nehren)

Imazighen is tribal (Gellner 1970) and can be described as patriarchal society
(Belahsen et al. 2017) with most of the tribes having male leaders—although there
were also a number of female leaders in the Middle Ages. Traditionally, women and
men share complementary roles in agriculture and food preparation: men are mostly
responsible for the livestock, while women take care of the family, work in agriculture, and produce handicrafts. However, societal transformation has changed this
distribution of roles and nowadays men are generally responsible for the income
and women take care of the family and the home and are still engaged in handicraft
or artistic production, such as pottery and artful weaving of textiles, in particular
kilims (tapestry) (Sadiqi 2007).

The domestic unit consists of a nuclear family. In marriages, the man selects the
woman or the family takes the decision. Traditionally men have important status in
community and family organization, which authorizes them to sit in the *djemââ*—
the tribal council that debates on community issues (Belahsen et al. 2017). Until the
colonization this was the political body to organize the entire village life, which

Fig. 15.3 (**a**) Restored Kasbah in Agdz; (**b**) Decayed earthen buildings in rural areas (Photo credits: M. Maurissens)

Fig. 15.4 Ahwash performance in Agdz (Photo credit: N. Dennig)

included the external and internal security, creation of solutions for land and water conflicts inside the community as well as irrigation cycles outside the community, organization of collective working tasks (Rademacher-Schulz 2014). Normally, the dominant ethnic groups had the monopoly over water and land rights and tried to maintain this power (Rademacher-Schulz 2014).

A historically important shift for the *Amazigh* groups was the change from semi-nomadism to permanent settlement with the beginning of the French protectorate (Belahsen et al. 2017). Land ownership can be divided into communal land owned by the tribe, private agricultural land owned by religious brotherhoods or mosques, and land owned by private individuals (Rademacher-Schulz 2014). To this day, land ownership rights are passed on within the family, with men generally being the landowners (Hoffman 2008). The status of women with regard to inheritance is regulated by the religious law (Montagne 1973). Although today men have often emigrated to the cities and women take responsibility for the children, the fields, and the harvest, men's property is maintained (Hoffman 2008). After the independence colonial district leaders were called *caids.* They were distinguished from the *anghar* who were elected by the tribe (Gellner 1970). Usually the *caids* were living in *kasbahs* inside the *ksur* (see Fig. 15.3b).

Project Design and Teaching/Learning Concept

In the initial project workshop, the German and Moroccan project partners together with representatives from the community and the involved NGOs decided to work on four specific areas, which are called workshops. The four defined workshops are related to (1) music and dance, (2) natural resources and development, (3) traditional earthen architecture, and (4) society and intercultural dialogue (Fig. 15.5).

Against the background of the objectives defined in the introduction of this volume (Huber-Sannwald et al. this volume), the project partners have defined three key terms that form the theoretical foundation of the project: (1) *transformation,* understood as the processes in the socio-ecological system within the different spheres and the system as a whole; (2) *transdisciplinary,* understood as a research approach and learning design that relates to the transformation processes and related challenges; and (3) *The Third Space,* which goes back to works by Bhabha (1994), Soja (1996), and other scientists. The Third Space—the "in-between"—can be understood as an ambiguous (fictive, metaphorical, or geographical) space in which

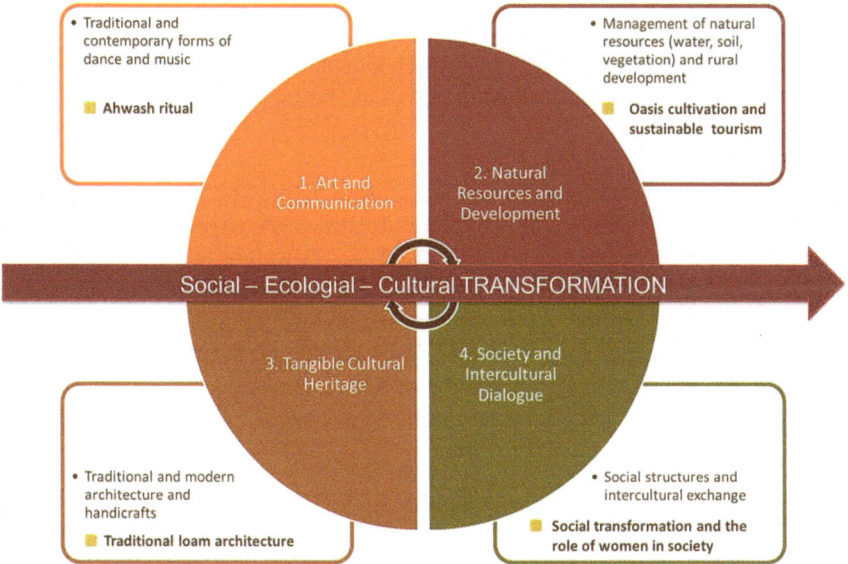

Fig. 15.5 The four workshops of the project

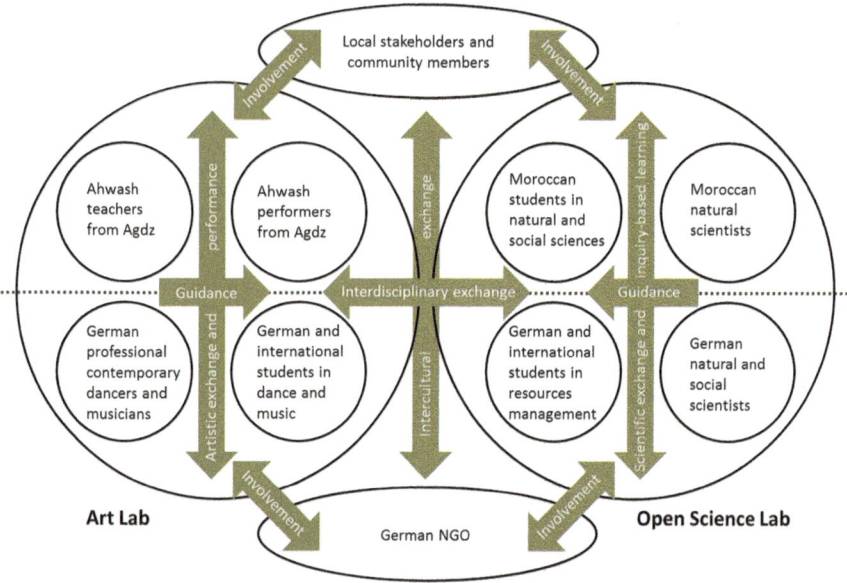

Fig. 15.6 Art Lab and Open Science Lab: activities and interaction

two or more individuals, cultures, or cultural paradigms interact to form "hybrids," which comprise hybrid identities, hybrid ways of thinking, or hybrid cultures. In the context of the project, the complex theoretical foundation and post-colonial discourse of the *Third Space Theory* was translated into an applicable approach by creating space for interaction, exchange, discussion, and reflection within and beyond cultural boundaries to foster the intercultural dialogue.

In the model region of Agdz, an Art Lab and an Open Science Lab were established to conduct artistic and scientific activities and support intercultural and interdisciplinary exchange. Both labs closely interacted with community members, local stakeholders, and the involved NGOs (Fig. 15.6).

The learning and teaching concept was guided by the Sustainable Development Goal 4, Quality education, in particular focusing on target 7, Education for Sustainable Development and Global Citizenship[1] as well as four specific program targets: (a) modernizing higher education teaching, (b) fostering academic junior scientists and gender equality, (c) establishing regional networks for teaching and research, and (d) contributing to social development and the creation of a knowledge society. Against this background we chose constructive alignment (Biggs 1996) as the guiding principle for teaching and learning activities and addressed the two basic concepts behind it as follows:

1. We provided students a deliberate alignment between planned learning activities and intended learning outcomes and defined clear criteria for the feedback process;

---

[1] https://en.unesco.org/education2030-sdg4/targets.

Fig. 15.7 Learning environments (**a**) courtyard of Kasbah Caid Ali; (**b**) oasis of Agdz

2. We supported students in linking their newly acquired knowledge with experiences from their own living environment and cultural background to promote abstraction of basic principles through self-reflection.

Furthermore, we applied the concept of inquiry-based learning, which is considered as a suitable concept to implement teaching and learning approaches within an active transdisciplinary learning environment (Dostál 2015). The concept is based on an explorative way of gaining new knowledge, for instance, by posing questions, analyzing problems, or discussing scenarios (Bell et al. 2010). To foster active learning, intercultural communication, and interdisciplinary exchange, we created spaces for discussion and reflection as well as for artistic performance (Fig. 15.7). Most activities were carried out in small, mixed-gender, and mixed-nationality groups, but there were also meetings and events where all participants came together. Another important element was the provision of "real-world experience" for the students. This was promoted through exchange with the local actors, field visits, surveys, and workshops.

In the Open Science Lab, the scientists acted as moderators and mediators providing the students professional and organizational support. They introduced the theoretical, methodological, and conceptual frameworks related to sustainable development. Fundamental concepts were discussed with the students to foster holistic and critical thinking; these include carrying capacity (*among others* Hardin 1986; Daily and Ehrlich 1992), common pool resources management (*among others* Ostrom 1990), socio-ecological systems (*among others* Berkes et al. 2001, Chapin III et al. 2009, Huber-Sannwald et al. 2012), socio-ecological resilience (among others Adger 2000, Folke 2006), as well as panarchy and adaptive cycles (Gunderson and Holling 2002). The learning outcomes for the Open Science Lab were defined as follows:

1. Analysis of natural, social, and cultural capitals and socio-ecological and cultural transformation processes in the Middle Draa valley and Agdz;
2. Investigation of interrelations between environment, resources, tangible and intangible cultural heritage;
3. Comparison of natural resources management in higher education in Morocco and Germany.

The Art Lab followed a more experience and practice driven approach, where the artists defined themselves as ambassadors for their own discipline and cultural practice and also being a guest and learner in the other project disciplines and cultural practices. Artists made use of individual artistic tools for an intercultural and interdisciplinary dialogue. At the same time, they had to perform in an artistic work setting in a non-familiar cultural context, which was seen as a major inspiration and challenge. The performative and scientific approach from the German artist side included artistic methods like interdisciplinary and transdisciplinary improvisation-techniques and choreographic methods (a more detailed presentation of the artistic approaches and methods would lead too far at this point). An adapted and both-directional use of "bi-musicality" (Hood 1960) was used as an intercultural communication tool. Further methods and tools were the "Tuning in" and "Drift 'n' Dialogue," as described in Chap. 7.

The artists defined their learning objectives as follows:

1. Incorporation of new impulses and influences to the own artistic practice while referring to own traditions;
2. Looking for new intercultural artistic and social practices and forms of nonverbal artistic communication;
3. Developing new methodological approaches, concepts, and perspectives in the light of the individual (academic) background and vita;
4. Learning to change roles from being guest in Morocco and being facilitator in Germany (and vice versa);
5. Incorporating non-artistic approaches from the scientific part to the own artistic practice.

Implementation of the Open Science and Art Labs

The project was launched in April 2016 with a kick-off event in Agdz where the university partners, together with involved stakeholders, jointly discussed the activities and expected outcomes. In each of the three project years, two main events took place, one in Morocco and one in Germany, each consisting of an Art Lab and an Open Science Lab. The main events lasted 8–10 days, which made them very intense, since the artistic and scientific laboratories were not only individually planned and carried out, but also space was planned for interactions between them. In the Art Lab, the main events in Morocco were carried out in the Kasbah Caid Ali in Agdz, where Moroccan *Ahwash* practitioners and German dancers and musicians worked up to an open and transdisciplinary "one group" setting. A key group of *Ahwash* performers was invited to three workshops in Germany, where they got insights into forms of contemporary dance and music, worked together with the German and international artists and students on a common choreography, made studio recordings, and practiced various forms and formats of artistic communication.

In the Open Science Lab, the main events were embedded in the format of a "Joint student project," which allows students from different cultural backgrounds to work in small groups of 4–6 participants on a specific topic. Since there was no master's program with a focus on natural resources management at the Moroccan partner universities, the students were selected from bachelor's programs in natural and social sciences. The joint activities were carried out during the main events and included among others field visits, interviews, and community workshops. In the Art Lab, the problem arose that there were no comparable study programs in Morocco in the field of dance and music at bachelor and master level. For this reason, the music and dance students worked directly with *Ahwash* artists from Agdz.

The research objectives were defined together with all partners at an early stage of the project. On the side of the Open Science Lab, students were asked to develop a conceptual basis and examine individual questions in more detail. The results were presented as reports and short scientific papers. They form the basis for the observed transformation processes presented in Chap. 7. In addition, the students developed a short educational film and contributed to a project exhibition as part of the final workshops in Rabat and Cologne. On side of the Art Lab, the main events formed the basis for the artistic cooperation, communication, and exchange. The artistic activities were accompanied by a filmmaker and photographer, who developed materials, which were partly used for project documentation in class, and which were also essential elements of the scientific–artistic events in Rabat and Cologne. Above all, the Moroccan partners placed particular emphasis on the fact that the final event in Rabat was to reach a wider public and, in this way, also spread the cultural heritage of the *Amazigh*. The core concepts of transformation, transdisciplinarity, and the Third Space were equally addressed in both events.

Finally, a self-evaluation of the scientists, artists, and students involved in the project took place with regard to the implementation of the transdisciplinary approach and the intercultural experiences and learning processes.

Main Outcomes of the Project

Transformation

Agdz and its surrounding villages can be seen as a historically grown socio-ecological system, well adapted to the harsh environmental conditions. The permanent availability of water in the Draa valley was a decisive factor for the sedentarization of the former nomads and semi-nomads and the development of a complex oasis culture. At the same time water was—and still is—a natural capital that is subject to major natural seasonal and long-term fluctuations in terms of its availability. In view of this predicament and the associated uncertainties and risks, the settlers were forced to develop strategies to secure a permanent and equitable division of water.

These challenges were met with technical, social, and agro-ecological developments. Technical measures include the *Khettara* system to bring water into the oasis with as little evaporation loss as possible as well as irrigation systems with *Saguias* (*or Sagias*) to distribute the water inside the oasis. Around the resource water a communal management system was formed, which regulates the water supply, distribution, and hydraulic maintenance. In addition, land ownership, rights of use, and inheritance were also regulated (see Chap. 4). From an agro-ecological point of view, the three strata cultivation system ensured optimal use of water and soil resources, created a comparably humid microclimate, and provided a habitat for insects, birds, and other organisms that perform important ecosystem functions such as pollination or pest control.

House construction was also optimally adapted to the environmental conditions. Locally available resources such as loam, stones, gravel, and plants were used, which were fed into the agro-ecosystem without non-degradable residues during the decay process. The complex social structures of the *Amazigh* and their past as nomads and semi-nomads have created the necessary social cohesion and knowledge about their living environment, which not only ensured survival under barren and unsafe conditions, but also enabled cultural progress. For example, earthen architecture techniques and the art of weaving could develop further, and the *Ahwash* could establish itself as an identity-forming ritual, which strengthens the cohesion of the community and thus contributes to the immediate resolution of conflicts, for example, related to water distribution.

Both from an ecological and social perspective, the oasis can be described as a fertile and resilient island within a hostile environment that is capable of buffering shocks such as droughts. However, due to its location on important trade routes of the caravans (see Chaps. 2 and 3) the system is not self-sufficient or even isolated. Like many rural regions in the world's drylands, Agdz is today confronted with the effects of global change, such as rural exodus and loss of traditions, ecological problems due to overexploitation of natural resources, and the impacts of climate change. However, in contrast to other rural towns in Morocco's hinterland, Agdz is characterized by population increase and economic growth and there are many signs of modernization and change (see Chap. 4). New housing estates, new and extended roads, concreted sewage systems, and an improved energy supply are visible in the urban area. There are also developments in the agricultural sector, such as the increasing use of drip irrigation and the concrete lining of irrigation canals. However, the rural exodus takes place in the small surrounding villages, which are characterized by an aging population and a weak infrastructure. Here the decay of the old earthen buildings is particularly striking.

Now the question arises as to what material and immaterial losses are associated with this modernization process and how they should be assessed. From an architectural and cultural-historical point of view, the decay of ancient *kasbahs* and *ksur* is a severe loss, especially when considering that in addition to the building deterioration, the intangible heritage of the traditional building technique is also lost. However, these earthen buildings have mostly lost their original function, so that there is no cogent incentive to preserve them since modern steel skeleton buildings promise greater living comfort, represent progress, and comply with today's building regulations. Therefore, it is mostly traditionalists, parts of the social upper class, and foreigners, such as the NGO Dindum, who are committed to their preservation. The renovated buildings are then often used for tourism, museums, or sometimes for handicraft cooperatives, which generates income in rural areas.

In the field of agriculture material losses and changes are also noticeable, such as a partial concrete lining of old water channels, a partially more intensive management with drip irrigation, as well as agricultural intensification within the traditional oasis system. Although the oasis system is still largely intact, the question arises how further intensification will affect the availability and distribution of water resources and thus also traditional water management.

Changes also occur in the *Ahwash* ritual. Undoubtedly, the *Ahwash* still has a very great spiritual, community-building and identity-forming significance. However, the ritual is also marketed for tourism and its musical elements are incorporated into popular music. Although this brings money into the rural area and increases awareness, there are also voices that are critical of this commercialization and fear a simplification of the *Ahwash* and thus a decoupling from its original ritual meaning. Of course, the Ahwash in Agdz was also influenced by the encounter with Western artists within the framework of the project. In contrast to the changes brought about by commercialization, however, these changes took place in a participatory approach in which all participants had the choice of whether and to what extent new ideas are integrated into the *Ahwash*.

The examples of earthen architecture, oasis agriculture, and the *Ahwash* ritual reflect the great changes that are currently taking place. It is very clear that the individual cultural elements are closely intertwined and should therefore be seen as part of the socio-ecological system and not isolated. The question now arises as to what impact the outlined changes could have on the future development of the region. The transdisciplinary approach was chosen to better understand these developments and interrelationships.

Transdisciplinarity

The collaboration of scientists from different disciplines with Moroccan and international artists allowed for observations from various viewpoints and the joined development of teaching and learning formats. Through the involvement of local actors and practitioners from the field of earthen architecture, interfaces to the local community and to practice were established. The transdisciplinary approach has made it possible to identify system contexts that would hardly have been recognizable from a purely disciplinary perspective. From this, possible changes in the socio-ecological system could be estimated and possible options for action developed. Some major cause-and-effect relationships and possible scenarios and leveling points are outlined in Table 15.2.

Table 15.2 Cause-and-effect relationships exemplified for earthen architecture, oasis agriculture and the *Ahwash* ritual

Observed change	Impacts on the environment	Physical and socio-economic impacts
Earthen architecture Traditional earthen buildings are increasingly being replaced by modern steel skeleton constructions (Fig. 15.8a); in the past, local clay deposits were used, today large quantities of sand are required for the concrete	Since desert sand is not suitable for the production of concrete and the availability of river sand is very limited, sand from coastal areas is increasingly being used The exploitation of sand on Moroccan coasts has led to the degradation of beaches, dunes, and marine ecosystems (Moussaid et al. 2015) Modern construction methods produce residues (steel, concrete) that are difficult to degrade	Modern design is accompanied by deterioration of thermal comfort (Baglioni 2010) Earthen buildings that are historically significant and attractive to tourism are lost; in this way, the region reduces its tourist potential and thus important sources of income Knowledge about earthen architecture, which represents an important intangible cultural heritage, is lost Scenery for the *Ahwash* ritual is impaired which might affect its spirituality
Oasis agriculture Decreasing number of people working in agriculture; fewer people farming larger areas more intensively (Ait el Kaid; personal communication)	Land use intensification increases pressure on water resources; construction of deeper wells leads to lowering of the groundwater level, so that during dry periods, water scarcity will become more severe Further land use intensification will probably result in a deterioration of the microclimate, soil properties, and agrodiversity According to climate models (Karmaoui et al. 2014, 2016), it is likely that climatic extreme events such as droughts will increase Modernization of the traditional water system (concrete canals, drip irrigation; Fig. 15.8b), leaving behind heavily degradable residues	Traditional knowledge on collective water management is being lost It is likely that agriculture will continue to develop towards intensification and market orientation; this would mean a stronger concentration on a few producers and possibly a stronger involvement of external actors (Ait el Kaid; personal communication) In the worst this development could weaken the community and cause conflicts over the scarce water resource

(continued)

Table 15.2 (continued)

Observed change	Impacts on the environment	Physical and socio-economic impacts
Ahwash ritual Commercialization and modernization	Classic frame drums and tambourines are replaced by western instruments (Fig. 15.8c), which has an impact on resource use in the region and in the countries of origin of the instruments	Commercialization and modernization entails the danger of "despiritualization," so that social functions and thus the cohesion of the community could be weakened Increased use of machine-made instruments will have a direct impact on local instrument making and the tradition of stretching the drumheads over open fire There is a danger that the younger generation will adopt an urban and western lifestyle, and that the *Ahwash* will thus be regarded as outdated; however, the current trend is more in the opposite direction

Fig. 15.8 Changes (**a**) steel skeleton buildings replace earthen architecture; (**b**) new irrigation channel made of concrete; (**c**) new western tambourine replaced handmade frame drum (Photo credits: (**a, b**) U. Nehren, (**c**) M. Maurissens)

Although the socio-ecological transformation processes with all causes, effects, and feedbacks cannot be comprehensively assessed, possible main threats to the stability of the socio-ecological system have been identified:

1. Socio-ecological threats: The overuse of water resources through land use intensification and the simultaneous effects of climate change can lead to water scarcity and cause conflicts over water resources.
2. Socio-cultural threats: Negative impacts of global change in Agdz can be perceived as future obstacles to local culture and social cohesion, which could lead to conflicts.

The Atlas Workshop project discussed with the participating scientists, artists, students, and local stakeholders how to counteract possible tipping points and thus promote sustainable development. The following points were highlighted:

- Strengthening the community and maintaining the traditional common pool resources management to avoid social conflicts;
- Maintaining sustainable oasis agriculture also under agricultural intensification in order to keep the socio-ecological system resilient;
- Implementing a circular economy to achieve a better resource efficiency and to avoid waste problems;
- Promoting sustainable tourism jointly with the local stakeholders in order to mitigate environmental and cultural problems;
- Searching for compromises between traditional and modern construction; as a suggestion, architects at Oldenburg University (Florian Schick, personal communication) propose hybrids that combine steel skeleton construction with earthen architecture;
- The *Ahwash* performers explicitly want to make the *Ahwash* known beyond the borders of the community and thus strengthen it as a cultural asset of the *Amazigh* in Morocco, and at the same time open it for new ideas.

The Third Space

The great diversity of the cultural backgrounds of the participants had a significant impact on the Atlas Workshop project. The Third Space is characterized on the one hand by demarcations and on the other hand by communication and hybridization of cultures. This semi-permeability includes not only a spatial, but also a temporal extension and can be understood as a process with an uncertain course for the participants. The challenge in such an arrangement is to examine the values of one's own culture, to question one's own cultural predisposition, and to integrate it as a participant into a common work. The cultural diversity and the artistic and scientific environment fostered an open exchange and resulted in (a) new methodological approaches, (b) synergies and hybrid artistic paths, and (c) a revitalization and valorization of tangible and intangible heritages.

New Methodological Approaches

The intercultural and transdisciplinary exchange showed the natural and social scientists that music and dance have very different approaches and methods than the ones they are familiar with. The artists invited all the participants to a "Tuning in" in the morning and to "Drift 'n' dialogues" during the day.

- "Tuning in" is a method of starting the day with somatics (according to Hanna 1986, soma is the unity of the conscious and unconscious sensory–motor functions) and/or bodypercussion, a kind of bodywork which forms part of the movement and sound studies and emphasizes on the internal physical perception and experience;

– "Drift 'n' dialogue" guided the participants into a dialogue during a walk in the garden. The practice of walking and thinking takes place simultaneously as a continuum between the dialogue partners. Both the "path of the walk" and the "path of thought" are open, without objective or fixed outcome. Participants are invited to drift and decide intuitively, without knowing exactly which of the two interlocutors will act when, because perhaps a way of thinking and walking only arises through joint negotiations that allow new thoughts beyond the main place of action.

In the artistic field, methods from the natural and social sciences were also taken up. Among other things, artistic reflections on space, society, and interculturality took place, and the performance also addressed the field of natural resources. For example, the percussionist played rhythms on the floor, thus creating an artistic connection to the resource soil.

Intercultural Artistic Experiences and Hybrid Formats

The intercultural exchange and the joint artistic work were perceived by the musicians and dancers from Agdz and Cologne as very inspiring. In addition to the experience gained, hybrid artistic formats were also created through the acquisition of techniques and forms of expression from the other cultural environment (Fig. 15.9). The common artistic path was initially characterized by curiosity, which was followed by an increasing exchange and finally by a common artistic

Fig. 15.9 Ahwash artist from Agdz in dialogue with contemporary dancer (Photo credit: M. Maurissens)

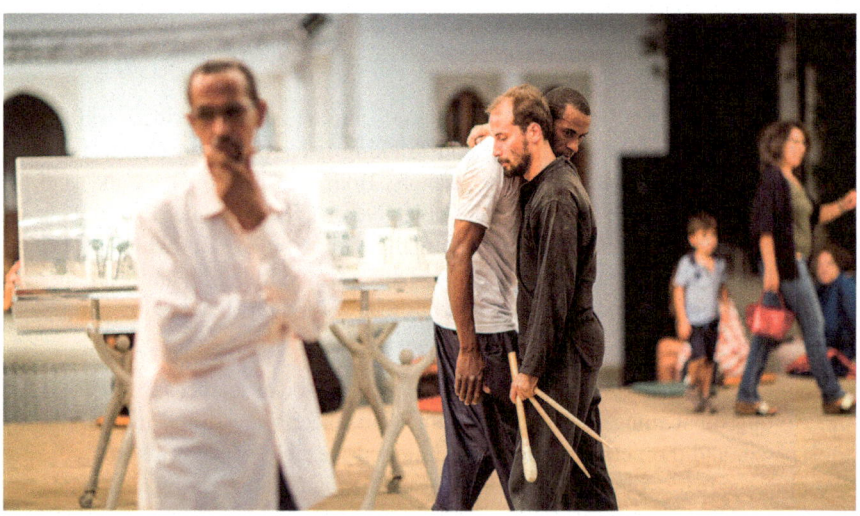

Fig. 15.10 Choreography as part of the final workshop in Rabat (Photo credit: M. Maurissens)

performance. The barriers between artists from different cultures were much smaller than one could have expected when a traditional, popular, and highly sophisticated ritual met a modern, less accessible form of art. Instead of a clash between the *Ahwash* tradition and contemporary dance and music, a continuous flow between the expressive forms could be observed.

The scientific–artistic outcome of this intercultural and transdisciplinary encounter was a jointly developed choreography for the two closing events in Rabat and Cologne, where transformation, transdisciplinarity, and the Third Space were staged in the form of an artistic performance and an interactive parkours (Fig. 15.10).

Revitalization and Valorization of the Ahwash

The intercultural, joint, and equal artistic work led to a strengthening of the *Ahwash* by recognizing it as an art form and thus giving this tradition an appreciation as an important cultural heritage of the *Amazigh*. During the 3 years of the project all artistic workshops in Agdz took place in the inner courtyard of the Kasbah Caid Ali in an environment of traditional earthen architecture. The interplay of space and performance was especially emphasized by the international participants of the project as very inspiring. This also enhanced the environment as a stage for the *Ahwash*. Initiated by the artistic exchange, the *Ahwash* teachers from Agdz introduced an *Ahwash* for children with an annual festival. Thus, the *Ahwash* received a further recognition and anchorage in the school education.

Conclusions and the Way Forward

Global change has caused rapid and profound changes in the rural areas of southern Morocco. Understanding these changes requires a transdisciplinary approach and new forms of learning, in which the acquisition of experiences in a "real-world environment" and the intercultural exchange are of particular importance. The Atlas Workshops project has met these challenges and directly contributed to the internationalization of the participating universities. It even went beyond this by implementing the concept of transdisciplinarity with the involvement of communities and local stakeholders, a concept that hardly exists in traditional university programs. At the Moroccan partner universities, research and teaching take place primarily in classical disciplines, while the potential for interdisciplinary activities and modern formats such as inquiry-based learning is not yet fully tapped. This way the project has triggered important impulses for a better understanding of complex socio-ecological contexts, including the relationships between resources, culture, and art. In the field of music and dance there are no study programs in Morocco yet. The *Ahwash* performer Boubker Marrouk was the first to receive an official certificate as an *Ahwash* teacher from the Polydisciplinary Faculty in Ouarzazate. This could be a first step towards anchoring *Ahwash* as a cultural asset and art form at university level.

The approach was very positively received by the students, although there were also points of friction, uncertainties, and obstacles to overcome. However, these were part of the learning process and therefore beneficial. In the labs in Agdz, German and international students and scientists were given a direct insight into *Amazigh* culture and oasis management, while in Cologne, the Moroccan participants were given insights into the German higher education system, teaching concepts and contemporary dance and music. In this way a mutual appreciation, learning, and understanding developed. The intercultural communication was significantly promoted by music and movement, whereby the outwardly open and inviting *Ahwash* ritual contributed in particular to overcoming cultural and language barriers.

Challenges to consider in the future are: More time is needed for field work and artistic rehearsals. The short-joined working phases created pressure and left too little room for reflection. Moreover, science–art interfaces need to be further strengthened. In most cases, the students of resource management and those of dance and music worked separately in their labs. In the joint presence phases, the art students reflected on the socio-ecological and cultural environments in which their art is performed; however, there were few interfaces with students from natural resources management in practical work. On the other hand, most students from natural and social sciences found access to contemporary dance and music difficult, while the *Ahwash* was perceived as open and integrative. Furthermore, the social sciences were too weakly represented in the project, so that not all potentials could be exploited. Another important point is the cooperation between the German and Moroccan partners. Cooperation is still characterized by a dominance of Western approaches. This "traditional" imbalance cannot be overcome in a short-term project, but the project has sought to achieve a better balance. An important aspect is certainly project funding from the German side, which is accompanied by certain framework conditions. In the course of the project, some of these conditions were changed, resulting in a better inclusion of local stakeholders to enable the transdisciplinary approach. This can give impetus to work towards an even better balance in the future.

Lastly, it can be said that the intercultural framework of the project, coupled with a respectful and appreciative exchange, has created a space of trust, curiosity, and mutual inspiration, and that the inclusion of music and dance as an integral part of the overall transdisciplinary project environment offered excellent opportunities for eye-to-eye encounters and nonverbal communication to overcome language and cultural barriers while at the same time opening a space for a different kind of interculturality and hybridity.

Acknowledgements We would like to express our sincerest thanks to the German Academic Exchange Service (DAAD) and the German Federal Foreign Office for funding the Atlas Workshops project. Furthermore, we are very grateful to Mr. M'Barek Ait El Kaid for providing us the space in his Kasbah and his valuable time for scientific discussion and practical support. Our heartfelt thanks go to the amazing Ahwash ensemble from Agdz, particularly to Boubker Marrouk, Mohamed Belahcen, and Abdessamad Belahcen, as well as to Prof. Vera Sander who played a decisive role in shaping the artistic part of the project. Furthermore, we would like to thank Prof. Dr. Youness Belahsen and Prof. Dr. Ahmed Ait Hou for their great support. We further thank the community of Agdz with the associations Assalam and the German NGO Dindum who supported the project activities. We are also very grateful to the Villa des Arts, the association La Paix/Salaam, and the Goethe-Institut for supporting the closing event in Rabat. Finally, we would like to thank Michael Maurissens (artist and film maker), Nezha Rhondali (dancer), Tom Dams (sound engineer), Justyna Niznik (violin player), Manfred Fahnert, Marc Haering, Jana Nolting, Florian Schick, Roman Seliger (scientific support), Belaid Zouhair (logistic support), and the students from science and arts: Oumaima Ait El Kaid, Francesco d'Amelio, Elsa Artmann, Margherita Dello Sbarba, Houda Dribila, Silvia Enis Duarte, Mohamed El-Hodaiby, Maximiliano Estudios, Mariana Garcia de la Torre, Leonie Gembler, Katharina Gübel, Diego Alexander Guarin Cifuentes, Karina Klein, Oumaima Kharkhach, Mihyun Ko, Linda Madio Ngankeng Epse Kemeni, Wyam Mardoukh, Alaa Obeid, Brahim Oumahrir, Maria Isabel Meza Rodríguez, Raphael Pak Lau, Anne-Lene Nöldner, Kenechukwu Albert Okoye, Gonzalo Rodriguez, Jannis Sicker, Marcella Ulrike Sobisch, Svenja Speen, Thea Soti, Seyedeh Taraneh Mousavi, José Carlos Tello Valle Hiriart, Dimitrios Thanos, Yannick Tiemann, and Moritz Wesp.

References

Adeel Z, Bigas H, Schuster B (eds) (2008) What makes traditional technologies tick? A review of traditional approaches for water management in drylands. The United Nations University, Hamilton

Adger N (2000) Social and ecological resilience: are they related? Prog Hum Geogr 24:347–364

Almasude E (2014) Amazighité and secularism: rethinking religious-secular divisions in the Amazigh political imagination. Decolonization: Indigeneity, Educ Soc 3(2):131–151

Baglioni E (2010) Sustainable Vernacular Architecture: the Case of the Draa Valley Ksur (Morocco). In: Proceedings: Sustainable Architecture and Urban Development, 12–14 July, Amman, Jordan, vol IV. pp 227–243

Baglioni E, Rovero L, Tonietti U (2015) Draa valley earthen architecture: construction techniques, pathology and intervention criteria. Conference paper, Restapia Congreso Internacional sobre Restauración de Tapia, Vol. 'Rammed Earth Conservation', Valencia, Spain

Belahsen R, Naciri K, El Ibrahimi A (2017) Food security and women's roles in Moroccan Berber (Amazigh) society today. Mater Child Nutr 13(Suppl 3):e12562. https://doi.org/10.1111/mcn.12562

Bell T, Urhahne D, Schanze S, Ploetzner R (2010) Collaborative inquiry learning: models, tools, and challenges. Int J Sci Educ 3(1):349–377

Berkes F, Colding J, Folke C (2001) Linking social-ecological systems. Cambridge University Press, Cambridge

Bhabha HK (1994) The location of culture. Routledge, New York

Biggs J (1996) Enhancing teaching through constructive alignment. High Educ 32:347–364

Boudraa N, Krause J (2009) North African mosaic: a cultural reappraisal of ethnic and religious minorities. Cambridge Scholars Publishing, Newcastle-upon-Tyne

Cardinale BJ, Duffy JE, Gonzalez A et al (2012) Biodiversity loss and its impact on humanity. Nature 486(7401):59–67

Chapin FS III, Kofinas GP, Folke C (eds) (2009) Principles of ecosystem stewardship -resilience-based natural resource management in a changing world. Springer Science & Business Media, Berlin

Daily GC, Ehrlich PR (1992) Population, sustainability, and earth's carrying capacity: a framework for estimating population sizes and lifestyles that could be sustained without undermining future generations. Bioscience 42(10):761–771

Dostál J (2015) Inquiry-based instruction: concept, essence, importance and contribution. Palacký University, Olomouc

Folke C (2006) Resilience: the emergence of a perspective for social-ecological systems analysis. Glob Environ Chang 16:253–267

Ennaji M (1997) The sociology of Berber: change and continuity. Int J Sociol Lang 123:23–40

Gellner E (1970) Saints of the atlas. The Trinity Press, Worcester

Graf K (2010) Drinking water supply in the Middle Draa Valley, South Morocco - options for action in the context of water scarcity and institutional constraints. Kölner Ethnologische Beiträge, Köln, p 34

Gunderson LH, Holling CS (2002) Panarchy: understanding transformations in human and natural systems. Island Press, Washington

Hanna T (1986) What is somatics? J Behav Optom 2(2):31–35

Hardin G (1986) Cultural carrying capacity: a biological approach to human problems. Bioscience 36:599–606

Haut-Commissariat au Plan (2019) Population légale des régions, provinces, préfectures, municipalités, arrondissements et communes du Royaume d'après les résultats du RGPH 2014, Rabat. Last Accessed 10 Jan 2019

Heidecke C (2009) Economic analysis of water use and management in the Middle Draa valley in Morocco. Dissertation, University of Bonn, Germany

Heidecke C, Heckelei T (2010) Impacts of changing water inflow distributions on irrigation and farm income along the Drâa River in Morocco. Agric Econ 41(2):135–149

Hoffman KE (2008) We share walls. Language, land, and gender in Berber Morocco. Blackwell Publishing, Malden

Hood M (1960) The challenge of bi-musicality. Ethnomusicology 4(2):55–59

Elfasi M, Hrbek I (eds) (1988) General history of Africa III, Africa from the seventh to the eleventh century. UNESCO Publishing, Paris

Huber-Sannwald E, Ribeiro Palacios M, Arredondo Moreno JT et al (2012) Navigating challenges and opportunities of land degradation and sustainable livelihood development in dryland social-ecological systems: a case study from Mexico. Philos Trans R Soc Lond Ser B Biol Sci 367(1606):3158–3177

Ilahiane H (2006) Historical dictionary of the Berbers (Imazighen). Rowman & Littlefield, Lanham

IPCC (2014) Climate change 2014: impacts, adaptation, and vulnerability. Part A: global and sectoral aspects. In: Contribution of Working Group II to the Fifth Assessment Report of the intergovernmental panel on climate change. Cambridge University Press, Cambridge

Jay C (2016) Playing the 'Berber': the performance of Amazigh identities in contemporary Morocco. J North Afr Stud 21(1):68–80

Johannsen IM, Hengst JC, Goll A et al (2016) Future of water supply and demand in the Middle Draa Valley, Morocco, under climate and land use change. Water 8(8):313

Karmaoui A, Messouli M, Yacoubi Khebiza M, Ifaadassan I (2014) Environmental vulnerability to climate change and anthropogenic impacts in dryland (Pilot Study: Middle Draa Valley, South Morocco). J Earth Sci Clim Chan S11:1

Karmaoui A, Ifaadassan I, Babqiqi A et al (2016) Analysis of the water supply-demand relationship in the Middle Draa Valley, Morocco, under climate change and socio-economic scenarios. J Sci Res Rep 9:1–10

Keenan J (2002) Tourism, development and conservation: a Saharan perspective. In: Mattingly DJ et al (eds) Natural Resources and Cultural Heritage of the Libyan Desert: Proceedings of a Conference held in Libya 14–21

Lafkioui, MB (2018) Berber languages and linguistics. Oxord bibliographies. https://doi.org/10.1093/OBO/9780199772810-0219

Maddy-Weitzman B (2001) Contested identities: Berbers, 'Berberism' and the state in North Africa. J North Afr Stud 6(3):23–47

Maddy-Weitzman B (2006) Ethno-politics and globalisation in North Africa: the Berber culture movement. J North Afr Stud 11(1):71–84

Maddy-Weitzman B (2011) The Berber identity movement and the challenge to North African states, 1st edn. University of Texas Press, Austin

Mileto C, Vegas F, Cristini V (eds) (2012) Rammed earth conservation. International Conference on Rammed Earth Conservation; Restapia. CRC Press, Boca Raton

MA (2005) Millennium ecosystem assessment MEA - ecosystems and human well-being: synthesis. Island Press, Washington

Montagne R (1973) The Berbers: their social and political organisation. T. & A. Constable LTD, Edinburgh

Moussaid J, Fora AA, Zourarah B et al (2015) Using automatic computation to analyze the rate of shoreline change on the Kenitra coast, Morocco. Ocean Eng 102:71–77

Ostrom E (1990) Governing the commons. The evolution of institutions for collective action. Cambridge University Press, Cambridge

Rademacher-Schulz C (2014) The making of the social order – migration, resource and power conflicts in the Moroccan Draa Valley. Erdkunde 68(3):173–183

Rockström J, Steffen W, Noone K et al (2009) Planetary boundaries: exploring the safe operating space for humanity. Ecol Soc 14(2):32

Sadiqi F (2007) The role of Moroccan women in preserving Amazigh language and culture. Mus Int 59(4):26–33

Schuyler PD (1979) Rwais and ahwash: opposing tendencies in Moroccan Berber music and society. World Music 21(1):65–80

Shoup JA (2011) Ethnic groups of Africa and the Middle East: an encyclopedia. ABC-CLIO, Santa Barbara

Soja EW (1996) Third space: journeys to Los Angeles and other real-and-imagined places. Blackwell, Malden

United Nations Convention to Combat Desertification (2017) The global land outlook, 1st edn. UNCCD, Bonn

United Nations, Department of Economic and Social Affairs, Population Division (2014) World urbanization prospects: the 2014 revision, highlights (ST/ESA/SER.A/352), New York

United Nations, Department of Economic and Social Affairs, Population Division (2018) World urbanization prospects: the 2018 revision, methodology, working paper No. ESA/P/WP.252, New York

Vorlaufer K (1999) Tourismus und Kulturwandel auf Bali. Geogr Z 87(1):29–45

Part IV
The Governance of Drylands

Chapter 16
Drylands, Aridification, and Land Governance in Latin America: A Regional Geospatial Perspective

E. Nickl, M. Millones, B. Parmentier, S. Lucatello, and A. Trejo

Abstract In this chapter, we evaluate whether drylands have expanded and become increasingly arid over time using geographic information systems (GIS) and climate data across 33 Latin American country boundaries and across two property regime types in Mexico. In all cases, we evaluate and identify changes using the United Nations Environment Programme (UNEP) Aridity Index (AI) that measures the annual ratio of potential evapotranspiration to precipitation. Annual fields of land surface precipitation and air temperature data are extracted from 1960 to 2017. We also compare aridity index summary statistics and trends between communal and non-communal land property regimes in Mexico as a proxy of communal environmental governance. Our results show that: (1) with some exceptions, most Latin American countries have experienced aridification from the 1961–1990 period to the 1991–2017 period, and the trend is for this pattern to continue; (2) in Mexico, communal lands experience slightly lower levels of aridification than non-communal lands. However, because the difference in aridity index values between periods is very small, the degree to which that difference is significant needs further research. If this difference were significant, it mean that communal land holders are at lower risk than non-communal land holders. At this point, however, we cannot claim that land regime practices have or have had any connection with the process of aridification itself.

E. Nickl
University of Delaware, Newark, DE, USA

M. Millones (✉)
Department of Geography, University of Mary Washington, Fredericksburg, VA, USA
e-mail: mmillone@umw.edu

B. Parmentier
SESYNC, Annapolis, MD, USA

S. Lucatello
Estudios Ambientales y Territoriales, Instituto Mora, Mexico City, Mexico

A. Trejo
Universidad Nacional Autónoma de México, Mexico City, Mexico

© Springer Nature Switzerland AG 2020   281
S. Lucatello et al. (eds.), *Stewardship of Future Drylands and Climate Change in the Global South*, Springer Climate,
https://doi.org/10.1007/978-3-030-22464-6_16

Keywords Aridification · Environmental governance · Geospatial analysis · Communal and non-communal lands

Introduction

Aridification of drylands and the conversion of landscapes into increasingly dry ones is a problem identified by the scientific and international policy community as well as the sustainable development goals. Societal responses to aridification, just like many other environmental change processes, are mediated by the institutional arrangements present in different geographic/historic contexts, and at different geographic scales. Nation-state boundaries, when used as a broad proxy variable for historical/and political trajectories, can provide insights about the interaction between countrywide institutions and environmental changes, such as aridification. Within subnational scales, analyzing aridification through the lens of land based institutions such as land tenure regimes, can provide further detailed insights. For example, we can explore the allocation lands (i.e., who received drier lands) and speculate about which type of property regimes was and will be most affected by aridification. In addition, analyzing land tenure regimes can potentially hint at links between resource management approaches (e.g., common property regimes) and the observed patterns and trends of aridification.

In this exploratory study, we address the degree to which drylands have increased and become increasingly arid over time using geographic information systems (GIS) and climate data across 33 Latin American country boundaries. In addition, we evaluate aridity trends across two property regime types in Mexico. In order to undertake these tasks, we assess and identify changes using the United Nations Environment Programme (UNEP) Aridity Index (AI) that measures the annual ratio of potential evapotranspiration to precipitation. Annual fields of land surface precipitation and air temperature data are extracted from 1960 to 2017 over a 0.5° grid resolution provided by the University of Delaware. In addition, we compare AI summary statistics and trends between communal and non-communal land property regimes in Mexico as a proxy of communal environmental governance.

Aridity Index (AI) in Latin America and Drylands Definitions

Recent research about aridification based on predictions of different climate models suggests that the aridity index (AI) is projected to decrease in the next coming decades. Counterintuitively, this decrease in AI means that most areas of the world—including Latin America as whole—will become drier as a consequence of anthropogenic climate change (Huang et al. 2016). These conditions would make desertification worse and make already vulnerable populations even more vulnerable.

Different aridity indices have been used to define drylands. Most of them are based on precipitation and temperature. In 1977 the United Nations Environment Programme (UNEP) adopted an aridity index (AI) based on the ratio between average annual precipitation and potential evapotranspiration (UNEP Index). This definition conceptualizes drylands as lands with an AI of less than 0.65. Using this index, drylands are further divided into hyper-arid lands, arid lands, semi-arid lands, and dry-subhumid lands. AI was used to prepare the Map of World Distribution of Arid Regions (UNESCO 1979) and the Map of Drylands by the UN Environment World Conservation Monitoring Centre based on the UNEP Index produced in 1991 (UNEP-WCMC 2007). Penman's formula which is based on solar radiation, atmospheric humidity, and wind was used to estimate potential evapotranspiration. In addition to global efforts to define arid regions, several region-specific climate classifications that include arid categories have been devised. For example, in Mexico the Koppen Classification modified by García (2004) is commonly used.

Environmental Governance, National Policies, and Land Property Regimes

Environmental governance has become an important analytical category for understanding the institutional and planning systems through which natural resources are accessed and managed in Latin America. According to Paavola (2007) environmental governance is best understood as the establishment, reaffirmation, or change of institutions to resolve conflicts over environmental resources. Although influenced by international agreements and pressures, national state governments are ultimately responsible and sovereign in the adoption and implementation of the array of policies available for the use and conservation of natural resources, as well as the measures to prevent, mitigate, and adapt to environmental change and natural disasters. It is true that many countries in Latin America are large enough to host multiple environmental contexts (e.g., ecoregions). It is also true that many countries in the region hold forms of government that allow for differing degrees of regional independence (e.g., Mexico or Brazil). Both of these realities could and do sometimes result in differentiated institutional arrangements of environmental governance within the boundaries of each country.

However, the power held by central governments in establishing environmental and land policy, as well as in empowering or weakening related institutions and norms, is consistently superior to those at regional scales, and arguably, this power is rarely contested. This is why we believe that for a continent-wide comparative exploration of aridity trends and patterns, a country boundary is a reasonable unit of analysis (Assies 2008).

Beyond subnational governments (states or regions), within country differentiation of environmental governance institutions and their potential associated outcomes can be analyzed through sectoral lenses. Existing forms of environmental governance are based on organized forms of land control such as in the case of forests (forests governance within local communities) or social movements against multinational exploitation and land control and/or institutional re-scalings based on huge development projects (such as public or private infrastructures). However, state strategies to govern environmental resources and reorder land spaces at local levels are commonly embedded within practices of both farmers and state actors. The interplay between the two often produces outputs where the re-shaping of the environment is constitutive of transformation processes at local level. (Barnes 2014) Actual places, peoples, and practices interact to make up the new landscape and new tenure regimes. In this chapter, we will focus on land policy and regime as proxy for institutional arrangements of environmental government. We will analyze communal lands in Mexico as a case study.

The Case of Communal Lands in Mexico

Land tenure and tenure regimes in Mexico date back to Spanish conquest in the fifteenth century and their evolution can be analyzed in the context of changing power relations and accumulation regimes over the Mexican economic and political history. From the economy of the colonial power to the *hacienda* system of late nineteenth century, several patterns of land tenure emerged in the country. During the second half of the nineteenth century, under an increasingly agro-export oriented economy, a liberal legislation of tenure has dominated the use of land in Mexico and by the early twentieth century, during the Mexican Revolution, a process of substantial redistribution of land was put in place (Assies 2008). In contrast to most other Latin American countries, the agrarian sector was assigned a privileged role in this new model and received extensive government support. Land redistribution peaked under the Cárdenas government (1940s), which consolidated the *ejido* as a form of tenure and experimented with collective *ejidos*.

The Mexican economic crisis of the 1980 and 1990s brought a new wave of liberal and neoliberal reforms that affected land tenure. The opening up of the economy, the NAFTA agreement and aggressive trade policies, reshaped the former land tenure system by opening the way to privatization of lands in the country. Under the Salinas government (88–94), many fundamental legal restrictions on ejidos were lifted. The legal reforms allowed to empower the owners of ejidos to change their tenure regime to private property (*dominio pleno*). The Ejidos vary considerably across the different regions of Mexico (Haenn 2006) and it is difficult to generalize given this heterogeneity.

Land units in Mexico are commonly known as *ejidos*. The *ejidal* lands are defined by the Agrarian law into: (a) Lands for human settlement; they integrate the area for the development of community life of the *ejido*. They are made up of the land in urban zones and where legal estates both private and public are located;

(b) lands of common use. They constitute the economic backbone of the community life within the *ejido*. They represent that area that has not been assigned by the *ejido* assembly for human settlement or plots. It can be defined also as public land, though there are several variations of "common use" within the *ejidos*; and (c) parceled land. These are lands owned by individual *ejidatarios*, or several *ejidatarios* as a whole, who have the right to use, usufruct, and sell the land (Jones and Ward 1998).

Given the array of land tenure regimes in Mexico, it can be argued that these regimes are an implicit form of environmental governance at local level. This is because Mexican states have fewer powers related to land use than the national government and the country's constitution itself specifies that municipalities are the planning authorities. They can decide on land use as long as they take other constitutional provisions and guidelines of higher levels of government into account. To do so, they rely upon several planning instruments which goes from developing land-use environmental and economic plans that control land-use changes and decide, for example, whether or not to issue building permits or allow the construction of landfills. Exceptions are limited to the mining and water extraction sectors that are currently regulated by the national governments. Municipalities are also responsible for land administration within their jurisdiction. Furthermore, they can set property taxes and are responsible for the provision of public services and infrastructure (Barnes 2014).

The relationship between the spatial patterns and the temporal trends of aridity and the two land property regimes analyzed in Mexico can be interpreted in at least two ways. First, in terms of the allocation of land during the key historical moments of land redistribution described above. In other words: were the majority of lands that were allocated to communal lands more arid than those allocated to non-communal lands? Second, the relationship could be interpreted in terms of outcomes of differentiated land governance institutional arrangements. In other words: are communal property regimes better at reducing aridification or lands? The environmental outcomes of communal land regimes in general have been analyzed extensively in the literature. Yet there is still no consensus on whether these outcomes are better than those of other land regime arrangements (Ojanen et al. 2017). Most research in Mexico has focused on deforestation. Less systematic research exists on the environmental outcomes and realities of common property institutions in arid lands.

In this chapter we first calculated and mapped the aridity index, its changes and trends based on climate data for two different time periods (1961–1990 and 1991–2017) for the 33 Latin American country members of the United Nations. We then analyzed the difference of aridity patterns between common property lands and non-common property lands in Mexico.

Data and Methods

Estimation of Aridity Index for Latin America, Changes, and Trends

For this research, an aridity index was estimated using the UNEP Index P/PET, where P is the annual average precipitation and PET is the annual potential evapotranspiration. Gridded precipitation and surface air temperature datasets were obtained from the University of Delaware at 0.5° grid resolution for the period 1960–2017. Surface air temperature was used to estimate potential evapotranspiration. Potential evapotranspiration was estimated based on the Thornthwaite method, which relates PET with temperature. For regions with average temperatures less than 0 °C ETP is zero. For regions with average annual temperatures greater than zero and lower than 26.5 °C, the ETP is 0.16× temperature. For regions with annual temperatures greater than 26.5 °C the PET is −415.85 + 32.24 T −0.43 T2.

In order to assess changes in drylands, wetlands, and to analyze possible extent growth of dry areas in the last years, averages of annual precipitation and temperature were estimated for two periods: 1961–1990 and 1991–2017. Aridity indices were calculated based on these two period averages and the differences (or changes) of aridity were estimated. In addition, annual aridity indices were estimated for the full study period 1961–2017 and a simple linear regression model was applied. This model allowed us to measure the temporal trends of aridity index for the whole 56-year data time series as well as their associated spatial patterns. Figure 16.1 shows the aridity index map for the period 1961–1990. Most areas of Argentina are semi-arid and arid and most of central eastern area of Brazil is semi-arid. The western regions of Peru, Ecuador, and Chile show more areas of hyper-arid indices, especially in Chile. Mexico has semi-arid, arid, and small areas of hyper-arid indices; the drier areas are in Baja California Peninsula.

Figure 16.2 exhibits the aridity index map for the period 1991–2017. Arid regions in Argentina, and in western areas of Peru, Ecuador, and Chile show similar and lower aridity indices compared to period 1961–1990. In Mexico, there are more areas of Arid indices, especially for the central region of the country.

Changes of aridity indices areas were calculated by subtracting first the image of the 1961–1990 period (Fig. 16.1) minus the image corresponding to the 1991–2017 period (Fig. 16.2). In addition a linear regression model was applied to estimate the trend (slope of the model) in aridification for each grid cell. Resulting maps can be seen in the results section of this chapter.

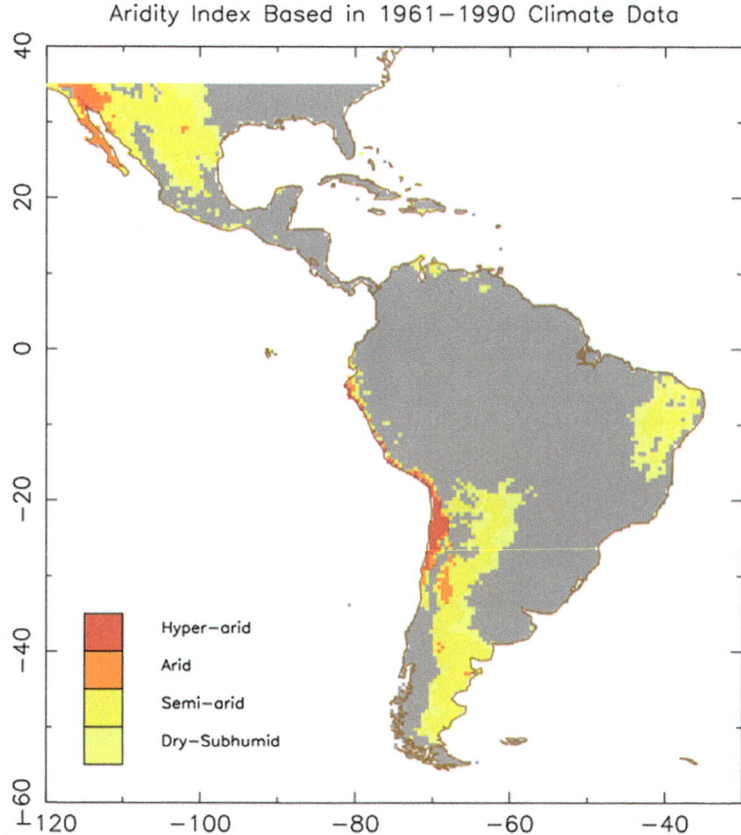

Fig. 16.1 Aridity index for Latin America based on 1961–1990 climate data

Communal Property Lands in Mexico

Communal land areas for Mexico were obtained from the Registro Agrario Nacional (RAN) which is part of the Secretaria de Desarrollo Agrario, Territorial y Urbano (SEDATU). These data consisted of two different GIS files in shapefile format. The first file consists of polygons that denote the perimeters of certified agrarian nuclei or *ejidos* (*perimetrales de nucleos agrarios certificados*). In this chapter we refer to these polygons as RAN. The second file consists of polygons within the RAN polygons that depict lands that are used for productive communal land uses within the *ejidos*. In this chapter, we call these second type of polygons TUC (*Tierra de uso comun*). While RAN polygons include the population centers (human settlement built up lands) as well as parceled land for "private use" within the *ejidos*, TUC polygons/lands do not. TUC lands constitute the economic source of communal life in each *ejido*. We used the most updated version (2018) of both TUC and RAN lands (polygons), since we are interested on the status of those communities in the

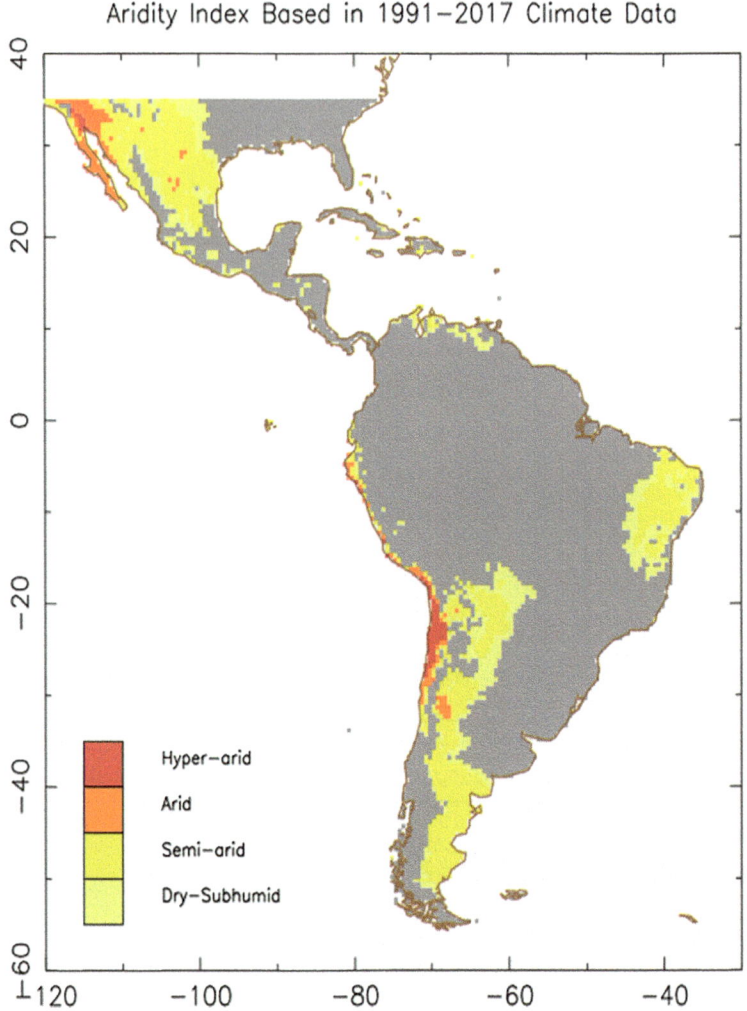

Fig. 16.2 Aridity index for Latin America based on 1991–2017 climate data

present, irrespective of when they were formalized. The idea behind the use of both types of polygons was to have an expansive or liberal definition of communal *ejido* lands (in the case of RAN), and a more restrictive/conservative definition of communal lands (TUC).

In order to extract summary aridity metrics, and assign them to each Latin American country, we overlaid the aridity maps (e.g., Figs. 16.1 and 16.2 coverted into XY points) over the country polygons, and extracted each characteristic via a point-in-polygon procedure. This resulted in summary tables by country (33 observations in total) which contained all aridity measures calculated: AI for each of the two multi-year perdiods, AI change or difference between the two periods, and AI 1991–2017 trends.

The multiple land categories of the original TUC and RAN polygons were reclassified into two: communal and non-communal lands, and then merged with the boundary outline of Mexico to create two binary polygon files, each with two categories. The two potential outcomes for each file were COMUNAL_RAN or REST for RAN *ejido* polygons; and COMMUNAL_TUC or REST for TUC communal lands within *ejidos*. Lands that did not fall under either of those categories were given the category REST. REST includes lands that are owned by either private landholders, the federal, or the state governments Fig. 16.3). The procedure to combine TUC and RAN polygons with the aridity maps measurements was similar to the one described above. A point-in-polygon operation resulted in summary tables that aggregated the 1464 aridity measures corresponding to Mexico into RAN_Communal, RAN_REST, TUC, Communal, and REST (see tables in Results section).

In addition, to make the interpretation of the aridity indices more intuitive, we classified index values into the categories shown in Figs. 16.1, 16.2, and 16.3, which were guided by the following ranges in AI:

Hyper-aridity: ≤ 0.05
Arid: 0.05–0.2
Sub-arid: 0.2–0.5
Subhumid: 0.5–0.65
Not Applicable (NA) ≥ 0.65

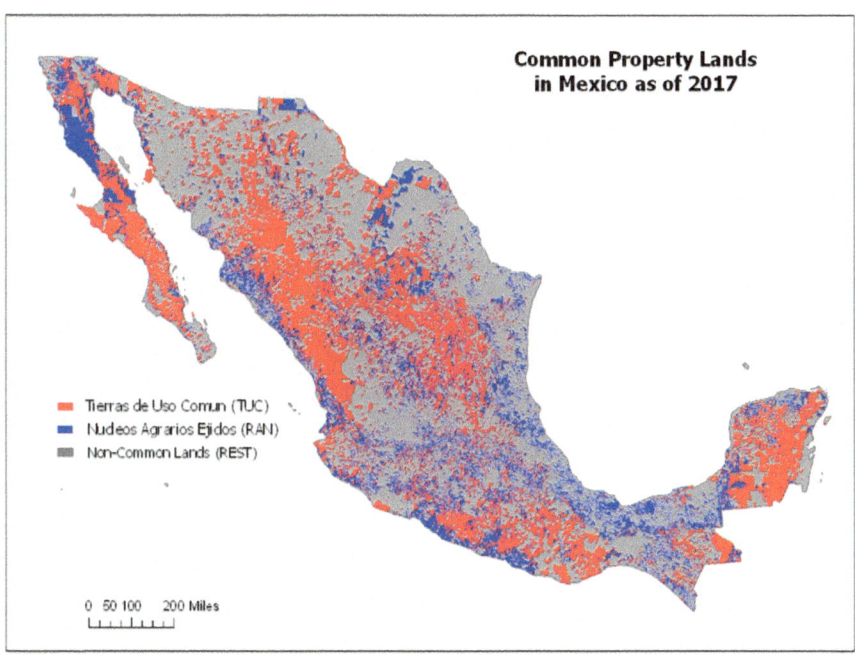

Fig. 16.3 Common property lands in Mexico: tierras de uso común (TUC) in red, and núcleos agrarios—*Ejidos* (RAN) in blue and red. RAN areas include TUC areas

NA stands for not applicable since it is not technically a category within the aridity spectrum (i.e., NA ranges are all in the humid spectrum), but it is important to note that our analysis shows locations that transitioned from NA into the arid spectrum within the analyzed time period of 1961–2017.

Results

Aridity Indices, Change/Difference, and Trends in Latin America Between 1961 and 2017

Figure 16.4 compares the aridity indices between the periods 1960–1991 and 1991–2017. Negative values (red colors) show a decrease in aridity index (drier conditions), while positive values (blue colors) indicate an increase in aridity index (wetter conditions). Most of arid regions shown in Figs. 16.1 and 16.2 remain the same (gray colors) or present lower aridity conditions (blue colors). Such is the case in Argentina, western areas of Ecuador, Peru, and Chile. In Mexico drying up is evident in most areas. When we observe the rest of the areas (not considered arid regions), we can see negative values (red colors) in most of the regions, which means a decrease in aridity index or aridification. Although these regions are not considered arid zones, they could convert in arid areas with time.

Aridity Indices, Change/Difference, and Trends by Country Between 1961 and 2017

The results for the aridity index averages for Latin America shown in the maps above (Figs. 16.4 and 16.5) are summarized by country and presented in the Appendix Table 16.3. The tables show a consistent pattern. All but three countries (Guyana, Peru, and Uruguay) show on average a decrease in the aridity index, a negative change difference between the two time periods analyzed (1961–1990 to 1991–2017), and a negative trend sign (the slope of a linear regression for the time series from 1961 to 2017). This means that overall the Latin American region is getting drier. The decrease in the index is the largest in El Salvador, Nicaragua, and Mexico. However, it must be noted that if we disaggregate trends geographically by regions within each country, there are areas that go counter the national average (see Figs. 16.4 and 16.5). For example, the south region of Chile presents a moderate increase of aridity. Most of Argentina has a slight decrease of aridity. We can observe slight to moderate increase of aridity in most regions of Brazil, Colombia, Ecuador, and Venezuela. In Peru, there are wetter conditions along the central and south coast, while the Central and Southern Andes show drier conditions.

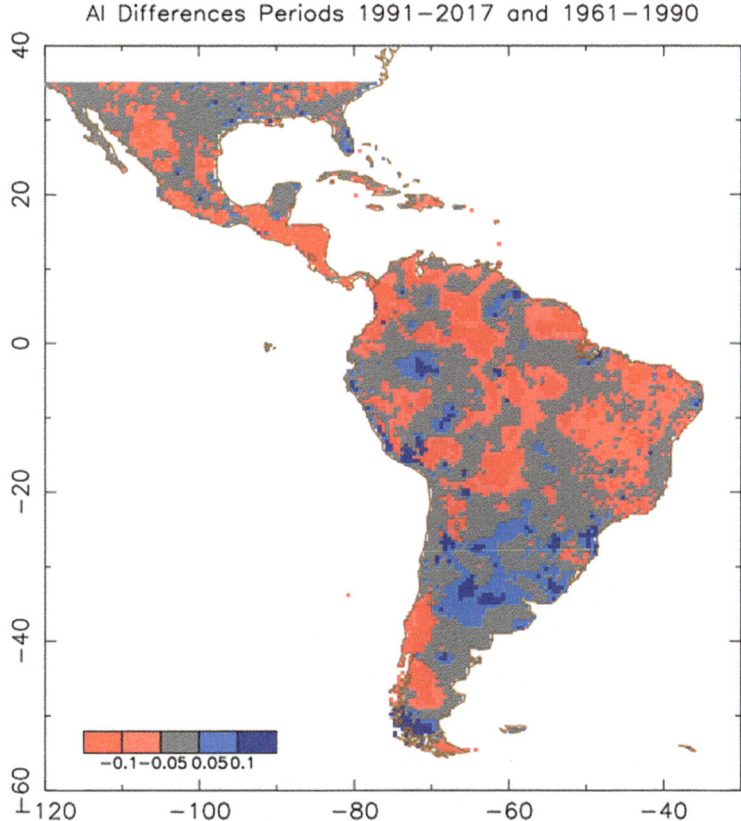

Fig. 16.4 Map of change/differences of aridity index between periods 1991–2017 and 1960–1991

Furthermore, if we analyze the results (in Table 16.1) by the ordinal subcategories of aridity: hyper-arid, arid, semi-arid, dry-subhumid, and NA (representing all humid categories) we see a more complex story. The dry spectrum of the categories in the aridity scale such as arid and hyper-arid remains more or less stable or even increases (becomes more humid) in some cases (for example, in Chile, Mexico, and Argentina). However, it is the categories in the more humid end of the spectrum such as dry semi-humid and especially NA (all humid categories) that seem to have experienced the largest decreases in the index (i.e., became drier) over the 1961–1997 time period.

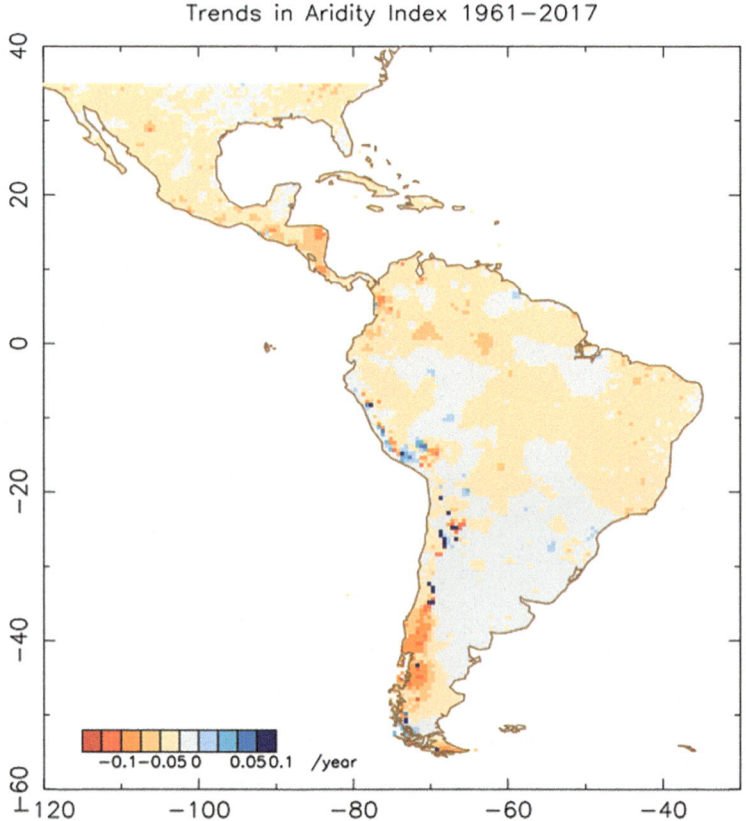

Fig. 16.5 Annual trends in aridity index. Yellow to red colors represent decrease in aridity indices (areas are getting drier), while blue colors show an increase of aridity index (areas are getting wetter)

Aridity Indices: Change/Difference and Trends In and Out of Communal Lands in Mexico Between 1961 and 2017

The results for the aridity index averages for communal versus non-communal lands Mexico are presented in Tables 16.1 and 16.2. Table 16.1 (RAN) shows the more liberal of the definitions of communal lands (Communal_RAN, which includes settlements, and "privatized" parceled lands), whereas Table 16.2 only includes a subset of RAN, composed of lands allocated to communal use within an *ejido* (i.e., tierras de uso común, Communal_TUC). The lands that are not communal or part of *ejidos* are labeled as REST in both cases. REST includes both government and private lands. The analysis that follows compares the aridity metrics on the communal fields (RAN or TUC) versus those in REST.

The results show that there has been a decrease in the aridity index from the 1961–1990 period to the 1991–2017 in all categories shown in both tables (total

Table 16.1 Average aridity indices for two periods, difference between periods, and trends for Mexico by type of land. The bold rows in the table have the purpose of highlighting the overall patterns and trends for the whole analysis. All the previous rows (not highlighted) are a disaggreation of the bold rows by different subcategories)

Aridity statistics	COMUNAL_RAN	REST	Grand total
Average of Index 1961–1990			
Arid	0.145	0.151	0.148
Dry-subhumid	0.629	0.626	0.627
Hyper-arid	0.053	0.053	0.053
NA	1.040	1.053	1.048
Semi-arid	0.405	0.399	0.401
Average of Index 1991–2017			
Arid	0.135	0.140	0.138
Dry-subhumid	0.577	0.575	0.576
Hyper-arid	0.045	0.045	0.045
NA	0.988	1.001	0.996
Semi-arid	0.361	0.349	0.353
Average of difference			
Arid	−0.010	−0.011	−0.010
Dry-subhumid	−0.053	−0.051	−0.052
Hyper-arid	−0.008	−0.008	−0.008
NA	−0.052	−0.051	−0.052
Semi-arid	−0.044	−0.049	−0.048
Average of trend			
Arid	0.000	0.000	0.000
Dry-subhumid	−0.002	−0.002	−0.002
Hyper-arid	0.000	0.000	0.000
NA	−0.002	−0.002	−0.002
Semi-arid	−0.001	−0.001	−0.001
Total average of Index6190	**0.657**	**0.655**	**0.656**
Total average of Index9117	**0.613**	**0.609**	**0.611**
Total average of difference	**−0.043**	**−0.046**	**−0.045**
Total average of trend	**−0.001**	**−0.001**	**−0.001**

RAN = *Ejido* lands in general, including settlement and "private" parcels, REST = Any land that does not include RAN or TUC in Mexico

average of difference). However, the change in lands labeled as REST is slightly higher than those labeled as communal (RAN or TUC). This result suggests that there is—on average—either no difference or only a very small difference in aridity between communal lands and non-communal lands in Mexico; however, they are defined. If the small difference is significant, the non-communal lands have suffered higher rates of aridity increase than communal lands.

When we disaggregate the analysis by the aridity subcategories on tables we see a similar trend to the one observed for Latin America as a whole and by country. There is an overall decrease in aridity index and the trend is negative (i.e., it is becoming drier). However, the largest decreases occur in the humid (wetter) end of

Table 16.2 Average aridity indices for two periods, difference between periods, and trends for Mexico by type of land

Aridity statistics	COMUNAL_TUC	REST	Grand total
Average of Index6190			
Arid	0.137	0.154	0.148
Dry-subhumid	0.621	0.630	0.627
Hyper-arid		0.053	0.053
NA	1.026	1.058	1.048
Semi-arid	0.410	0.397	0.401
Average of Index 9117			
Arid	0.129	0.143	0.138
Dry-subhumid	0.578	0.574	0.576
Hyper-arid		0.045	0.045
NA	0.973	1.007	0.996
Semi-arid	0.364	0.350	0.353
Average of difference			
Arid	−0.008	−0.011	−0.010
Dry-subhumid	−0.043	−0.055	−0.052
Hyper-arid		−0.008	−0.008
NA	−0.054	−0.051	−0.052
Semi-arid	−0.047	−0.048	−0.048
Average of trend			
Arid	0.000	0.000	0.000
Dry-subhumid	−0.001	−0.002	−0.002
Hyper-arid		0.000	0.000
NA	−0.002	−0.002	−0.002
Semi-arid	−0.001	−0.001	−0.001
Total average of Index6190	**0.661**	**0.653**	**0.656**
Total average of Index9117	**0.617**	**0.608**	**0.611**
Total average of difference	**−0.044**	**−0.046**	**−0.045**
Total average of trend	**−0.001**	**−0.001**	**−0.001**

Communal_TUC = communal lands within Ejidos, REST = Any land that does not include RAN or TUC in Mexico

the spectrum. This suggests that we might see an increase in areas newly categorized as arid in regions where humid climates predominated in the past (e.g., NA areas converting to dry-subhumid).

Discussion and Conclusions

The analyses presented in this chapter allow us to explore several dimensions and scales of aridification in Latin America. We first analyzed the changes in aridity as measured by the aridity index over two periods of time (1961–1990 and

Table 16.3 Aridity index in two time periods (1961–1990 and 1991–2017), difference or change between periods, and trend (based on the slope of a simple linear regression)

Country	Average of Index6190	Average of Index9117	Average of Difference	Average of Trend 1961–2017
Argentina	**0.900**	**0.874**	**−0.026**	**0.034**
Arid	0.161	0.194	0.034	0.001
Dry-subhumid	0.570	0.607	0.037	0.001
NA	1.440	1.354	−0.086	0.071
Semi-arid	0.360	0.387	0.026	0.001
Belize	**1.422**	**1.373**	**−0.048**	**−0.001**
NA	1.422	1.373	−0.048	−0.001
Bolivia	**0.927**	**0.871**	**−0.056**	**0.004**
Dry-subhumid	0.589	0.521	−0.068	−0.002
NA	1.039	0.982	−0.058	0.005
Semi-arid	0.418	0.386	−0.032	0.000
Brazil	**1.245**	**1.208**	**−0.037**	**−0.001**
Dry-subhumid	0.573	0.520	−0.053	−0.002
NA	1.319	1.283	−0.036	−0.001
Semi-arid	0.411	0.375	−0.036	−0.001
Chile	**52.949**	**52.877**	**−0.072**	**0.092**
Arid	0.117	0.132	0.015	0.000
Dry-subhumid	0.576	0.544	−0.032	−0.001
Hyper-arid	0.025	0.024	−0.001	0.000
Arid	0.048	0.063	0.015	0.000
Hyper-arid	0.022	0.018	−0.003	0.000
NA	78.034	77.927	−0.107	0.136
Semi-arid	0.327	0.335	0.009	0.000
Colombia	**1.760**	**1.702**	**−0.058**	**−0.002**
Dry-subhumid	0.608	0.525	−0.083	−0.003
NA	1.788	1.729	−0.058	−0.002
Semi-arid	0.418	0.363	−0.055	−0.002
Costa Rica	**2.748**	**2.476**	**−0.271**	**−0.008**
NA	2.748	2.476	−0.271	−0.008
Cuba	**0.885**	**0.844**	**−0.040**	**−0.001**
Dry-subhumid	0.585	0.565	−0.020	−0.001
NA	0.901	0.859	−0.041	−0.001
Dominican Republic	**0.974**	**0.895**	**−0.079**	**−0.001**
Dry-subhumid	0.540	0.580	0.040	0.001
NA	1.003	0.916	−0.087	−0.001
Ecuador	**1.525**	**1.496**	**−0.028**	**−0.002**
Arid	0.170	0.190	0.020	0.001
Dry-subhumid	0.564	0.574	0.010	0.000
NA	1.735	1.698	−0.037	−0.002
Semi-arid	0.398	0.419	0.021	0.001

(continued)

Table 16.3 (continued)

Country	Average of Index6190	Average of Index9117	Average of Difference	Average of Trend 1961–2017
El Salvador	**1.288**	**1.079**	**−0.209**	**−0.006**
NA	1.288	1.079	−0.209	−0.006
Guatemala	**1.702**	**1.561**	**−0.141**	**−0.004**
Dry-subhumid	0.560	0.370	−0.190	−0.006
NA	1.732	1.593	−0.139	−0.004
Guyana	**1.420**	**1.433**	**0.013**	**0.001**
NA	1.420	1.433	0.013	0.001
Haiti	**0.880**	**0.870**	**−0.010**	**−0.001**
Dry-subhumid	0.580	0.500	−0.080	−0.003
NA	0.944	0.935	−0.008	−0.001
Semi-arid	0.480	0.520	0.040	0.002
Honduras	**1.258**	**1.028**	**−0.230**	**−0.006**
NA	1.258	1.028	−0.230	−0.006
Jamaica	**1.328**	**1.245**	**−0.083**	**−0.003**
NA	1.328	1.245	−0.083	−0.003
Mexico	**0.654**	**0.607**	**−0.047**	**−0.001**
Arid	0.132	0.136	0.003	0.000
Dry-subhumid	0.578	0.523	−0.055	−0.002
Hyper-arid	0.050	0.043	−0.007	0.000
NA	1.018	0.958	−0.060	−0.002
Semi-arid	0.365	0.326	−0.039	−0.001
Nicaragua	**1.543**	**1.251**	**−0.292**	**−0.007**
NA	1.543	1.251	−0.292	−0.007
Panama	**1.752**	**1.689**	**−0.063**	**−0.002**
NA	1.752	1.689	−0.063	−0.002
Paraguay	**0.832**	**0.842**	**0.011**	**0.001**
Dry-subhumid	0.585	0.561	−0.024	0.000
NA	1.014	1.049	0.035	0.002
Semi-arid	0.441	0.402	−0.039	0.000
Peru	**8.743**	**8.751**	**0.008**	**0.045**
Arid	0.112	0.169	0.057	0.002
Dry-subhumid	0.564	0.603	0.039	0.002
Hyper-arid	0.032	0.059	0.026	0.001
Arid	0.038	0.076	0.038	0.001
Hyper-arid	0.024	0.034	0.010	0.000
NA	10.119	10.120	0.001	0.052
Semi-arid	0.353	0.429	0.077	0.003
St. Vincent and the Grenadines	**1.400**	**1.210**	**−0.190**	**−0.004**
NA	1.400	1.210	−0.190	−0.004
Suriname	1.473	1.402	−0.071	−0.001

(continued)

Table 16.3 (continued)

Country	Average of Index6190	Average of Index9117	Average of Difference	Average of Trend 1961–2017
NA	1.473	1.402	−0.071	−0.001
The Bahamas	**0.718**	**0.715**	**−0.003**	**0.000**
Dry-subhumid	0.520	0.520	0.000	0.000
NA	0.783	0.780	−0.003	0.000
Trinidad and Tobago	**1.260**	**1.150**	**−0.110**	**−0.004**
NA	1.260	1.150	−0.110	−0.004
Uruguay	**1.139**	**1.204**	**0.066**	**0.002**
NA	1.139	1.204	0.066	0.002
Venezuela	**1.354**	**1.294**	**−0.060**	**−0.002**
Dry-subhumid	0.596	0.554	−0.041	−0.002
NA	1.421	1.359	−0.062	−0.002
Semi-arid	0.401	0.386	−0.015	−0.001
Grand Total	**3.768**	**3.728**	**−0.040**	**0.011**

Hyper-arididy: Index ≤0.05; arid: 0.05–0.2; semi-arid: 0.2–0.5; dry-subhumid: 0.5–0.65; Not Applicable (NA) ≥0.65; NA is not applicable since it is not technically a category within the aridity spectrum, (i.e., they are humid). But it is important to note that our analysis shows locations that transitioned from NA into the arid spectrum within the analyzed time period of 1961–2017

1991–2017). Then we disaggregated, described, and briefly interpreted the results of those changes by country and then by land property regime type for Mexico. Given the results of both analyzes we come to the following findings and preliminary conclusions. We also outline an agenda for future work.

1. Taking into account changes in aridity indices and trends by country (Appendix Table 16.3), there is a clear pattern of drying in Latin America. All but 4 countries, Peru, Nicaragua, Paraguay, and Guyana, have negative differences, between the two time periods analyzed. El Salvador, Mexico, and Nicaragua have on average the larges decreases in the aridity index, which suggests that on average, as a whole country, they are getting drier.
2. The patterns described above are more complex if we disaggregate them within each country, there are regions that have the opposite trend than the national average presented in Appendix Table 16.3. For example, in Peru, the Central and Southern Peruvian Andes showed drier conditions (Fig. 16.5).
3. The areas that are becoming drier are not necessarily the areas considered arid (aridity index lower than 0.65), although that is the case sometimes. For the most part, it is wetter, more humid areas (classified as NA in the tables and displayed as grey areas in Fig. 16.1) that are becoming drier—or less wet. This finding suggests that we might see a geographic increase in areas categorized as arid in the future. This pattern would become stronger taking into account the future scenarios of lower precipitation for some regions and higher global temperatures for the next decades (Huang et al. 2016).

4. There is a virtually no difference between the aridification of communal lands and non-communal lands in Mexico (the negative change and trend in Tables 16.1 and 16.2). Non-communal lands have seen a slightly larger drop in the aridity index (i.e., they are getting drier) than communal lands. That difference should be taken with caution for several reasons. First, because it is a very small difference, and second, because communal lands were compared to all non-communal lands, without differentiating between them. Federal lands could behave very differently than private lands, for example. The trend values in both cases is negative and the same.

5. If the difference were meaningful, it would suggests that communal lands are drying less and less rapidly than non-communal lands. The interpretation of this statement is tricky because it could be the result of two processes that are not mutually exclusive: a) the allocation of common lands in the large land redistribution processes in Mexico could have disproportionally placed arid lands under non-common land tenure regimes; or b) communal land practices and the institutions that govern them could have created the conditions that prevented or reduced the levels of aridification in them. At this point, we are not in position to make either claims. We especially warn against drawing causal conclusions with regard to the favorability of environmental governance practices of communal lands and institutions.

Future research will address many of the questions that arouse from this early analysis of the data. We would like to look at how the different trajectories of aridity metrics of Latin American countries observed in this chapter match each country's environmental policy, including natural resource management, climate change, and natural disaster policies. Ideally we would attempt to establish potential links between policies and aridity trends and further analyze emerging patterns observed in this chapter. We also would like to disaggregate the analysis into more property regimes (including private, state and federal owned), as well as to include examples of countries other than Mexico. Finally, we would like to further develop our methodological framework to include statistical hypothesis testing to evaluate potential links between land property regimes and aridification. In doing so, we think that it would be important to account and control for environmental differences (e.g., land cover types) within each reagion in order to produce a more robust analysis.

Appendix

Data Source References

1. Clasificación de Koppen modificada por Enriqueta García, para adaptarlo a las condiciones de la República Mexicana
2. http://www.igeograf.unam.mx/sigg/utilidades/docs/pdfs/publicaciones/geo_siglo21/serie_lib/modific_al_sis.pdf

3. http://smn.cna.gob.mx/es/component/content/article?id=13:lop&catid=1:gene ral
4. Datos geográficos de las tierras de uso común, por estado - Formato SHAPE.
5. https://datos.gob.mx/busca/dataset/datos-geograficos-de-las-tierras-de-uso-comun-por-estado%2D%2Dformato-shape
6. Diccionario de datos http://datos.ran.gob.mx/filesdd/ran_dd_dgcat_poligonos_tierras_uso_comun.pdf
7. Números 11 al 17
8. http://datos.ran.gob.mx/conjuntoDatosPublico.php
9. Sistema de Información Geoespacial del Catastro Rural
10. http://www.ran.gob.mx/ran/index.php/sistemas-de-consulta/sistema-de-informacion-geoespacial
11. Listado de títulos de concesiones mineras
12. https://datos.gob.mx/busca/dataset/cartografia-minera-de-se/resource/0f34b008-cc81-478d-ba9b-dbce1873d6a9
13. Consulta de la Cartografía Minera Digitalizada
14. http://www.cartografia.economia.gob.mx/cartografia/

References

Assies W (2008) Land tenure and tenure regimes in Mexico: an overview. J Agrar Chang 8(1):33–63
Barnes G (2014) Land Administration of Communal Land: Lessons from the *Ejidos* in Mexico, (7299) FIG Congress 2014. Engaging the Challenges – Enhancing the Relevance. Kuala Lumpur, Malaysia
García E (2004) Modificaciones al Sistema de Clasificación Climática de Kopen. (original de 1964). Instituto de Geografía de La Universidad Nacional Autónoma de México. Serie Libros Num. 6. México
Huang J, Yu H, Guan X, Wang G, Guo R (2016) Accelerated dryland expansion underclimate change. Nat Clim Chang 6:166–171
Haenn N (2006) The changing and enduring ejido: a state and regional examination of Mexico's land tenure counter-reforms. Land Use Policy 23(2):136–146
Jones AG, Ward P (1998) Privatizing the commons: reforming the Ejido and urban development in Mexico. Int J Urban Reg Res 22(1):76–93
Ojanen M, Zhou W, Miller DC, Nieto SH, Mshale B, Petrokofsky G (2017) What are the environmental impacts of property rights regimes in forests, fisheries and rangelands? Environmental Evidence 6(1):12
Paavola J (2007) Institutions and environmental governance: a reconceptualization. Ecol Econ 63(1):93–103
UNEP World Conservation Monitoring Centre (2007) A spatial analysis approach to the global delineation of dryland areas of relevance to the CBD programme of work on dry and subhumid lands, Cambridge
UNESCO (1979) Map of the world distribution of arid regions: explanatory note. UNESCO, Paris

Chapter 17
Vulnerability to the Effects of Climate Change: Future Aridness and Present Governance in the Coastal Municipalities of Mexico

G. Seingier, O. Jiménez-Orocio, and I. Espejel

Abstract We estimated vulnerability to climate change in the coastal municipalities of Mexico through an interdisciplinary approach using an index model with three components: (1) Exposure to dryness and climate change (Lang's dryness index), (2) socioeconomic sensitivity, and (3) adaptation capacity. Data input were national census data and general circulation model outputs (2045–2069 scenario). Scenarios were compared to reference climatology through Lang's aridity index, which was found to be practical as an aridity indicator and foresee its future change in value. This methodological approach allowed to set priorities by identifying groups of more threatened and less prepared municipalities in the territories, which show more differences between the present climatological parameter values and the values predicted assuming future climatic scenarios, have less climate change related institutional instruments, present social weaknesses (poverty and inequality), or show environmental degradation. Our results showed the presence of *arid* zones in 43% of Mexico's 266 coastal municipalities, comprising more than half of the total coastal population. Comparing present and future scenarios allows differentiating regions already *arid* that will remain in this situation (Northern Pacific) from regions that are currently humid but will shift to an *arid* (Caribbean Sea) or *desert* category (Northern Gulf of Mexico). Both used models predicted a future shift of all categories towards dryness and a worrisome number that will turn to *desert* and *arid* categories. We found regional heterogeneity and a complex contribution of each subindex to the total coastal vulnerability index. We concluded that our results can be used as an input to guide the current coastal policies to promote a development model that decreases the vulnerability of coastal municipalities and

G. Seingier (✉) · O. Jiménez-Orocio
Facultad de Ciencias Marinas, Universidad Autónoma de Baja California,
Ensenada, BC, Mexico
e-mail: georges@uabc.edu.mx

I. Espejel
Facultad de Ciencias, Universidad Autónoma de Baja California UABC, Ensenada, BC, Mexico

© Springer Nature Switzerland AG 2020   301
S. Lucatello et al. (eds.), *Stewardship of Future Drylands and Climate Change in the Global South*, Springer Climate,
https://doi.org/10.1007/978-3-030-22464-6_17

advance towards a model that includes the complexities of the Mexican coastal dryland socio-ecological systems, its challenges, and its opportunities.

Keywords Coastal vulnerability · Coastal sensitivity · Coastal exposure · Lang's index, Dryness · Adaptation capacity · Governance · Coastal municipalities

Introduction

This study is based, first, on the definition of climate change vulnerability used worldwide by the Intergovernmental Panel on Climate Change (IPCC 2007) as "the degree of susceptibility or inability of a system to face the adverse effects of climate change, and in particular the variability of climate and extreme events. Vulnerability will depend on the nature, magnitude and speed of climate change to which a system is exposed, and on its sensitivity and adaptive capacity." Second, it is based on the definition of climate change as any modification in the climate produced during the course of time, either due to natural variability or to human activity (IPCC 2007).

Conceptually, vulnerability is divided into three components: (1) Exposure, which refers to the degree to which a system is exposed to external stimuli that act on it (IPCC 2007; Hinkel 2011). In the case of climate change in coastal municipalities the stimuli that might affect their territory more than that from inland municipalities include: increase in ocean or continental temperature, the sea level rise, floods, and the higher frequency of hydrometeorological events among others. (2) Sensitivity, which is the degree to which the system is affected by climate change, both positively (opportunities) and negatively (challenges; IPCC 2007). It can be estimated through land use change, population density, population concentration in the coastal plain, municipality's coastal length, and other indicators. (3) Adaptation capacity refers to the natural or human systems' capability for adjustment in response to actual or expected climatic stimuli, or to their effects, mitigating the harmful effects or exploiting the beneficial opportunities (IPCC 2007). It estimates how much a coastal municipality is prepared to face and manage its "exposure" and "sensitivity" as part of a governance process. In general, the municipality's economic, political, cultural, social, and technological resources are evaluated. Vulnerability increases with exposure and sensitivity, and will be reduced when adaptation capacity increases.

Vulnerability is assessed from different risk perspectives in particular cases but always aiming to evaluate the response of the system (Adger 2006; Füsel 2007), that is, the response to a natural or anthropic disturbance of a social system, an ecosystem, or a socio-ecological system.

These components were adapted to this chapter's scope by seeing the coastal region as a socio-ecological complex system at the municipal administrative level. In particular, in this chapter we focus on the exposure to dryness that climate change scenarios predict and on the capacity to adapt considering municipal policy indicators from an interdisciplinary perspective.

In many national economies, the social importance of coastal zones is due to the numerous social and economic activities concentrated near the shoreline. As Nguyen et al. (2016) assert, the importance of these zones will be further intensified in the future because of the ever-increasing number of people inhabiting them. Adger et al. (2005) state that 1.2 billion people—which accounts for 23% of the world's population—are now living within 100 km of the coast and about 50% of the world's population is likely to do so by 2030 (Neumann et al. 2015). In coastal countries characterized by having a long heterogeneous coastline as Mexico, to be able to precisely assess the vulnerability to the stress factors associated with climate change in coastal municipalities is a great challenge, since it involves taking into account social and ecological components, as well as multiple temporal and spatial interactions.

In the complex socio-ecological system of the coastal fringe in Mexico, the traditional—agriculture, fisheries, and forestry activities—and the growing new economic sectors—tourism, urbanization, aquaculture, and mariculture—compete along the 11,000 km long coastline, where, according to Lang's index, 36% of the 271 coastal municipalities are presently *arid* or *desert*; a condition that is apparently increasing its surface according to the general climate change models that were to be used in our study. The methodological approach put forward in the present work allows to set priorities by identifying groups of more threatened and less prepared municipalities: those which show more differences between present climatological parameter values and predicted future climatic scenarios values (through an aridity indicator), have less climate change related institutional instruments (adaptation capacity building), present social weaknesses (poverty and inequality), or show environmental degradation. Results will be usefull as input for promoting public policies and constructing governance in the socio-economic-political context of an emerging country immersed in an increasingly globalized world.

Synthesis of the Major Conceptual Advances Made During the Past 25 Years

Indicators of Vulnerability to Climate Change Used for the World's Coasts

The Organization for Economic Cooperation and Development (OECD) defines an indicator as: "a parameter (property that is measured and observed), or value derived from other parameters, aimed at providing information and describing the state of a phenomenon, environment or area, with an added meaning greater than that directly associated with its own value (Nardo et al. 2005; OECD 2000, 2003)."

Indicators are a support for the evaluation and monitoring of the dynamic and complex ecosystems and their economic development. These indicators contribute to the development of governmental proposals, guiding policies and actions of society, and provide timely information to decision makers (Azuz-Adeath et al. 2010a).

Indicators are inputs to improve governance. Nowadays, it is necessary to sensitize people about the possible phenomena, processes, and environmental problems that could affect them. For an indicator to be useful it must comply with certain characteristics such as being simple, relevant, adequate, meaningful, and having a large-scale and long-term vision. This means that indicators must be sufficient, have a methodology of transparent construction, prove their usefulness for making evident decisions, and be able to show changes (sensitive) or to measure progress clearly (Spangenberg and Bonniot 1998).

At the global level, Nguyen et al. (2016) reviewed 53 published studies assessing the impact of climate change on the world's coasts with different vulnerability indicators. They concluded that a consensus of the best indicators has not been reached, since there are discrepancies in the spatial and temporal scale at which the evaluation is made and in the choice of indicators and their measurement ranges (Table 17.1).

Table 17.1 Review of indicators used to assess vulnerability to global climate change (modified from Nguyen et al. 2016)

Indicators	Number of studies	Global	National	Regional	Local
Exposure indicators					
Temperature	8			4	4
Sea level rise	7		1	6	
Precipitation	4	1		2	1
Hurricanes	3			3	
Flood	4			3	1
Drought	2			2	
Others	15	1	3	5	6
Not added	9				
Total	52	3	4	25	12
Sensitivity indicators					
Population density	13		2	10	1
Population	10	2	1	4	3
Land use	8			4	4
Others	18	1	3	11	3
Not added	3				
Total	52	3	6	29	11
Adaptation capacity indicators					
Economic indicators	10		5	4	1
Government institutions	6	1	3	2	–
Administrative units	2	1	1	–	–
Management strategies	1	1	–	5	–
Others	22	1	5	13	5
Not added	12	–	–	–	–
Total	53	4	14	17	6

Table 17.2 Physical variables used for exposure studies in coastal areas (modified from Nguyen et al. 2016)

Variables	Authors
Land relief (m)	Gornitz (1991)
Sea level rise (mm/years)	Gornitz (1991), Özyurt and Ergin (2010)
Tidal range (average) (m)	Gornitz (1991), Özyurt and Ergin (2010)
Height wave (máximo) (m)	Gornitz (1991)
Flood depth (m)	Kafle et al. (2007), Bormudoi et al. (2008), Le et al. (2009), Dang et al. (2011), Özyurt and Ergin (2010), Mackey and Russell (2011), Dinh et al. (2012), Balica et al. (2013), Tingsanchali and Karim (2005)
Salinity (ppt)	Grattan et al. (2002), Mackey and Russell (2011), Hoang et al. (2012), Le (2003)
Shoreline changes (m/years)	Gornitz and Kanciruk (1989), Gornitz (1991), Gornitz et al. (1994), Pham and Nguyen (2005), Dwarakish et al. (2009), Pendleton et al. (2010), Abuodha and Woodroffe (2010b), Nguyen (2012)

Table 17.3 Social variables used for sensitivity studies in coastal areas (modified from Nguyen et al. 2016)

Variables	Authors
Population density (people/km^2)	Kafle et al. (2007), Dang et al. (2011), Mackey and Russell (2011)
Land use	Preston et al. (2008), Özyurt and Ergin (2010), McLaughlin and Cooper (2010), Liu (1996), Huang et al. (2012), Yin et al. (2012)
Local income level (mil/per capita/years)	Dang et al. (2011)

According to the results of Nguyen et al. (2016), 87% publications contained indicators of exposure, 85% of sensitivity, and 74% of adaptation capacity. Most of the indicators in the reviewed literature were of regional level—especially for exposure and sensitivity—but only a few had a local (except for exposure indicators), national (except adaptive capacity), or global scope. The global scale studies of coastal vulnerability to climate change (Abuodha and Woodroffe 2006; Eakin and Luers 2006; Nicholls et al. 2008) were focused on measuring geophysical processes (coastal geomorphology) or direct physical impacts (increase in mean sea level, floods, etc., Table 17.2). Nguyen et al. (2016) highlight the evident lack of coastal vulnerability to climate change studies that included socioeconomic parameters in their models, seven indicators were based on physical variables and only three used socioeconomic variables (Table 17.2 and Table 17.3); for example, one as important as income level was only mentioned in a single reference and only partially because the indicator was assessed by the income level (Dang et al. 2011).

Indicators of Vulnerability to Climate Change Proposed for the Coastal Zone of Mexico

In a national recompilation effort, Azuz-Adeath et al. (2010b) listed 50 indicators to assess the vulnerability to climate change of the coastal zone of Mexico. The authors mention that each model must select indicators that, first, are available and, second, are not redundant. Their proposal is to use indicators to model the baseline of the coastal zones of the country in order to elaborate scenarios of their possible evolution and of the changes that the coasts would experience in response to different actions aiming at mitigating—or adapting—to the impacts of climate change.

In Fig. 17.1, all 12 identified exposure indicators correspond to the ecological category and are related to exposure to coastal processes or physical phenomena typical of the coast, for example, the frequency and intensity of hydrometeorological events like hurricanes, but also of precipitation and temperature. Coastal land use, demographic variables, and heritage indicators can be identified as related to the sensitivity component. As for adaptation capacity components, one-third of the indicators were related to man power and two-thirds to governance. Seingier et al. (2011a, b) did not mention if such indicators were applied at all scales but agree that most of them have a local and regional scope, only a few being available at the national scale.

In the past 25 years, the outputs of numerous general circulation models at different time scales and with different emission scenarios have been made available to the general public and have become tools allowing to foresee the possible future

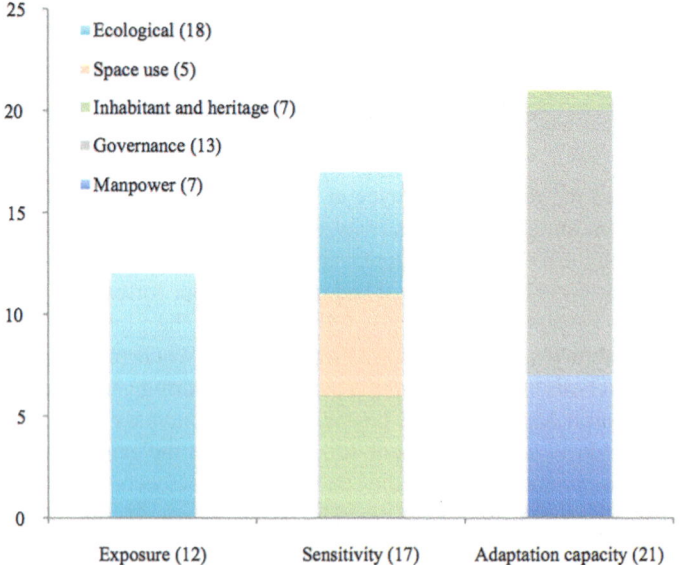

Fig. 17.1 Number and category of indicators used to assess vulnerability to climate change in the coastal zone of Mexico (modified from Azuz-Adeath et al. 2010b)

with different degree of likelihood. Future values of temperature and precipitation allow to calculate the degree of dryness or aridity in the medium or long term and, if compared to the present data, also allow to identify changes. Indeed, aridity can be quantified using Lang' index—which is a ratio between precipitation in mm and the temperature in °C—that can help us to delimit and characterize arid and semi-arid zones (Hubálek and Horáková 1988; Sánchez-Torres et al. 2011).

Description of the Broader Societal Implications of Conceptual Advancement

The conceptual advancements with possible implications for society can be summarized by saying that there is a tendency towards an interdisciplinary perspective that assumes the system of interest not as being social (sectorial) or ecological but as a socio-ecological complex system. Baseline data are being generated and monitoring must continue and be itself evaluated. Integration of the conceptual and methodological models as well as of the data needs to be made, and the steps to reach this goal need to be fastened up.

These interdisciplinary efforts will help to set priorities for promoting sound public policies and construct governance within the socio-ecological context of an emerging economy country immersed in a globalized world.

Description of Coastal Mexico: A Case Study

Using both ecological and social indicators, an estimation of the present and future vulnerability to dryness was made for Mexico's coastal municipalities, a territory under growing pressure. We used indicators and methodologies that have been suggested for regional scales at the international (Abuodha and Woodroffe 2006, 2010a; McFadden 2007; Harvey and Woodroffe 2008; McLeod et al. 2010) and national scales, in the case of Mexico, through a selection of the indicators listed by Azuz-Adeath et al. (2010b) evaluated for each component of climate change vulnerability in the country's coastal municipalities.

Our territorial units of evaluation were the 271 Mexican coastal municipalities as defined by the Program for the Sustainable Development of Oceans and Coasts of Mexico (PANDSOC by its Spanish acronym; SEMARNAT 2006) and the current Policy of Seas and Coasts (SEGOB 2018) which corresponds to 17 coastal states (Fig. 17.2).

The applied model of coastal vulnerability to the effects of climate change integrated three types of indicators, related to exposure, sensitivity, and adaptation capacity, as defined by (IPCC 2007; Fig. 17.3). The corresponding maps and results shown here allowed the interpretation of the main issue we wanted to focus on in this chapter, the present and future dryness, and the capacity for adaptation to climate change effects.

Fig. 17.2 Mexican coastal socio-ecological system: five regions, 17 coastal states, and 266 municipalities (divided in 271 due to the size of five municipalities in Baja California Peninsula)

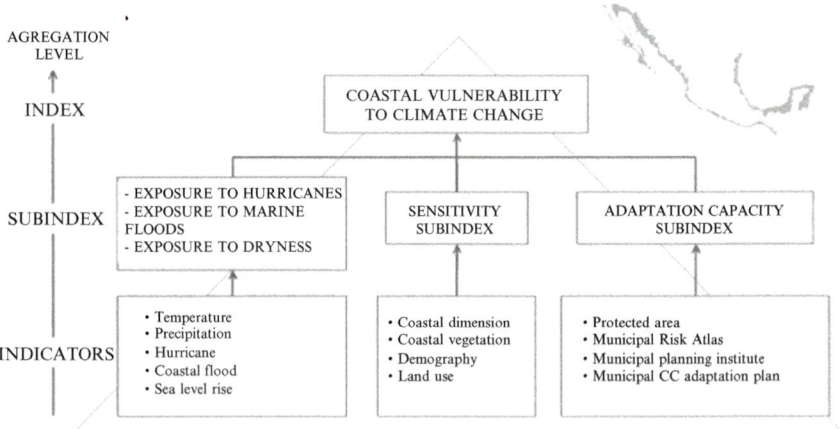

Fig. 17.3 Climate change coastal vulnerability model based on the selected indicators

Exposure to Dryness Subindex

Dryness is related to aridity and availability of water resources and is an essential phenomenon and an indicator for determining the negative impacts of climate change. Variations in temperature (such as the capacity of the climate to evaporation) and the decrease in rainfall (as a source of water) have repercussions in various economic sectors going from agriculture to public health. Aridity indices are based on the precipitation occurred throughout the year (as a source of water) and on the temperatures (as an indicator of the evaporation capacity of the climate). For example, Lang's index (Hubálek and Horáková 1988) is an explicit and easily applicable measurement for estimating the magnitude of dryness that allows an evaluation of the changes in the distribution of drought and—with temperature and precipitation outputs from general circulation models of climate change—allows to calculate its future projections (Sánchez-Torres et al. 2011). Consequently, the dryness exposure index was calculated through the Lang's drought index as described below, both for the current situation (reference climatology) as for one and two horizon models (GFDL-CM3 and MPI-ESM-LR climate model outputs for medium term [2015–2069] RCP 8.5 scenarios). Form then on the significance of GFDL_CM3 and MPI_ESM_LR terms would be explained[1].

Present and Future Scenarios

Different methods for the delimitation and characterization of arid and semi-arid zones have been proposed and used for the numerical quantification of drought and for its application in the interpretation of the aridity process, one of the alternatives applied in this type of studies being the calculation of the *PP/Tm* ratio (Amador et al. 2011), published and referred to as Lang's index (Hubálek and Horáková 1988; Sánchez-Torres et al. 2011). We proposed to use Lang's index, which, as mentioned above, is a thermo-pluviometric index calculated by dividing the average annual precipitation (in mm) by the average annual temperature (in ° C):

$$Pf = PP / Tm,$$

where PP is the annual precipitation in mm; Tm is the average temperature in °C.

Climatological data were used for the current situation and Lang's index was calculated with the temperature and precipitation values corresponding to the outputs of the different models for the future projections and the selected climate change scenarios. Municipalities were classified according to the standard categories proposed by Troyo-Diéguez et al. (2014, Table 17.4). To quantify the changes of exposure to dryness along the coasts, Lang's index values and categories were

---

[1] Climate change projections have been developed at a global scale considering different scenarios. General Circulation Models (GCM) and are used to predict climate change for different future time periods under different CO2 emission scenarios and related atmospheric heat transfer processes. In this work we choose the highest emissions pathways (Representative Concentration Pathways RCP 8.5), which assumes that greenhouse gas emissions will continue to rise throughout the current century, so that we could distinguish larger change. Two models from the world available GCM considered to represent the best climate regime for Mexico were analyzed: GFDL_CM3 (developed by the Geophysical Fluid Dynamics Laboratory), and MPI_ESM_LR (developed by the Max-Plank Institute).

Table 17.4 Classification according to the Lang's index value (Troyo-Diéguez et al. 2014)

Value of Pf	Category
0–20	Desert
20–40	Arid
40–60	Humid of steppe and savanna
60–100	Humid of clear forests
100–160	Humid of large forests
>160	Perhumid with meadows and tundras

Table 17.5 Current and future (GFDL_CM3 and MPI_ESM_LR) dryness categories for the 271 coastal municipalities of Mexico according to Lang's index value

		Desert	Arid	Humid of steppe and savanna	Humid of clear forests	Humid of large forests	Total
Number of municipalities	Current climate	31	67	121	45	7	271
	GFDL_CM3	35	105	89	40	2	271
	MPI_ESM_LR	37	118	80	34	2	271
	Average future	36	111.5	84.5	37	2	
% of total 271 municipalities	*Current climate*	*11*	*25*	*45*	*17*	*3*	
	GFDL_CM3	*13*	*39*	*33*	*15*	*1*	
	MPI_ESM_LR	*14*	*44*	*30*	*13*	*1*	
% change in future number of municipalities			*16*	*66*	*−30*	*−18*	*−71*

compared to determine the regions with the greatest impact as those regions having larger differences between the reference climatological value and the two models' output values. The higher the differences in the index are, the greater will be the future challenge to build capacity and to strengthen the institutions and the interactions among actors to be prepared to cope with a municipality transiting towards aridity. The current situation and the scenarios of future projection of exposure to dryness for coastal municipalities are shown in Tables 17.5 and 17.6, and in Fig. 17.4.

Currently, *humid of steppe and savanna* is the most numerous category, followed by *arid*, and at the medium horizon, the *arid* category predominantes. The two models are very similar, MPI giving a result more towards aridity, due to forecasting greater decrease in precipitation. The largest difference between the present and future Lang's index value was observed between the *humid of steppe and savanna*

Table 17.6 Regional redistribution of aridity

Region/Aridity change	Northern Pacific	Southern Pacific	Gulf of California	Gulf of México	Caribbean Sea	Total
x < 0	3	0	0	0	0	3
0 < x < 5	2	36	27	0	0	65
5 < x < 10	2	38	13	39	63	155
10 < x < 15	0	6	0	37	5	48
Total	7	80	40	76	68	271

Change (x) in Lang's index between the present and future scenarios (considering the maximum between MPI and GFDL output calculations, representing worst scenario) by number of municipalities in the coastal regions of Mexico

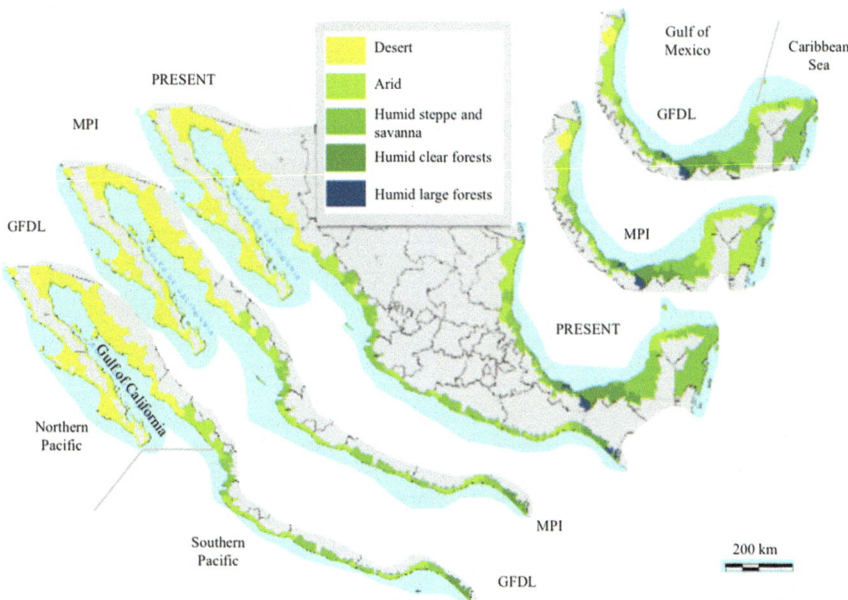

Fig. 17.4 Current exposure to drought subindex. Medium term exposure to drought for reference climatology (present) and future GFDL_CM3 and MPI_CM3 models output

and the *arid* categories. Regarding the number of municipalities by categories, the three humid categories are predicted to be found in less municipalities in the future, while the number of *arid* municipalities will increase 66% (Table 17.5).

Regions with more challenges due to climate changes are at the bottom line corresponding mainly to municipalities in the Gulf of Mexico, followed by those in the Caribbean and the Southern Pacific regions. The importance of comparing present and future dates is to differentiate regions that are already *arid* (for example, Northern Pacific), and that will stay in that situation, from regions that are currently humid and will shift to the *arid* (Caribbean Sea) or the *desert* categories (Northern Gulf of Mexico; Fig. 17.4).

Coastal Sensitivity

A sensitivity index was calculated by the combination of four indicators: littoral ratio, demography, land use, and vegetation (not thoroughly described in this chapter). Table 17.7 and Fig. 17.5 show that the municipalities of the Southern Pacific and Gulf of Mexico regions have very high sensitivity values and that the high values were concentrated in the Caribbean Sea as well as in the Southern Pacific regions. Regarding coastal dimension, the comparison of municipal length of the coast and its land surface shows that there is a greater proportion of coastline in the municipalities of both regions of the Pacific and in the state of Quintana Roo in the Caribbean Sea region.

Although the population in municipalities has grown in recent years, the coast is not yet densely populated. The largest populations are clearly concentrated, for

Table 17.7 Sensitivity index summary by region related to Fig. 17.4

Region/class	Northern Pacific	Southern Pacific	Gulf of California	Gulf of México	Caribbean Sea	Total
Low	0	0	0	0	1	1
Medium	1	2	0	2	6	11
High	2	33	19	19	41	114
Very high	4	45	21	55	20	145
Total	7	80	40	76	68	271

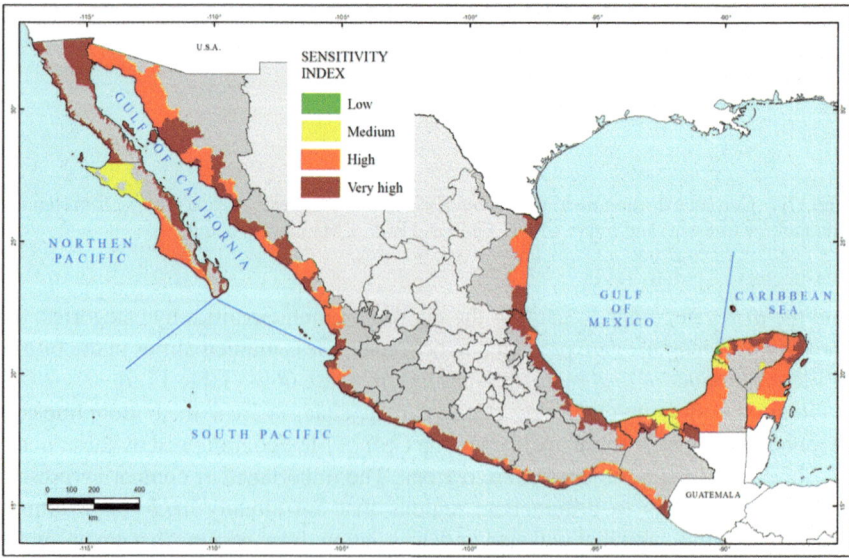

Fig. 17.5 Sensitivity subindex (SI) of the coastal municipalities of Mexico

example, in cities of the municipalities of Tijuana and Mexicali in the state of Baja California, of Peñasco in the state of Sonora, and of Benito Juárez in the state of Quintana Roo. Most coastal cities have expanded to promote tourism projects, but some do so centered in the marine zones like in Cancun, Los Cabos, and Vallarta, and others focusing in industrial zones, as in the case of the oil producing cities of Coatzacoalcos and Minatitlán in the state of Veracruz. Land use change reflects the results obtained with the previous indicator but it is agricultural use which has transformed most of the coastal natural vegetation. Forests transformed into pastures and coconut plantations are the dominant land use changes in the coastal areas of the municipalities of the Southern Pacific and the Gulf of Mexico regions. The coastal vegetation indicator measures the proportion of coastal vegetation (wetlands, mangroves, and dunes), and reaches very high values in most municipalities.

Coastal Adaptation Capacity Subindex

The capacity of adaptation to the effects of climate change of the country's coastal municipalities was estimated as a measure of institutional governance. The basic premise was that a coastal municipality has a better capacity to adapt if it has enough planning policy instruments for the attention of emergencies caused by the effects of climate change. At the municipal level, four indicators to estimate this subindex are: municipal risk atlas, action plans for the effects of climate change, municipal research and planning institutes, and environmental protection instruments like protected areas.

The presence of a municipal risk atlas is an indicator that reflects the existence of an official instrument generated by the Mexican government (SEGOB 2019) to provide local knowledge about exposure to events and the sensitivity of the population to them, which is the basis for adaptation. The Mexican government also has a program to develop municipal actions for the effects of climate change (SEDESOL 2012). This indicator reflects a step further in the capacity of the municipalities to adapt. There is an association of organisms or institutes for the integral planning of municipalities committed to the comprehensive development planning that promotes the collection of information in long-term planning to give continuity to plans and projects, and to favor citizen participation through public policies having an effect on the local and regional culture towards sustainability (AMIMP 2003). Therefore, we selected the presence of such planning institutes as an indicator reflecting the capacity of the municipality to carry out planning tasks and establish continuity in the processes with a vision of sustainable development, elements conducive to an adaptation to the impacts of climate change.

The values of these three indicators were scored as presence or absence (risk atlas, climate change actions plan, and planning institutes), since we could not find performance evaluations of such instruments for each municipality at the national scale. Another indicator was added, the indicator of protected areas, defined as the ratio of protected areas recognized by the National Commission of Natural Protected

Areas (CONANP by its Spanish acronym) in the municipality or in its neighborhood coastal and marine area. Most protected areas have action plans for the adaptation to climate change effects. We consider this a good governance (institutional) indicator since these plans "seek to identify, sustain and guide the implementation of adaptation measures to reduce the vulnerability of socio-ecosystems and achieve articulation with key actors. These instruments integrate information on climate scenarios and their possible effects on conservation objectives and rural productive activities" (CONANP 2017).

The results of the subindex of coastal adaptation capacity are shown in Table 17.8 and Fig. 17.6. Contrary to the sensitivity subindex, most of the coastal municipalities of the country (red color in Fig. 17.6) have a low adaptation capacity. These municipalities lack the four public policy instruments that allow them to be prepared to face the impacts of climate change on their coasts. In addition, for the cases

Table 17.8 Summary of the adaptation capacity index (ACI) by region related to Fig. 17.5

Region/class	North Pacific	South Pacific	Gulf of California	Gulf of México	Caribbean Sea	Total
Low	0	53	12	53	45	163
Medium	1	17	9	16	15	58
High	0	7	5	2	3	17
Very high	6	3	14	5	5	33
Total	7	80	40	76	68	271

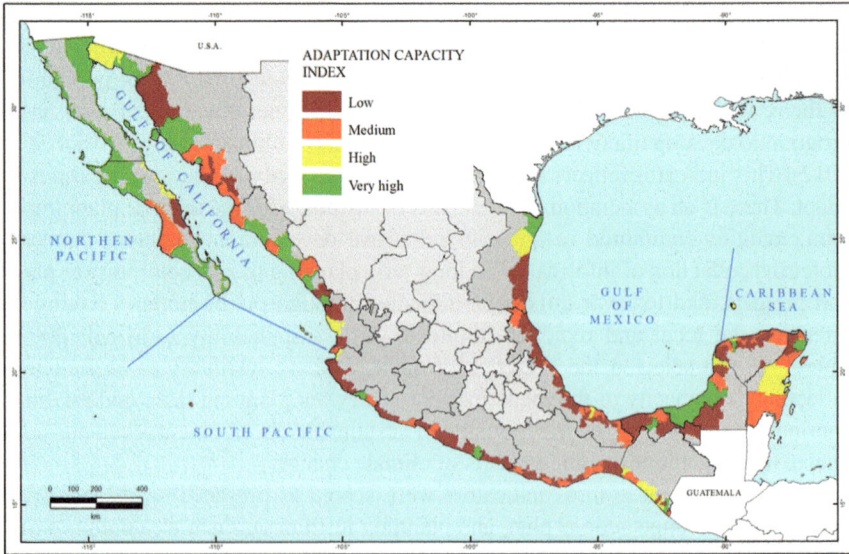

Fig. 17.6 Adaptation capacity subindex (ACI). Low corresponds to no instruments and very high to the presence of the four instruments

of the Gulf of Mexico and Southern Pacific regions, the lack of legal protection instruments (Protected Natural Areas ratio) further decreases their possibilities to adapt to the effects of climate change. There are three major new marine reserves in the Caribbean and in the deep Pacific and its islands. The Northern Pacific and Gulf of California regions are more prepared because they have legal instruments, protection instruments, or institutions like municipal planning and research institutes that are responsible for preparing local development plans and adaptation strategies to the effects of climate change.

Recommendations and Future Perspectives

Climate change effects on coastal areas have been one of the major issues addressed in scientific journals and media. Sea level rise and erosion have been most mentioned as the major threats. Floods associated with hurricane and resilience related issues are historical priorities to which economic and human resources have been assigned, maybe because of the strength and visual impact and the sudden nature of these events. Nevertheless, other climate changes are equivalently important, even if their more medium to long term character is perceived only at a longer time scale (or are unseen). The results of this preliminary study show a country where the majority of municipalities will have dryer climates in the medium term, combined with an average low governance. Resilience to dryness associated with climate changes must be constructed from now on, in order to reach a stable and locally adaptable governance and to count with tools to face the different present and future coastal aridity situations characteristic of coastal Mexico.

More detailed analysis of these three components will allow to group municipalities in clusters with similar characteristics in terms of vulnerability to dryness as we determined it. This will be the first step to assign priorities and to identify the component to be attended (sensitivity or adaptation capacity, for example). The analysis represented in Fig. 17.7 provides more information like, for example, that *desert* categories are those with more adaptation capacities, compared to the more humid categories that will suffer a greater shift towards aridity (larger circles representing greater than 10 change in Lang's index) and that have a maximum of two adaptation instruments. Plotting a similar graph with sensitivity index values, or running statistical analysis (cluster analysis), might aid us to identify these homogeneous groups of planning units.

In this chapter, we used indicators and methodologies suggested for regional scales at the international (Abuodha and Woodroffe 2006, 2010a; McFadden 2007; Harvey and Woodroffe 2008; McLeod et al. 2010) and national level through a selection from the list of indicators for each coastal municipalities climate change vulnerability components proposed by Azuz-Adeath et al. (2010b) for the case of Mexico. The next step will be to downscale the analysis towards local issues. An aspect of the complexities of dryland socio-ecological systems is the administrative unit at which it must be attended; if future exposition data are available at a global

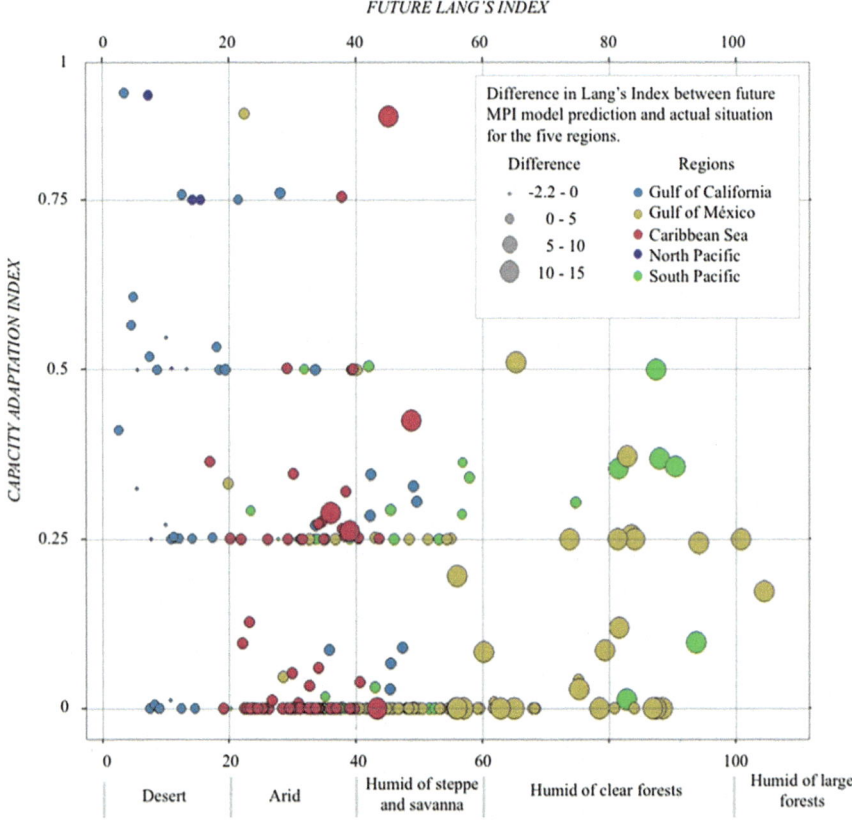

Fig. 17.7 Adaptation capacity index (0–1 normalized upwards) vs. future Lang's index (more arid towards left). Symbol size represents the change between actual value and its future prediction. Symbol position is worse (more arid) future situation. Each color represents a region

or regional scale, sensitivity and capacity adaptation in general will need to be attended at a more local scale. This implies adjusting the indicators to the new scale, along with facing the corresponding data availability issues. As for future data, nesting of general circulation models to obtain more accurate data is needed, together with the use of other models to define the likelihood of these predictions.

Obtaining a value for the future adaptation capacity subindex is difficult to achieve in a model for any of the horizons considered in the climate models. Although it is desirable that all coastal municipalities have at least the four abovementioned preventive policies (indicators), the future existence of all these instruments is not predictable. Moreover, other indicators of governance should be included for constructing a more complex and complete subindex, especially those describing the communities' capacity of adaptation, for example, active NGOs programs, environmental education levels, participation, and social interactions, among others. The problem with these indicators is their unavailability for the whole coastal municipalities of the country.

On the other hand, predicting the future value of sensitivity indicators (as is done for temperature and rainfall with global circulation models) is also complicated, but consulting, for example, future regional projects and development plans can give us a hint of the direction towards which development is heading, and help us predict social and ecological implications.

This situation is alarming because it shows that, as a coastal country, much has to be implemented in the near future to protect the coastal inhabitants and infrastructure in Mexico. Also, Mexico has to change its coastal policies to promote a type of coastal development model that decreases the coastal vulnerability of the municipalities and that takes into account the potential impacts of a shift towards dryness of this complex coastal socio-ecological system.

Acknowledgments The present research is part of the characterization and regionalization of the coastal zones of Mexico that includes methods of Geographic Information Systems and biophysical and socioeconomic statistics in current conditions and with climate change project of the 2013–2016 Collaborative Platform on Climate Change and Green Growth between Canada and Mexico.

References

Abuodha P, Woodroffe CD (2006) International assessments of the vulnerability of the coastal zone to climate change, including an Australian perspective. Australian Greenhouse Office, Department of the Environment and Heritage, Australia

Abuodha P, Woodroffe CD (2010a) Vulnerability assessment. In: Green DR (ed) Coastal zone management. Thomas Telford, London, pp 262–290

Abuodha P, Woodroffe CD (2010b) Assessing vulnerability to sea-level rise using a coastal sensitivity index: a case study from Southeast Australia. J Coast Conserv 14:189–205

Adger WN (2006) Vulnerability. Glob Environ Chang 16:268–281

Adger WN, Arnell NW, Tompkins EL (2005) Successful adaptation to climate change across scales. Glob Environ Chang 12(2):77–86

Amador GA, López G, Mendoza ME (2011) Three approaches to the assessment of spatio-temporal distribution of the water balance: the case of the Cuitzeo basin, Michoacán, México. Investigaciones Geográficas, Instituto de Geografía, UNAM, México 76:34–55

Asociación Mexicana de Institutos Municipales de Planeación (AMIMP) (2003) https://www.amimp.org.mx/. Accessed 26 Apr 2019

Azuz-Adeath I, Arredondo-García MC, Espejel I, Rivera-Arriaga E, Seingier G, Fermán JL (2010a) Referentes internacionales sobre indicadores e índices. Historia y estado del arte. In: Rivera-Arriaga E, Azuz-Adeath I, Alpuche GL, Villalobos-Zapata GJ (eds) Cambio climático en México un enfoque costero-marino. Universidad Autónoma de Campeche, Cetys-Universidad, Gobierno del Estado de Campeche, pp 845–858

Azuz-Adeath I, Arredondo-García MC, Espejel I, Rivera-Arriaga E, Seingier G, Fermán JL (2010b) Propuesta de indicadores de la Red Mexicana de Manejo Integrado Costero-Marino. In: Rivera-Arriaga E, Azuz-Adeath I, Alpuche GL, Villalobos-Zapata GJ (eds) Cambio climático en México un enfoque costero-marino. Universidad Autónoma de Campeche, Cetys-Universidad, Gobierno del Estado de Campeche, pp 901–940

Balica SF, Popescu I, Beevers L, Wright NG (2013) Parametric and physically based modelling techniques for flood risk and vulnerability assessment: a comparison. Environ Model Softw 41:84–92

Bormudoi A, Hazarika MK, Samarakoon L, Phosalath S, Sengtianthr V (2008) Flood hazard in Savannakhet Province, Lao PDR mapping using HEC-RAS, remote sensing and GIS. In: 29th Asian conference on remote sensing. Colombo, Sri Lanka

Comisión Nacional de Áreas Naturales Protegidas (CONANP) (2017) Programas de Adaptación al Cambio Climático en Áreas Naturales Protegidas. CONANP. https://www.gob.mx/conanp/documentos/programas-de-adaptacion-al-cambio-climatico-en-areas-naturales-protegidas

Dang NM, Babel MS, Huynh TL (2011) Evaluation of food risk parameters in the day river flood diversion area, Red River Delta, Vietnam. Nat Hazards 56:169–194

Dwarakish GS, Vinay SA, Natesan U, Asano T, Kakimuna T, Venkataramana K, Jagadeesha Pai B, Babita MK (2009) Coastal vulnerability assessment of the future sea level rise in Udupi coastal zone of Karnataka state, west coast of India. Ocean Coast Manag 52:467–478

Eakin H, Luers AL (2006) Assessing the vulnerability of socio-environmental systems. Annu Rev Environ Resour 31:365–394

Füsel HM (2007) Vulnerability: a generally applicable conceptual framework for climate change research. Glob Environ Chang 17:155–167

Gornitz VM, Kanciruk P (1989) Assessment of global coastal hazards from sea level rise. In: Proceedings of the 6th symposium on coastal and ocean management. American Society of Civil Engineers, Charleston, SC

Gornitz VM, Daniels RC, White TW, Birdwell KR (1994) The development of a coastal risk assessment database: vulnerability to sea level rise in the US southeast. J Coast Res 12:327–338

Grattan SR, Zeng L, Shamnon MC, Robert SR (2002) Rice is more sensitive than previous thought. Calif Agric 56:189–195

Harvey N, Woodroffe CD (2008) Australian approaches to coastal vulnerability assessment. Sustain Sci 3:67–87

Hinkel J (2011) Indicators of vulnerability and adaptive capacity: towards a clarification of the science-policy interface. Glob Environ Chang 21:198–208

Hoang HN, Huynh HK, Nguyen TH (2012) Simulation of salinity intrusion in the context of the Mekong Delta Region (Viet Nam). IEEE, pp 1–4

Huang Y, Li F, Bai X, Cui S (2012) Comparing vulnerability of coastal communities to land use change: analytical framework and a case study in China. Environ Sci Pol 23:133–143

Hubálek Z, Horáková M (1988) Evaluation of climatic similarity between areas in biogeography. J Biogeogr 15:409–418

IPCC (2007) Summary for policymakers. In: Palutikof J, van der Linden P, Hanson C (eds) Climate change 2007: impacts, adaptation and vulnerability. Contribution of working group II to the fourth assessment report of the Intergovernmental Panel on Climate Change. Cambridge University Press, Cambridge, pp 7–22

Kafle TP, Hazarika MK, Samarakoon L (2007) Flood risk assessment in the flood plain of Bagmati river in Nepal

Le S (2003) The restructuring of production in coastal areas in the Mekong River Delta. J Agric Rural Dev 5

Le QT, Nguyen HT, Le AT (2009) Climate change impacts and vulnerabilities assessment for Can Tho City. Asian Cities Climate Change Resilience Network (ACCCRN) program. DRAGON-Mekong-CTU, Can Tho, Vietnam

Liu J (1996) Macro-scale survey and dynamic study of natural resources and environment of China by remote sensing. Press of Science and Technology of China, Beijing

Mackey P, Russell M (2011) Climate change scenarios, sea level rise for Ca Mau, Kien Giang- climate change impact and adaptation study in the Mekong Delta. Asian Development Bank, TA 7377 e VIE. Sinclair Knight Merz (SKM), Vietnam Institute of Meteorology, Hydrology, and Environment (IMHEN), y the Kien Giang Peoples Committee, Melbourne VIC 8009 Australia

McFadden L (2007) Vulnerability analysis: a useful concept for coastal management? In: McFadden L, Nicholls RJ, Penning-Rowsell E (eds) Managing coastal vulnerability. Elsevier, Amsterdam, pp 15–28

McLaughlin S, Cooper JAG (2010) A multi-scale coastal vulnerability index: a tool for coastal managers? Environ Hazards 9:233–248

McLeod E, Poulter B, Hinkel J, Reyes E, Salm R (2010) Sea-level rise impact models and environmental conservation: a review of models and their applications. Ocean Coast Manag 53:507–517

Nardo M, Saisana M, Saltelli A, Tarantola S (2005) Handbook on constructing composite indicators: methodology and user guide. Organisation for Economic Co-operation and Development, Paris, p 207

Neumann B, Vafeidis AT, Zimmermann J, Nicholss RJ (2015) Future coastal population growth and exposure to sea-level rise and coastal flooding - a global assessment. PLoS One 10(3):e2371

Nguyen NT (2012) Assessing the vulnerability of coastal Phu Quoc Island, Kien Giang in terms of sea-level rise. Unpublished Master of environmental sciences Master degree. University of Natural Sciences, National University of Ho Chi Minh City, Ho Chi Minh, Vietnam

Nguyen NT, Bonetti J, Rogers K, Woodroffe CD (2016) Indicator-based assessment of climate-change impacts on coasts: a review of concepts, methodological approaches and vulnerability indices. Ocean Coast Manag 123:18–43

Nicholls RJ, Wong PP, Burkett V, Woodroffe CD, Hay J (2008) Climate change and coastal vulnerability assessment: scenarios for integrated assessment. Sustain Sci 3:89–102

Organisation for Economic Co-operation and Development (OECD) (2000) Frameworks to measure sustainable development. Organisation for Economic Co-operation and Development, París, p 164

Organisation for Economic Co-operation and Development (OECD) (2003) OECD environmental indicators development, measurement and use. Organisation for Economic Co-operation and Development, París, p 37

Özyurt G, Ergin A (2010) Improving coastal vulnerability assessments to sea-level rise: a new indicator-based methodology for decision makers. J Coast Res 26:265–273

Pendleton EA, Thieler ER, Williams SJ (2010) Importance of coastal change variables in determining vulnerability to sea- and lake-level change. J Coast Res 26:176–183

Pham HT, Nguyen VC (2005) Forecasting the erosion and sedimentation in the coastal and river mouth areas and preventive measures. State level research project. Hanoi. 407p

Preston BL, Smith TF, Brooke C, Gorddard R, Measham TG, Withycombe G, Beveridge B, Morrison C, McInnes K, Abbs D (2008) Mapping climate change vulnerability in the Sydney Coastal Councils Group. Report prepared for the Sydney Coastal Councils Group. Sydney Coastal Council Groups. 117 p

Sánchez-Torres G, Jospina-Noreña JE, Gay-García C, Conde C (2011) Vulnerability of water resources to climate change scenarios. Impacts on the irrigation districts in the Guayalejo–Tamesí river basin, Tamaulipas, México. Atmósfera 24(1):141–155

Secretaria de Desarrollo Social (SEDESOL) (2012) Guía Municipal de Acciones frente al Cambio Climático, Con énfasis en desarrollo urbano y ordenamiento territorial. http://www.inafed.gob.mx/work/models/inafed/Resource/330/1/images/

Secretaria de Gobernación (SEGOB) (2018) Acuerdo mediante el cual se expide la Política Nacional de Mares y Costas de México. Diario Oficial de la Federación. https://dof.gob.mx/nota_detalle.php?codigo=5545511&fecha=30/11/2018

Secretaria de Gobernación (SEGOB) (2019) Cobertura de Atlas Municipales. http://www.atlasnacionalderiesgos.gob.mx/archivo/cob-atlas-municipales.html. Accessed 26 Apr 2019

Secretaria de Medio Ambiente y Recursos Naturales (SEMARNAT) (2006) Política Nacional de Mares y Costas. https://www.biodiversidad.gob.mx/pais/mares/pdf/A4_PNMC_actualizada_dic2015.pdf

Seingier G, Espejel I, Ferman JL, Montaño G, Azuz I, Aramburo G (2011a) Halfway to sustainability. Ocean Coast Manag 54(2):123–128

Seingier G, Espejel I, Fermán JL, Montaño G, Azuz I, Aramburo G (2011b) Design of an inte-grated coastal orientation index. Cross-comparison of Mexican municipalities. Ecol Indic 11(2):633–642

Spangenberg JH, Bonniot O (1998) Sustainability indicators. A compass on the road towards sus-tainability. Wuppertal papers. http://nbn-resolving.de/urn:nbn:de:bsz:wup4-opus-7218

Tingsanchali T, Karim MF (2005) Flood hazard and risk analysis in the southwest region of Bangladesh. Hydrol Process 19:2055–2069

Troyo-Diéguez E, Mercado-Mancera G, Cruz-Falcón A, Nieto AG, Valdez-Cepeda RD, García-Hernández JL, Murillo-Amador B (2014) Análisis de la sequía y desertificación mediante índi-ces de aridez y estimación de la brecha hídrica en Baja California Sur, noroeste de México. Invest Geogr 85:66–81. https://doi.org/10.14350/rig.32404

Yin J, Yin Z, Wang J, Xu S (2012) National assessment of coastal vulnerability to sea-level rise for the Chinese coast. J Coast Conserv 16:123–133

Chapter 18
Social Cohesion and Environmental Governance Among the Comcaac of Northern Mexico

N. Martínez-Tagüeña and R. F. Rentería-Valencia

Abstract The Comcaac have inhabited the central coast of the Sonoran Desert in Northern Mexico since time immemorial. Acknowledging the value of their continuous presence and the adaptations it has generated, scholars have documented for decades the intricacies of their environmental knowledge—a complex corpus of socio-ecological relations in constant refinement and transformation. Yet, a crucial point missing within these efforts is the recognition of the ways in which the colonial encounter and the eventual incorporation of this indigenous people into a market economy in the twentieth century drastically re-organized the ways knowledge and power flux locally—an acknowledgement that consequently challenges scholarly understandings of traditional knowledge as extemporal. As old system of reciprocity and collective accountability transformed under new forms of social organization, the individualistic inclinations that characterize the Comcaac society were drastically exacerbated by capitalist logics, producing in turn new forms of power and governance that stand at odds with previous social logics and balances. The present chapter sheds light into the existing tensions that define Comcaac livelihoods in order to better understand the social creation and transformation of environmental knowledge while reflecting upon the vulnerability and resilience that characterizes the different governance systems of the Global South dryland regions.

Keywords Social cohesion · Governance · Capitalism · Indigenous peoples · Sonoran Desert

N. Martínez-Tagüeña (✉)
Cátedra CONACYT, Consortium for Research, Innovation and Development of Drylands,
Instituto Potosino de Investigación Científica y Tecnológica, San Luis Potosi,
San Luis Potosí, Mexico
e-mail: natalia.martinez@ipicyt.edu.mx

R. F. Rentería-Valencia
Anthropology and Museum Studies in Central Washington University, Ellensburg, WA, USA

Introduction

As established in the introduction of this book, the successful governance of coupled social-environmental systems is at the heart of the conceptual considerations
articulated in the 2030 Sustainable Development Agenda. The creation of stable
governance institutions with clearly defined equitable rules, and openness for shared
learning as a basis for collective action related to sustainable resource use is essential for the conservation of the biotic and cultural diversity (Ostrom 2000). Thus,
documenting and understanding the specific historical, cultural, economic, and
political unfolding of local communities can allow us to conduct interregional,
intercultural, and inter-policy comparisons among similar dryland socio-ecological
systems and their divergent responses to global environmental change drivers
(Chap. 1). In this context, the present chapter contributes to the advances on this
burgeoning literature by articulating the connection between transitional forms of
social cohesion and concomitant forms of environmental governance in a dry setting context among the Comcaac, an indigenous community of the coastal desert of
Northern Mexico.

The Comcaac are a small indigenous society of former hunter-gatherers living
next to the Gulf of California, in the Sonoran Desert of Mexico (Fig. 18.1). Residing
since time immemorial (Fig. 18.2) in small, mobile groups across ancestrally based
territories, by the turn of the twentieth century the Comcaac established in two

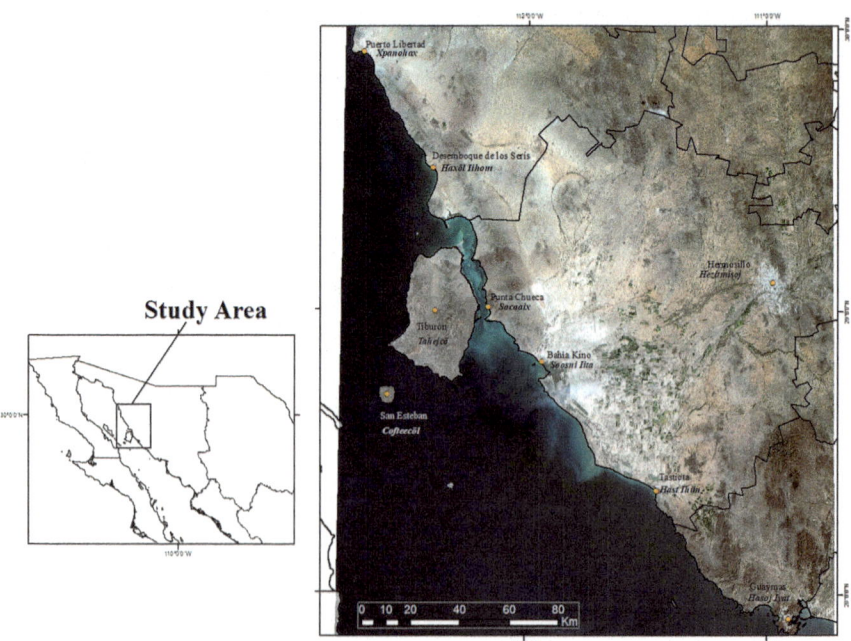

Fig. 18.1 Map indicating the Comcaac territory with some places mentioned throughout the text

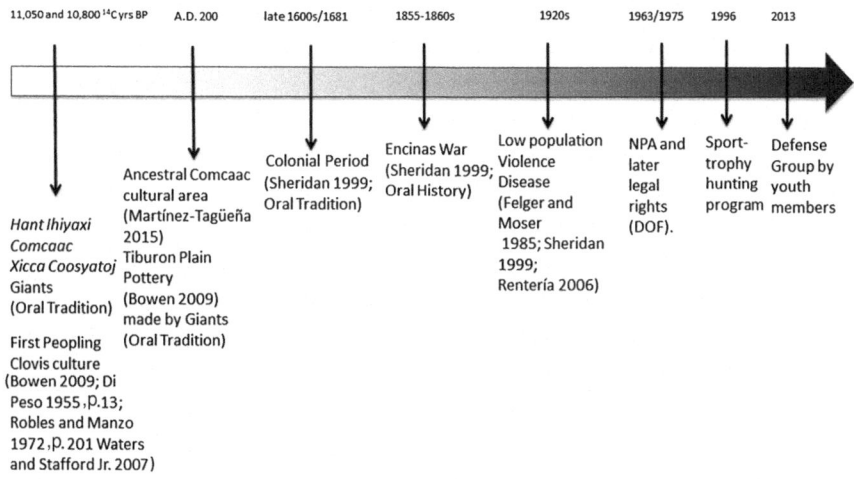

Fig. 18.2 Time line with relevant events for the proposed discussion

permanent settlements in the coast of Sonora: Punta Chueca (*Socaaix*) and Desemboque (*Haxöl Iihom*). Incorporated in the regional economies as small-scale commercial fishermen and wildlife managers, the Comcaac subsistence patterns and concomitant forms of social cohesion have experienced drastic transformations in the last century. Aiming to shed light on the implications that these changing conditions have had in the way this indigenous society approach and govern their environment, the present chapter broadly documents these transformations, emphasizing the relation between social cohesion, ecological knowledge, and environmental governance.

The objective of this ethnohistorical and ethnographic account is to shed light on the ways social cohesion and individualism have been negotiated throughout Comcaac history and the ways in which this negotiation has impacted the governance of their environment. Importantly, this is a story of radical tensions and transitions between common tenure of resources and their privatization or commodification; between metaphysical and institutional forms of power; between traditional forms of environmental knowledge and their academic translations; between malleable residential patterns of appropriating the territory and sedentarized lifestyles; and, inevitably, between indigenous and neoliberal forms of indigenous governance.

It could be argued that historically there is no aspect of the Comcaac form of life in the Sonoran Desert that has not been affected, first, by the colonial encounter and then by their incorporation into the Mexican state. And yet, the Comcaac retain important cultural principles that differ in fundamental ways from the ways of their contemporary Mexican and North American interlocutors; differences that hold fundamental implications in the way they govern and manage their environment.[1] In this context, our aim is to emphasize the plasticity of Comcaac social cohesion as it

---

[1] See Basurto (2005, 2006, 2008) for an analysis of how different land/maritime tenure systems produce radically different sustainable environmental management practices between Mestizo Mexicans and Comcaac people.

has adapted to different regimes of power. The ultimate goal being the analysis of the Comcaac case as a story of resilience, adaptability, and cultural creation in one of the most intriguing drylands of North America.

Social Cohesion and Governance Among the Comcaac

Comcaac leadership is an evasive topic of analysis; scant data and simplistic approximations permeate early heuristic efforts. Twenty-century scholars interested in understanding Comcaac settlement patterns and social organization focused their attention on analyzing band organization, their range and territory dynamics, and their descent system (Bahre 1980; Moser 1963; Sheridan 1996 1999) compiled an impressive ethnohistorical scope of the sources in search for subsistence information for, nearly all Colonial documentation deals with battles and expeditions to subdue the Comcaac into the mission life, leaving out the actual descriptions of the environment, people, and culture. This initial effort facilitated a later endeavor by Martínez-Tagüeña (2015) in which Comcaac oral accounts about the colonial Spaniards were combined with archival documents and with archaeological survey evidence from mentioned places. Comcaac accounts explain the relationship between leadership, prowess in warfare, and spiritual power.

While archival documents indicate that the Spaniards understood the role that Comcaac leaders play in their society, it is not until the actual voice of these actors is documented that a more complete historical narrative can be achieved—especially when we consider that spiritual power enhanced the capabilities of these leaders for survival against Spanish domination. The Comcaac favored certain qualities in their leaders: fast and efficient movement through the land, precise aim while shooting, the ability to innovate and strategize during difficult times, spiritual power, to be expert dancers, and possess a vast song-singing repertoire, not to mention strong personalities. Today, these characteristics are reflected in personalities idealized by the Comcaac—individualistic characters, opportunistic, and adaptable to the changing world that surrounds them (Martínez-Tagüeña et al. forthcoming).

Compiled evidence indicates that Comcaac bands had an extremely fluid social composition where individuals and families moved according to different needs in different pathways. Also, on occasions, several families congregated at different places for fiestas (traditional parties). Importantly, cultural and environmental reasons for movement cannot be dichotomized; they are interwoven on the land (for a detailed description, see Martínez-Tagüeña and Torres 2018) so this fission–fusion dynamic has to be understood as a complex socio-environmental engagement that was modified as a whole with the colonial encounter and their integration in a market economy.

Once existing in small groups, Comcaac families had recurrent grounds that composed an *ihiizitim*, an ancestral ground, and the fundamental unit of their territoriality. An extended family or group of families held priority in the use of a given *ihiizitim*. Ritual gestures like burying a newborn's placenta at a particular *ihiizitim* permanently established a relation between the individual to the land. Social interactions among the Comcaac were often determined by the possession of specific environmental knowledge, since resource availability and type varied from *ihiizitim* to *ihiizitim* and consequently, different kinds of environmental expertise developed along different familial lines. This condition was reinforced by social patterns of transmission of knowledge that emphasized individual creation over reproduction (Martínez-Tagüeña 2015; Alfredo Lopez, personal communication, June 2013). In turn, highly specialized knowledge "belonged" to just a few. Moser characterizes this:

> […] the hant iiha quimxoj or those who tell about the ancient ways, who in turned passed it on to the hant iiha cöhacomxoj or those who have been informed about the ancient ways. Today only a few members of the community are considered to be hant iiha cöhacomxoj, entrusted to be the carriers of knowledge to transmit it to future generations (Moser 2014, p. 39).

Nonetheless, while specialized knowledge and the resulting subsisting skills it created were considered a matter of individual achievement, the distribution of resources resulting from such specialized skills was not. Comcaac society enforced different modalities of reciprocity within and between different *ihiizitim*, generating a network that ensured resource distribution. The *Kimosin* represented one of such modalities:

> The *kimo?sími?a* represents a type of begging, somewhat akin to the 'palanqueo'. When food is being prepared, a person can approach another's house and announce his purpose, using the proper grammatical form of *kimo?sími?a*. If the family has a sufficient supply on hand or is of "good heart", the person will receive something to eat; the procedure indicates that one has come to ask for food because he is hungry—he has not come to visit or to trade. Today some people abuse this practice and go around saying "give me" but this is not the same as the *kimo?sími?a*; others, on the other hand, are not willing to give as freely as they did in the past (Griffen 1959, p. 33).

From this it should be evident that reciprocity and individuality as moral values played a key role in Comcaac subsistence. The first assured a relatively balanced distribution of resources across different families. The latter propitiated the constant updating of specialized knowledge and the mastering of basic subsistence skills. Together, they orchestrated power and expertise among the Comcaac—domains that were greatly affected not only by the colonial encounter but also by different waves of external influence during the twentieth century, namely the re-structuration of the many *ihiizitim* into a single "territory" owned as communal ejido land; the incorporation of the Comcaac as labor in the regional small-scale fishing industry of the region; the appropriation of new technologies and material goods; the establishment of permanent settlements; and the influence of Protestant logic all constitute processes that greatly altered said balance.

Comcaac Governance Through Time

The ancestral territory of the Comcaac extended along the coastal plains of what today is Sonora, including the islands Angel de la Guardia, San Esteban Island, and Tiburón (see Fig. 18.1). Shreve (1964) recognized seven subdivisions for the Sonoran Desert, and this area of study corresponds to the Central Gulf Coast subdivision characterized as Sonoran Desertscrub (Van Devender et al. 1994). This ecological-geographic typification is characterized by its extreme aridity and water scarcity. In the past, the Comcaac relied on springs and seasonal *tinajas* or bedrock tanks that collected rainwater which likely limited the size of encampments at different times. The absence of water in this sense can be considered a key factor in their fission dynamics. In contrast, their intertidal marine habitats are among the most diverse of the world (Álvarez-Borrego 2002) creating a tendency for social fusion during specific bonanzas. It is against the background of these contrasting realities of scarcity and abundance that the Comcaac people found in the Sonoran Desert their life-stage (Felger and Moser 1985).

Archaeological evidence places human presence in the ancestral Comcaac territory between 11,050 and 10,800 ^{14}C years BP (Waters and Stafford 2007). Regional material cultural differentiation is evident later in time with the development of pottery alongside the establishment of agricultural villages (Martínez-Tagüeña 2015). The Ancestral Comcaac cultural area is defined in specific by the presence of Tiburón Plain pottery type: a thin, light, and hard well-fired ware made without added temper (previously defined by Bowen (1976) as Central Coast Archaeological Tradition). Complementing these understandings, deep-time oral tradition of the Comcaac recognized two kinds of ancient predecessor conceived of as "Giants" (Felger and Moser 1985, p. 10; see Fig. 18.2 for a time line and Martínez-Tagüeña 2015 for an ethnohistorical discussion of these ancestors). Contemporary Comcaac speak Cmiique iitom, an isolated language (Marlett et al. 1998; O'Meara 2010) that corroborates a long history of occupation in the area (Marlett ms.26 quoted in O'Meara 2010, p. 16).

Initiating with the Colonial period and extending to Mexico's independence era, the Comcaac experienced drastic process of intercultural exchange that redefined their lifeways, settlement patterns, and social organization. Oral tradition and colonial archival documents indicate that the Comcaac lived in small, shifting, dispersed family groups that belonged to a number of subdivisions or bands reflecting geographic areas (Moser 1963; Smith 1947). The six subdivisions recognized by the Spaniards in late 1600s and early 1700s (Sheridan 1999) were: the *Xiica hai iic coii* or Tepocas or Salineros, the *Tahejcö Comcaac* or Tiburones or *Comcaacs*, *Heeno Comcaac* or Tiburones who lived in the central valley of *Tahejcö* or Tiburón Island, *Xnaa motat* or Guaymas and Upanguaymas, *Xiica xnaii iic coii* or Tastioteños, and *Xiica hast ano coii* or the people from San Esteban Island (Moser 1963). Each of this subdivision was at the same time composed by a small number of ihiizitim. Across time and space these bands had varying social interactions among themselves and also with their mobile and agricultural neighbors.

The European settlers were uninterested for the most part in the Comcaac territory. The lack of permanent or abundant water in the area, the perceived inadequacy of the land for agricultural production and more importantly, the often-impossible-to-understand cultural ways of the Comcaac and their unequivocal reluctance to become labor at the service of colonial expansion made the effort to missionized the area an unpleasant and low-priority task for the Crown (Sheridan 1999). It was until the first part of the Nineteenth century that the presence of European settlers was established well within the Comcaac rage. The radical confrontation of their cultural ways led to recurrent campaigns of extermination as the only stable form of expanding the colonial reach. The Comcaac resisted by guerrilla warfare, engaging and disappearing into the heart of their territory. In an extraordinary story of defiance, the Comcaac were considered, next to the Apache tribes the single most preoccupying presence for colonial peace (Sheridan 1999). But the reduction of the Comcaac territory had significantly started to take place.

After the Mexican Independence in 1821, mestizo inhabitants steadily encroached on Comcaac territory, populating large expanses of land with livestock. Pascual Encinas, owner of the "Costa Rica" ranch, carried out the most vicious of these campaigns. The "Encinas Wars" stretched from 1855 to the late 1860s, killing about half of the Comcaac population (McGee 1971/1898). The violence exerted during the colonial and postcolonial period left a small surviving population of ~130 Comcaac individuals by the early 1920s (estimated number reported in Felger and Moser 1985; see Sheridan 1999 and Rentería 2006 for details regarding this time period). This small group was soon incorporated into the emergent mestizo small-scale fisheries economy of the region, establishing two remote villages (Smith 1966) in the late 1940s became the only strategy to survive the violent relation with the Mexican population. The relative spatial and cultural isolation they enjoyed allowed them to slowly recover their numbers; nowadays, they represent a population of over a thousand individuals.

In 1963, Tiburón Island, the heart of the Comcaac territory, was designated as a natural protected area, resolution that henceforth prohibited this indigenous population from residing on the island. In a compensatory move, in 1975 the federal government granted communal legal rights to the Comcaac (via an "ejido"[2]) to part of their ancient territory, including Tiburón Island—although residing on the Island remains prohibited. The act created considerable legal confusion by imposing diverse legal frameworks that determine access and use of resources on the island, a situation that resulted in severe conflicts between the Comcaac people and a plethora

---

[2] After the Mexican revolution (1910–1917), the government established a collective land reform program in which lands were expropriated from large private owners and redistributed to landless peasants. Ejidos are a form of communal property in which land was distributed to a group of peasants (or fishermen in this case), where land ownership resides with the ejido community rather than the individual. In fact, the redistribution reform law stipulated that the redistributed lands remained the property of the federal government. The administration and management of these lands and their resources are collective. In 1991, however, the federal government passed neoliberal regulatory changes that allowed, among other changes, the sale of ejidos. Nonetheless, ejidos remain the second largest form of land tenure in Mexico (Valdez 2006).

of different Mexican interlocutors. In the late 1990s, the state-driven neoliberal conservation management programs that proliferated at the time in Mexico found in Tiburón Island their "golden goose." Bighorn sheep (*Ovis canadensis mexicana*) had been introduced to the island in 1975 as part of a repopulation program in Sonora. The growing numbers of the ram population in the following decades set the stage for a local NGO which in 1996 proposed the establishment of a sport trophy-hunting program among the Comcaac. The claims of success made by those associated with this program after it was implemented were largely based on the revenues that the Comcaac received for the hunting and extraction activities on the island, as well as their involvement in the management of the species (Renteria 2009, 2015).

The changes brought up during the twentieth century drastically changed Comcaac governance. The need to negotiate resources with Mexican villages and governmental institutions force the emergence of new figures of authority among the Comcaac—a governor, an elder's council, and ejido (communal land property) managers. These new figures, albeit crucial in their relation to Mexican state authority remain internally fragile and incohesive. While an interconnection between the different governance levels (local, regional, national, and international) is crucial for successful environmental stewardship and sustainable development, the Comcaac prefer to achieve self-management and independence.

The relation between the Comcaac and the Mexican mestizos has always been tense—intensely violent at times. The quotidian experiences of racism and the troubling history of genocide that mark their encounter are never minimized by this indigenous community. In this context, the Zapatista movement that erupted in 1994 in Southern Mexico to challenge the narratives and structures of a modern, neoliberal Mexico, a country that had symbolically erased and systematically oppressed its indigenous societies, empowered many of these groups to take a more defiant stance against processes that continue the colonial approach. The Comcaac were not the exception. Building on the singular territorial isolation that characterized their *ejido* (in comparison to many other indigenous communities in Mexico) paired with a strong sense of independence in their natural resources set the ground for a Comcaac nationalist narrative that portrayed them as an independent nation within Mexico—to a certain extent reproducing perceived logics of the reservation system in the USA. A "Comcaac Nation" flag and a national anthem—an ancient song of resistance dating back to the colonial encounter—became ubiquitous symbols of independence. More importantly, an armed youth, organized as a Comcaac Traditional Guard, organized to defend their territory and resources, a strategy that produced new clashes with the Mexican government and the mestizo communities that surround their territories but effectively empowering this indigenous community in terms of environmental governance.

In recent years the Comcaac have been threatened by drug trafficking related problems and different external initiatives that seek to conduct "development" activities in their territory: illegal mining; wildlife poaching; illegal fishing in their exclusive fishing zone; the establishments of a desalination plant along the Infiernillo Channel, among others. These threats pushed several young community members to organize as a defense group in 2013; branching out of the Comcaac Traditional Guard, these youths strongly relied on social media and external alliances to challenge these interventions. In addition, young artists, through music and poetry are participating in many national festivals and venues to promote the protection of their land and the sustainable development of their people. The reaction of Comcaac youth against these threats should not pass unnoticed for it represents, in first instance, a story of resilience in the face of social and environmental adversity. Yet, perhaps more importantly, it sheds light on the plasticity that Comcaac forms of social cohesion have in the production of new forms of environmental governance. And it is in this context, that the Comcaac have valuable lessons to contribute to the key conversations that compose this chapter.

Re-conceptualizing Comcaac Environmental Knowledge

The academic emphasis on (traditional) ecological knowledge has promoted international recognition of its potential applications in resource management practices and sustainable development. It is not only relevant to understand how some local communities have sustained resilient landscapes, but it is also essential for the successful stewardship of diverse socio-ecological systems (SES).[3] In the different studied contexts, it is primordial to understand the social creation and transformation of environmental knowledge that will thus shed light on the vulnerability and resilience that characterizes the different governance systems that communities' transition from or maintain in the dryland regions. Reynolds et al. (2007) emphasize the importance of system memory (local knowledge and social learning), as well as system legacy (socio-environmental history and path dependence), and their implications for SES stewardship. While the Comcaac are known for their high survival capacity and their adaptability to external change due to their vast traditional knowledge, they are in constant threat not only due to external pressures that are a menace to their territory and their resources, but also by internal divisions that limit their capacity to organize. And it is at the interphase that the role of scholars documenting and disseminating Comcaac ecological knowledge during the last century should be analyzed with a critical eye.

A plethora of researchers have documented Comcaac traditional knowledge[4] mostly through ethnobiological studies (Felger and Moser 1970, 1971, 1974a, b, 1985; Felger et al 2012; Malkin 1962; Moser 2014; Narchi et al. 2015; Hernández-

---

[3] Socio-ecological systems understood as complex adaptive systems, where the relationship between humans and nature is based on interconnections among system parts whose interlinkages and their dynamics create new, so-called emerging properties with synergistic effects compared to the original system elements (Berkes and Folke 1998) Linking Social and Ecological Systems. Cambridge University Press, Cambridge.

[4] Traditional knowledge understood as the result of countless observations and empiric experiments by nature observers that has been transmitted orally or through writing, which has value of its own, is contextual and historically variable (Berkes 2009).

Santana and Narchi 2018) but also through the compilation of rich oral history centered in their sophisticated views of the natural order expressed in stories, poetry, and songs (Di Peso and Matson 1965, p. 5; Felger and Moser 1985; Marlett et al. 1998; Martínez-Tagüeña 2015; Martínez-Tagüeña and Torres 2018; Moser and Marlett 2010; Renteria 2009, 2015). At large, it was not until the 1960s that scholars enjoyed systematic access to a more intimate view of the Comcaac universe through the pioneering language-based work of Edward and Rebecca Moser who documented the relationships between ecological traditional knowledge and material life (Felger and Moser listed above; Moser 1963). The vast research that resulted in the last part of the twentieth century has nonetheless uncritically generalized assertions in atemporal vacuums ("Comcaacs believe this or that") omitting in turn an examination of the enormous variability that characterizes Comcaac knowledge. It includes how their power structures affect and have affected its production, distribution, and implementation, conditions that reflect the vertiginous historical, cultural, and economic transformations described above.

More problematic, however, is the fact that until today and with the exception of a few Comcaac members, the community does not have access to much of this information (published in expensive volumes or erudite dissertations written mostly in English) nor they have been given credit and ownership for this knowledge. With a few exceptions were publications have been made in co-authorship, see Marlett et al. 1998; Martínez-Tagüeña and Torres 2018; Wilder et al. 2007, 2008a, b. While published studies provide ethnobiological information centered on cataloging lists of plants and animals with their associated uses, they do not describe how knowledge is generated, how it evolves and changes through time, and how it is distributed among community members (Narchi et al. 2015 explored the association between ethnomedicinal knowledge gender, age, and literacy but through the problematic lens of proficiency assessment).

Descriptions of management practices around useful/relevant plants and animal populations are scarce or non-existent nor an acknowledgment of these changing practices through time. In the same tenet, attention has failed to underpin the underlying processes of territorial and/or species-specific types of expertise that sustain differential types of expertise within this society and its implications in terms of social networks, reciprocity, and access to resources. Families possess differing knowledge about their land, reflecting the different territories of the Comcaac bands to which they belonged—affiliations that are still relevant to many of them. It is very important not to consider the Comcaac as a homogenous group. Even though contemporary Comcaac have lost much of their homeland, they relate to their ancestral territory through long-term continuity expressed in their vast knowledge (Martínez-Tagüeña 2015).

Furthermore, no information has been published about the local community members' perception on the changing contexts that they are living, on the one hand, by their contact with researchers that come to the community and document their traditional knowledge, and on the other hand, by their interactions with government agencies and extractive industries that are creating conflict over their resources and their territory and are promoting the inadequate appropriation of their patrimony.

Thus, its documentation can be a two-edge sword since it makes traditional knowledge available to be stolen through biopiracy[5] but it can also serve as a mechanism for indigenous communities to legitimize and establish traditional knowledge (and concomitant practices) as their own.[6]

In this chapter we conceptualize Comcaac environmental knowledge as a complex corpus of socio-ecological relations in constant refinement and transformation; a synchronic installment of known principles and facts to the new changing contexts of the everyday. We also recognize the ways in which the colonial encounter and the eventual incorporation of this indigenous society into a market economy in the twentieth century drastically re-organized the ways knowledge and power flux locally—an acknowledgement that consequently challenges scholarly understandings of traditional knowledge as extemporal, characterizing it instead as a diachronic process. Thus, this dynamic understanding of (ecological) knowledge recognizes the emergence of indigenous research. For instance, Humberto Romero-Morales a plant expert collaborated in the project on the Gulf of California islands flora (Felger et al. 2012; Wilder et al. 2007, 2008a, b). Young members have had their own monitoring organizations (the most enduring one being Grupo Tortuguero Comcaac, originally funded by the NGO Ocean Revolution[7]) and have had projects financed by the CONANP (the Federal Commission for the Natural Protected Areas) to establish bird diversity and migration patterns, sea turtle behavior and egg protection, bighorn sheep and sea mammals population estimates, plus trash collection, among others. These conservation efforts in some cases have been translated into bachelor's degrees earned by Comcaac youth in environmental law, biology and ecological engineering. This weaving of academic knowledge with traditional ecological knowledge in resource management and decision-making is key for environmental stewardship (Berkes and Seixas 2005).

On the Emergence of Comcaac Environmentality

The last two decades have characterized the Comcaac's subsistence mode as focused on small-scale fisheries and wildlife conservation and management. In 1998, the first wave of "Comcaac para-ecologists" was formalized as part of a multi-institutional effort to train Comcaac youth as stewards of their natural resources (Nabhan 2003). Some youth were trained to conduct basic monitoring with

---

[5] Biopiracy refers to the use of traditional knowledge without permission and the appropriation of genetic resources by individuals that seek monopoly control in the search for new bioresources (http://www.etcgroup.org/issues/patents-biopiracy).

[6] Both the World Intellectual Property Organization and the Nagoya Protocol that is part of the Convention on Biological Diversity employ published data as a source to verify the ownership of this knowledge by the different indigenous groups: https://www.wipo.int/tk/en/; https://www.cbd.int/abs/doc/protocol/nagoya-protocol-en.pdf.

[7] http://www.oceanrevolution.org/index.php?option=com_content&view=article&catid=35:OR%20Revolutionaries%20Entries&id=18.

ecological concepts. Since then, many pilot conservation programs among the Comcaac were implemented, relying on these trained individuals—ranging from trash collection to fostering the transmission of "Traditional Ecological Knowledge." The "Grupo Tortuguero Comcaac," previously mentioned, represents the most stable and successful of all these efforts where every year, Comcaac youth conduct surveys and tag resident sea turtles of various species and they also protect and carry for the eggs until they hatch and return to the ocean. Many young Comcaac, including boys and girls, are heavily invested in this group, to the point that it constitutes part of their identity.

As the Comcaac are incentivized by neoliberal conservation programs to learn and care for species highly valued by conservationists, their youths' relation to "nature" balances precariously on uneasy grounds, where pressing economic necessities, foreign ecological epistemologies, and new ethical stances have had a destabilizing effect on the constituency of their selves. Individuals, aiming to make sense of the often-contradictory forces that compose their quotidian subsistence, rely on narratives like the one described in this article. Thus, as symbols, moral stances, practices, and identities work in tandem across neoliberal landscapes, this article has drawn attention to the crucial role that individualism plays in the articulation of the antagonistic laminations of the self—and the concomitant moral tensions between individuality and reciprocity that it negotiates.

The above point has profound implications for anyone conducting research among the Comcaac—especially those that insist on treating them as a homogenous "community." There is an extraordinary variation in terms of how the Comcaac approach the world—something evident to anyone who has ever conducted fieldwork among them. Yet, in books and articles Comcaac plurality is erased, clumped together at best into conveniently cohesive narratives. These homogenizing narratives fail to represent the nuances, complexities, and contradictions that permeate their daily life. In addition to being an epistemological oversimplification that blatantly erases the brutal process of Spanish and Mexican colonization that forced a mobile, territorially subdivided and highly individualistic society into the sedentary indigenous settlements they form today. The theoretical and practical consequences that the colonial process exerted in the configuration of contemporary Comcaac life are evident at any level of analysis—at least when we choose to pay attention to it—the uneven distribution of knowledge, power, and wealth is a testament to it. In this sense, a new academic understanding of what being Comcaac means nowadays is needed; an attempt capable of recognizing the historical continuity of this tradition as well as the insurmountable losses and sociocultural (re)creations that co-constitute a definition of their current daily existence—their quotidian becoming.

Therefore, researchers working in the area face a true challenge for our theoretical and methodological attempts to document and explain their plurality. It is through true transdisciplinary endeavors that further promote the integration of knowledge systems that we will collaboratively co-produce knowledge (i.e., Johnson et al. 2016). Academics and other stakeholders working in the area need to have equitable collaborations with Comcaac members, where participation takes place throughout all the project's stages. Furthermore, the publications should provide information on how many persons shared a specific oral history and when relevant acknowledging the individual who recounted it. When complete oral stories are published, the Comcaac elder should be co-author of the publication with a complete description of his source of knowledge.

The Comcaac have much to teach about how to conceptualize the environment as part of us, they do not refer to nature as a resource to be used, extracted, or appropriated; moreover, their land is their ancestors' bodies and their oral stories reveal an entanglement between objects, people, nature phenomena, animals, and plants. As a millenary indigenous community with a vast socio-ecological memory, they have lived through change and uncertainty, developing the capacity to learn from crisis to respond and adapt to global socio-environmental changes. Therefore, the Comcaac are the key players to achieve integrated assessments, define conservation, restoration, management and development projects, and guide and inform adaptive management and policy development.

The different transformative waves experienced by the Comcaac people in this becoming have forged, out of different intercultural dialogues, a paradoxical, yet fascinating environmentality. Whereas the connection between environmental knowledge and territoriality was drastically altered as the ihiizitim systems collapsed and the relationship between Comcaac subsistence practices and their environmental resources was redrawn by the arrival of new technologies, the implementations of new forms of land tenure (ejidos) and their insertion in market-oriented economies (commercial fishing and wildlife management) changed the panorama. The emergent Comcaac leadership are renegotiating the different ecological and capitalist forces that determine their livelihoods by framing their decision-making processes in cultural logics and principles of appropriation of their environment embedded in their ancestral ecological knowledge; a set of principles and information crystalized in the hundred songs that all children, adults, and elders continue to sing at every fiesta, puberty ceremony, or new year celebration, but perhaps more importantly, every night, when the lights go out and the desert prepares to sleep.

Acknowledgments The authors acknowledge the many members of the Comcaac Indigenous Community that have supported and collaborated with them. They also thank their respective affiliations for their support. NMT thanks her Cátedra CONACYT ID Number 6133 as part of the project 615. And RRV appreciates the support provided by the Anthropology and Museum Studies at Central Washington University.

References

Álvarez-Borrego S (2002) Physical oceanography. In: Case TJ, Cody ML, Ezqurra E (eds) A new Island biogeography of the sea of Cortés. Oxford University Press, New York, pp 41–59
Bahre CJ (1980) Historic seri residence, range and sociopolitical structure. Kiva 45:197–209
Basurto X (2005) How locally designed access and use controls can prevent the tragedy of the commons in a Mexican small-scale fishing community. Soc Nat Resour 18(7):643–659
Basurto X (2006) Commercial diving and the Callo de Hacha fishery in Seri territory. J Southwest 48(2):189–209

Basurto X (2008) Biological and ecological mechanisms supporting marine self-governance: the Seri callo de hacha fishery in Mexico. Ecol Soc 13(2). https://doi.org/10.5751/ES-02587-130220

Berkes F, C Folke (1998) Linking Social and Ecological Systems: Management Practices and Social Mechanisms for Building Resilience. Cambridge University Press, Cambridge

Berkes F, Seixas CS (2005) Building resilience in lagoon social-ecological systems: a local-level perspective. Ecosystems 8(8):967–974

Berkes F (2010) Indigenous ways of knowing and the study of environmental change. J R Soc N Z 39(4):151–156

Bowen T (1976) Prehistory: the archaeology of Central Coast of Sonora, México. Anthropological Papers of the University of Arizona Number 27. Tucson

Di Peso C (1955) Two Cerro Guaymas Clovis fluted points from Sonora, México. Kiva 21(1–2): 13–15

Di Peso CC, Matson DS (1965) The Seri Indians in 1692: as described by Adamo Gilg, S.J. Arizona West 33(I):33–56

Felger RS, Moser MB (1970) Comcaac use of agave (century plant). Kiva 35(4):159–167

Felger RS, Moser MB (1971) Comcaac use of mesquite (*Prosopis glandulosa var. torreyana*). Kiva 37(3–4):53–60

Felger RS, Moser MB (1973) Eelgrass (*Zostera marina L.*) in the Gulf of California: discovery of its nutritional value by the Comcaac Indians. Science 181:355–356

Felger RS, Moser MB (1974a) Columnar cacti in Comcaac Indian culture. Kiva 39(3–4):257–275

Felger RS, Moser MB (1974b) Comcaac Indian pharmacopeia. Econ Bot 28(4):414–436

Felger RS, Moser MB (1985) People of the desert and sea. Ethnobotany of the Comcaac Indians. The University of Arizona Press, Tucson

Felger RS, Wilder BT, Romero-Morales H (2012) Plant life of a desert archipelago: flora of the Sonoran islands in the Gulf of California. University of Arizona Press, Tucson

Griffen WB (1959) Notes on Seri Indian Culture, Sonora, México. University of Florida Press, Gainesville

Hernández-Santana G, Narchi N (2018) The seri traditional food system: cultural heritage, dietary change, and the (Re) awakening of dietary resilience among coastal hunter-gatherers in the Mexican Northwest. In: Price LL (ed) NE Narchi coastal heritage and cultural resilience. Springer, New York, pp 135–182

Johnson JT, Howitt R, Cajete G, Berkes F (2016) Weaving Indigenous and sustainability sciences to diversify our methods. Sustain Sci 11:1–11

Malkin B (1962) Seri ethnozoology. Occasional papers of the Idaho state college museum, Pocatello

McGee WJ (1971/1898) The Seri Indians of Bahía Kino and Sonora, Mexico. Bureau of American Ethnology to the Secretary of the Smithsonian Institution 17th Annual Report, 1895–1896. The Río Grande Press, Glorieta

Marlett S (2007) Las relaciones entre las lenguas 'hokanas' en México: ¿cuál es la evidencia? In: Buenrostro C et al (eds) Memorias del III Coloquio Internacional de Lingüística Mauricio Swadesh. UNAM/INI, Ciudad de México, pp 165–192

Marlett SM, Astorga de Estrella ML, Moser BM, Nava F (1998) Las canciones Comcaac: una visión general. Cuarto Encuentro Internacional de Lingüística en el Noroeste 1(2):499–526

Martínez-Tagüeña N (2015) And the giants keep singing: Comcaac anthropology of meaningful places. Dissertation, The University of Arizona

Martínez-Tagüeña N, Torres LA (2018) Walking the desert, paddling the sea: Comcaac mobility in time. J Anthropol Archaeol 49:146–160

Moser EW (1963) Comcaac bands. Kiva 28(3):14–27

Moser C (2014) Shells on a Desert Shore. Mollusks in the seri world. The University of Arizona Press, Tucson

Moser EW, Marlett S (2010) Comcáac quih yaza quih hant ihíip hac: Diccionario seri- español-inglés. Plaza y Valdés Editores, Mexico City

Nabhan GP (2003) Singing the Turtles to Sea: The Comcáac (Seri) Art and Science of Reptiles. University of California Press, Berkeley

Narchi NE, LE Aguilar-Rosas, JJ Sánchez-Escalante, DO Waumann-Rojas (2015) An ethnome-dicinal study of the Seri people: a group of hunter-gatherers and fishers native to the Sonoran Desert. J Ethnobiol Ethnomed 11(1): 62
O'Meara C (2010) Comcaac Landscape Classification and Spatial Reference. Buffalo State University of New York, Dissertation
Ostrom E (2000) El gobierno de los bienes comunes. La evolución de las instituciones de acción colectiva. Fondo de Cultura Económica, Mexico City
Rentería R (2006) Los bordes indomables; ritualidad e identidad etnica entre los Comcaacs o Comcaac Bachelor Thesis. Escuela Nacional de Antropologia e Historia
Rentería R (2009) Hunting on the slopes of Tiburon: bighorn sheep's market-oriented conservation in Northern Mexico. MA thesis, The University of Arizona
Rentería R (2015) Ethics, Hunting tales and the multispecies debate: the entextualization of nonhu-man narratives. Vis Anthropol Rev 31(1):94–103
Reynolds JF, Stafford Smith DM, Lambin EF (2007) Global desertification: building a science for dryland development. Science 316:847–851
Sheridan TE (1996) The Comcáac (Comcaacs): people of the desert and sea. In: Sheridan TE, Perezo NJ (eds) Paths of Life. American Indians of the Southwest and Northern México. The University of Arizona Press, Tucson, pp 187–212
Sheridan TE (1999) Empire of Sand. The Comcaac Indians and the Struggle for Spanish Sonora, 1645–1803. The University of Arizona Press, Tucson
Shreve F (1964) Vegetation of the Sonoran Desert. In: Shreve F, Wiggins I (eds) Vegetation and Flora of the Sonoran Desert, vol 1. Stanford University Press, Stanford, pp 1–45
Smith WN (1947) Seri Field Notes, 1947. Unpublished manuscript filed at Tucson. the University of Arizona Libraries Special Collections
Smith WN (1966) Seri Field Notes, 1966. Unpublished manuscript filed at Tucson. the University of Arizona Libraries Special Collections
Valdez R, Guzmán-Aranda JC, Abarca FJ, Tarango-Arámbula LA, Sánchez FC (2006) Wildlife conservation and Management in Mexico. Wildl Soc Bull 34(2):270–282
Van Devender TR, Burgess TL, Piper JC, Turner RM (1994) Paleoclimatic Implications of Holocene Plant Remains from the Sierra Bacha, Sonora, México. Quatern Res 41:99–108
Waters MR, Stafford TW Jr (2007) Redefining the age of Clovis: implications for the peopling of the Americas. Science 315(5815):1122–1126
Wilder BT, Felger RS, Romero H (2008a) Succulent plant diversity of the Sonoran Islands, Gulf of California, Mexico. Haseltonia 14:128–161
Wilder BT, Felger RS, Van Devender TR, Romero H (2008b) *Canotia holacantha* on Isla Tiburón, Gulf of California, Mexico. Can Underwrit 4(1):1–7
Wilder BT, Felger RS, Romero H, Quijada-Mascareñas A (2007) New plant discoveries for Sonoran Islands, Gulf of California, Mexico. J Bot Res Inst Tex 1:1203–1227

Chapter 19
Governing Drylands Through Environmental Mainstreaming: How to Cope with Natural Resources Scarcity and Climate Change

M. Zortea and S. Lucatello

Abstract Since the original conception of "sustainability" as a guiding paradigm for development planning, it has been evident how necessary but also difficult can become the integration of social, political, ecological and economic aspects together, because of their complex interrelations and trade-offs. And in the specific case of drylands a transversal or mainstreaming approach is particularly recommendable. After framing the concept of environmental mainstreaming as integrated and cross-cutting approach, its theoretical functioning and the mechanism of practical application, the challenge of how to enhance basic skills, which are the main prerequisite for an effective application of the EM, will be explored. Then we will examine how to apply it concretely, through three distinct and consequential stages: to build-up policies, to construct the implementation paths, and to build constant monitoring and evaluation mechanisms. The final part of the chapter will be focusing on some experiences in Mexico and other countries of LAC.

Keywords Environmental governance · Environmental mainstreaming · Capacity building · Resources scarcity · Climate change · Drylands

M. Zortea (✉)
UNESCO Engineering for Human and Sustainable Development,
DICAM – Department of Civil, Environmental and Mechanical Engineering,
University of Trento, Trento, Italy
e-mail: massimo.zortea@unitn.it

S. Lucatello
Estudios Ambientales y Territoriales, Instituto Mora, Mexico City, Mexico

© Springer Nature Switzerland AG 2020   337
S. Lucatello et al. (eds.), *Stewardship of Future Drylands and Climate Change in the Global South*, Springer Climate,
https://doi.org/10.1007/978-3-030-22464-6_19

Introduction

In territories like drylands—already made more fragile by climate change and by their context, in itself more difficult—the progressive marginalization of populations from the attention of policy makers and from economic circuits has provoked a vicious circle: depopulation, abandonment of the territory, economic unproductivity, which multiplies the depopulation, and so on. In such a framework, it is evident that administrative planning plays a decisive role with a broad vision and capable of considering all the relevant social, economic, and environmental aspects (Zortea and Lucatello 2016).

A transversal approach is needed and expressed by the principle 4 of the Rio Declaration on Environment and Development (UNCED 1992) with an integration of the various environmental aspects and impacts within all the components of the administrative and political management of a territory that is broad and disrupted, by its nature. Figure 19.1 (cited from UNDP 2008) shows an excellent example of the need to consider the interdependence of factors.

This requires the introduction of the political and technical culture of environmental mainstreaming (EM) in decision-making processes and in the design and implementation both of wide-ranging routes and specific projects. We need to get out of the emergence logic and adopt long-term logic, precisely. This perspective implies strengthening capacities, training trainers, and introducing mandatory forms of learning for local administrators. On the other hand, since the original conception of "sustainability" as a guiding paradigm of development planning, it has become

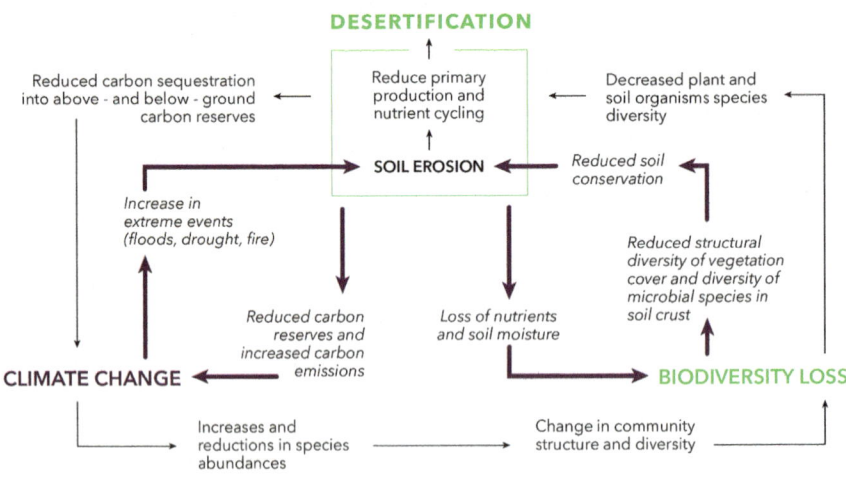

In green: major components of biodiversity involved in the linkages
Bolded: major services impacted by biodiversity losses

Fig. 19.1 Interdependence of factors, integration of environmental, socio-political, and administrative components

evident how necessary but also how difficult integrating social, political, ecological, and economic aspects is—because of their complex interrelations and trade-offs (Gibberd 2017). Furthermore, in the specific case of drylands, challenges are multiplied by the complexity of well managing the so-called fragile ecosystems, like deserts, and in the last decades, the situation has been worsened by climate changes. In many cases even accelerated by anthropogenic factors, desertification actually holds a significant weight in the framework of global human suffering: it affects around one-sixth of the world's population, seven-tenths of all drylands, and one-quarter of the total land area of the world, according to the most recent World Atlas of Desertification (Cherlet et al. 2018).

All phenomena related to this not encouraging picture are interlinked, as well as the underlying factors, both natural and human. Even within the 2030 Agenda for Sustainable Development this interdependence of factors has been made evident, by including them in several goals and targets, rather than locking all in a specific, separate goal.

Environmental Mainstreaming: Concept and Relevance for Drylands

Desertification can be defined by the process of land degradation in arid, semi-arid, and dry sub-humid areas resulting from various factors, including climatic variations and human activities. Land degradation is commonly described as the reduction or loss in arid, semi-arid, and dry sub-humid areas of the biological or economic productivity and complexity of rain-fed cropland, irrigated cropland, or range, pasture, forest, and woodlands (Cherlet et al. 2018).

Combating desertification and land degradation includes activities which are part of the integrated development of land in arid, semi-arid, and dry sub-humid areas for sustainable development; they are aimed at prevention and/or reduction of land degradation, rehabilitation of partly degraded land, and reclamation of desertified land, as remarked by UNDP (2008). Land degradation results from a process or combination of processes, including those arising from human activities and habitation patterns, including soil erosion caused by wind and/or water, deterioration of the physical, chemical, and biological or economic properties of soil and long-term loss of natural vegetation. The promotion of processes of sustainable development is already in itself very complex, in an increasingly globalized world where many of the factors favoring and disfavoring development have a transversal character and go beyond local and often national contexts (UNEP 2012). This requires an open, transversal, and above all interdisciplinary approach. The structuring of the new global agenda for development, the 2030 Agenda for Sustainable Development, is very eloquent in this regard.

Faced with these challenges, also political and technical responses must be inspired by greater transversality, inclusiveness, and transdisciplinarity. The mainstreaming approach, in particular the Environmental Mainstreaming (EM), represents an effective methodological setting, even if not always easy to apply (Dalal-Clayton and Bass 2009). According to UNDP, Environmental Mainstreaming Strategy 2004, EM refers to the integration of environmental policy considerations into core institutional thinking with other policies and related activities, as well as with coordination and harmonization to ensure policy coherence (Bass 2008; Dalal-Clayton and Bass 2011). To be successful therefore, environmental mainstreaming must be adopted as an institutional culture of doing business (Dalal-Clayton and Bass 2010). If mainstreaming is to feed into planning and decision-making, it should permeate all types of planning frameworks involved in the implementation of drylands issues, such as policies, laws, standards, institutions, technologies, curricula, funding mechanisms, programmes, projects, plans, etc. At the same time, EM permeates the different stages of the formulation of these frameworks: conceptualization and identification, design, appraisal, budgeting, implementation, and monitoring and evaluation (UNDP 2008).

Since the work of the World Commission of the Environment and Development, in 1980s, mainstreaming environment into strategic decision-making was perceived as an essential prerequisite of sustainable development, moving beyond the traditional idea of environmental policy being separate and discrete from other policy. Thus, the Commission states: "The ability to choose policy paths that are sustainable requires that the ecological dimensions of policy be considered at the same time as the economic, trade, energy, agricultural, industrial and other dimensions on the same agendas and in the same national and international institutions" (WCED 1987).

As well known among scholars studying the theme, there are at least three broad perspectives of mainstreaming: (1) procedural mainstreaming; (2) methodological mainstreaming, and (3) substantive mainstreaming. For an effective application of EM to drylands, each of those is to be studied and implemented, both at global and local level.

In turn, drylands themselves have been the subject of a mainstreaming, to integrate their consideration into all development policies and actions. For example, "Drylands mainstreaming is a systematic practice to integrate drylands issues in all decision-making processes, policies and laws, institutions, technologies, standards, planning frameworks, etc. and to ensure that they continue to be part of the agenda in subsequent decision-making processes, implementation and revision" (UNDP 2008).

Focusing specifically on the sustainable development and sustainability for drylands, the positive relationship between the services drylands provide and human well-being deserves to be remarked, but also the risk of negative serious consequences from drylands degradation. Drylands ecosystems directly contribute to basic materials for human use, security, and society cohesiveness as featured in Fig. 19.2. In parallel, the capacity of the ecosystems to sustain those functions can be undermined by natural, physical, and biological factors, both direct and indirect, e.g., high population growth rates, poor socio-political environment, cultural and religious barriers, market failures or market absence, flawed policies, and weak

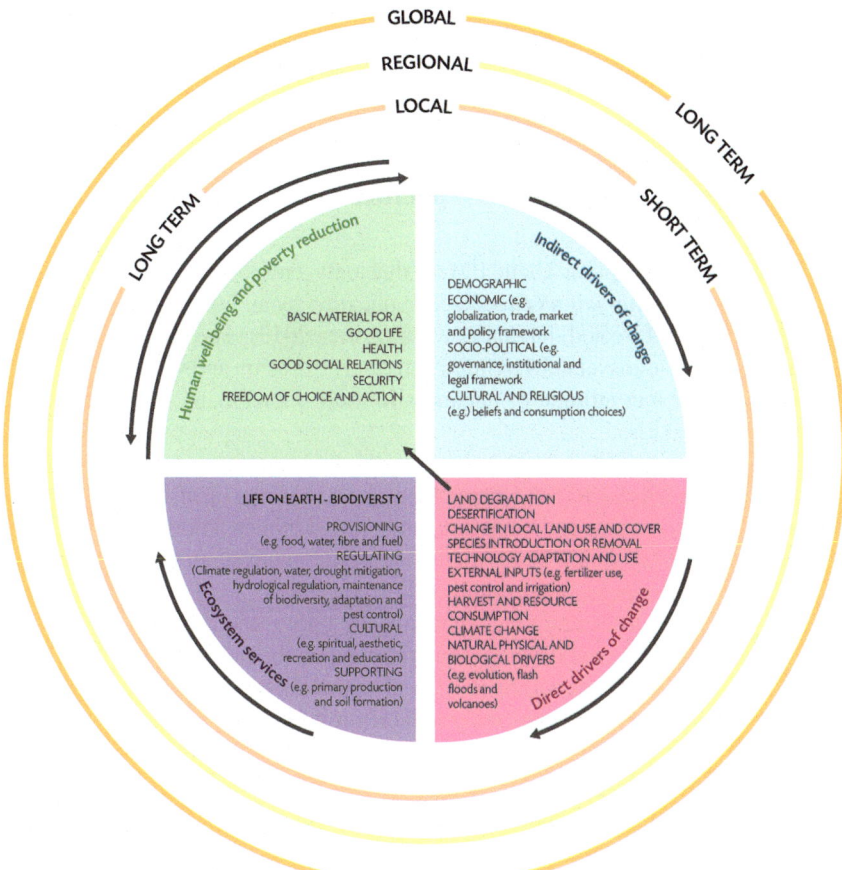

Fig. 19.2 Drylands ecosystems contribution to basic materials for human use, security, and society cohesiveness in long and short terms

institutional capacities (Bass et al. 2010). Acting on that tangle of factors requires a typical mainstreaming process.

According to World Desertification Atlas (Cherlet et al. 2018) underlying and familiar factors—some old, some new—are driving environmental change/land degradation at a global scale and some recurring global issues (such as surface and ground water) have an alarming urgency that could not be known 20 years ago. There is a growing confirmation of suspected global trends (such as a decline in productivity) that may impact sustainability. Some regional patterns of potential degradation are reconfirmed (south Asia; China) and some underlying causes are revealed (heavy fertilizer use and irrigation). Concerns emerge at the regional level that brings into question our ability to meet the demands of future populations, e.g., maintaining and increasing yields on high-density croplands.

The same 2030 Agenda contains several references to drylands contexts, even though not ever explicitly: in particular, Targets 1.5, 2.3, 2.4, 6.4 referring to water scarcity, 13.1, 13.2, 13.3, 13.a, 13.b, 15.1, 15.2, 15.3, 15.5, 15.8, 15.9, 17.18.

A Paramount Component: Capacities and Capacity Building

A common misconception about EM is that only environmental specialists can apply it. On the contrary, if a generalized application by all policy-makers and project managers is not treated, it doesn't work (Zortea 2013). The basic characteristic of the EM, i.e., its transversality, requires precisely diffusion and widespread knowledge and requires an attitude of openness to multi-, inter-, and trans-disciplinary knowledge and skills.

Since governance and institutional systems should be reformed in order to assure a more balanced approach to the three classical components of sustainable development—economic, social, and environmental—the attitudes, knowledge, and skills of the human capital should be re-oriented to accept mainstreaming as a culture of doing business as opposed to an additional responsibility.

It is therefore worth examining what capacity public and private policy-makers and project managers must develop. Much depend on the individual context, on the level of intervention (regional, national, and regional), and the objectives that are set. In this regard, it is interesting to remind the initiative of the Chinese government that has created a manual on traditional knowledge and practical techniques in the fight against desertification (UNDP 2008).

But in general, two aspects should be taken care of:

(a) Basic knowledge of the concept and the method;
(b) EM application tools, which are at least four different types: informative; educational; operational; evaluative (Zortea 2013).

To build these skills, it is useful to divide the learning paths into two levels: a **basic**, introductory, and suitable for all categories of trainees, and an **advanced**, focused on more targeted contexts and application objectives. For instance, in the advanced level it's recommendable to study the synergies among UNCCD and other conventions and initiatives for mitigating the effect of drought and climate change, such as the UNDP/GEF funded project on "Coping with drought and climate change: best use of climate information for reducing land degradation and conserving biodiversity." That project supports sustainable livelihoods of drylands populations, promoting an integrated ecosystems management approach that hinges on better use of local and scientific knowledge on climate in farming and herding. Given the complex but close linkages between droughts, land degradation, biodiversity, and climate change, the programme gives attention to the interface among the three Rio Conventions: UNCCD, CBD, and UNFCCC. A very useful tool is the *Sourcebook of opportunities for enhancing cooperation among the Biodiversity-related*

Conventions at national and regional levels (UNEP 2015), especially Chap. 5 on capacity building.

Another example is the Global Drylands Imperative (GDI), a collaboration of organizations involved in drylands development, and its tools, such as the challenge papers on these thematic areas such as Poverty and the Drylands; Strategies for the Sustainable Development of Drylands Areas; Biodiversity in the Drylands; Vulnerability and Adaptation to Climate Change in the Drylands. Furthermore, we can cite also UNDP-GEF's "Guidelines for developing GEF projects in the cross-cutting area of land degradation," as a tool suitable for advanced training.

Finally, the topic of resources mobilization deserves to be mentioned as a "learning must." In addition to UNDP core resources through the Drylands Development Centre a range of other funds and special programmes including the Global Environment Facility (GEF), GEF Small Grants Programme, Capacity 21 Programme, the United Nations Capital Development Fund (UNCDF), Africa 2000 Initiative, and the Special Unit for Technical Cooperation among Developing Countries (SU/TCDC) are available. All of them include drylands development objectives in their activities.

How to Do (1): Building Up Policies

In order to try to build the EM application processes in a solid and lasting way, it is essential to take care of three distinct levels of action:

1. Strategic planning;
2. Implementation;
3. Monitoring and evaluation.

For greater convenience it is worth examining them in separate paragraphs.

To apply the EM to the first level in a winning way, experience teaches that it is important keeping in mind and taking care of three methodological aspects: logical planning, participatory approach, and attention to results (Zortea 2013).

In other words, it is recommendable to adopt a strategic sequence, endowed with some characteristics, i.e.:

(a) Planning (set up according to criteria of linear design, with well-defined actions in a given space and time oriented to the achievement of definite objectives);
(b) Logic (rational, non-emotional, non-emergency, but far-sighted, with coherent policies);
(c) Transversal (able to consider the contexts in all their complexity of subjects, levels, interaction of problems, causes and effects, taking into consideration both the global dimension and the local dimension);

 – to involve all stakeholders (to make a careful mapping of stakeholders and their characteristics, both favorable and unfavorable to environmental

mainstreaming; to carefully plan their participation, enhancing in particular the inter- and trans-disciplinarity),
- to pay attention to the results from the outset (to define them, also setting indicators and parameters to monitor and evaluate them; to invest in monitoring and evaluation: see also paragraph 6).

The EM must be applied at the local (community), sub-national, national, regional, and global levels, otherwise the minimum scale required to significantly impact the livelihoods of people cannot be reached. At the same time, the optimal level for a best impact of mainstreaming depends on several factors. For instance, regional institutions can best handle issues of trans-boundary nature, such as regional conflict over natural resources and use of shared resources such as river basins and lakes, whereas nation-specific problems such as regulating irrigation practices in drylands or defining access to land can be handled at national level. Information flow among all the levels is critical to ensure congruence and consistency.

It is worth remembering that even UNDP adapted its global strategies in supporting UNCCD Global Mechanism, making them more transversal and adopting institutional and management changes to improve support to drylands development. In particular, it established a new Drylands Development Centre in Nairobi. The Centre has developed and will be coordinating a new UNDP integrated drylands development programme that focuses on issues critical to drylands development and the implementation of the UN Convention to Combat Desertification.

On the other hand, experiences from worldwide UNDP projects and from single countries highlight several common difficulties in applying EM. For example, National Action Programmes (NAPs) often are not integrated into the Poverty Reduction Strategy Paper (PRSP) process; reasons to mainstream environment and drylands issues are not clear to all actors; mainstreaming is an expensive and a time demanding process; budgets for mainstreaming are insufficient; a serious gap exist between planning and implementation; participatory approach matters; communication in the mainstreaming process has a critical role (UNCCD-UNDP-UNEP 2007).

Furthermore, policy advocacy and awareness raising have great importance, as well as raising awareness on desertification and its consequences. There is a clear need for more investments in scientific research to provide data/information and scientific evidence to decision makers, in order to ensure mainstreaming is taken up as an important issue. In this sense, technology plays a relevant role, in particular GIS for collecting and processing data/information on desertified areas, for planning strategies, and for giving decision makers scientific evidence to the importance of environment.

How to Do (2): Building Up Implementation Paths

Even in presence of an excellent strategic planning, the construction of suitable routes of implementation is not easy and almost always represents the real obstacle to overcome. It is not possible to provide universal recipes and certainly not successful. However, some recommendations can be identified, which help to set up solid paths. First of all, EM must be planning and far-sighted: to resist short-term temptations. Secondly, it is necessary to take proper care and not to leave critical aspects to the case: adequate timing, adequate scale (planning globally, building locally), broad coordination, open attitude to transversality. Thirdly, it is useful to deal with all the interlocutors, with a sustained effort of inclusiveness, not only in the most operative phase of the interventions but also in that planning phase.

Furthermore, it is advisable to implement the EM gradually, choosing first the themes and the tools that are most easily understood and visible to the various stakeholders, then moving on to topics, tools, and size of broader application. Also for this reason it is good to choose narrow pilot paths: either a theme (e.g., water resource management) or an area or a community.

Finally, it is necessary to manage the communication of paths to beneficiaries and also to promote the dissemination of the results.

The steps for mainstreaming may be structured in five phases: the assessment phase; the awareness raising, participation, and partnership building phase; the planning phase; the implementation phase; the evaluation phase (UNDP 2008).

A good methodology is to proceed with systematic and gradual steps; among other, the following are remarkable: identifying the environmental, economic, and social impacts; identifying and filling information gaps; assessing the legal, political, and institutional environment for mainstreaming, making a deep stakeholder analysis; carrying out capacity assessment and building; drawing up a communication and awareness creation strategy (Zortea and Lucatello 2016).

How to Do (3): Monitoring and Evaluating

As highlighted at the beginning of paragraph 4, it is essential to pay attention to the results of the mainstreaming process and therefore to address all the interventions according to the achievement of these results. That means, on one hand, to commit to accurately define these results and, on the other hand, to ensure adequate monitoring during the interventions and an in-depth evaluation, once concluded.

It is therefore necessary to plan and invest in monitoring and evaluating (M&E), even with adequate human and financial resources. Preliminarily, it requires a planning of the M&E as if it was a project in itself: it should not be neglected, as a merely secondary activity. But even when it has been properly planned, time, energy, and adequate resources must be invested in its execution. Finally, a further element often overlooked reveals all its importance: giving value and operational follow-up to M&E, adequately using the results in contemporary or subsequent communication and awareness activities. In particular, the impact of implemented activities on the well-being of the people and the effectiveness of mainstreaming processes must undergo periodic monitoring and evaluation with a view to: identifying barriers to addressing drylands issues and building on good practices in order to upscale and replicate (UNDP 2008).

Evaluation of the mainstreaming process can be carried out concomitantly with evaluation of the development programmes, projects, or initiatives. It should focus also on the impacts risen from implementing the mainstreamed programmes. However, how to link M&E with communication and awareness raising strategies?

Successful mainstreaming includes citizen participation by ensuring that they have sufficient knowledge about drylands issues and they are informed about the policy or plan being developed. Unfortunately, many factors act as barriers to communication, including the diversity of languages and dialects, the liberalization of the media, and poor infrastructure. For these reasons, a well-planned communication strategy is crucial. The strategy should pervade all processes in an iterative manner. Most importantly, it should target the stakeholders that were prioritized during stakeholder analysis. A key element for successful communication strategies is advocacy. That aims to bring to the forefront country-specific evidence needed to convince skeptical policy makers, economists, and planners of the need for drylands mainstreaming.

Some Experiences in Mexico and Latin American Countries

Throughout all Central and South America, there are numerous contexts of drylands where the opportunity of cross-cutting analysis and of strategic/operational approaches of EM emerges clearly. The areas of work or phases, which the method is applicable to, are many: national or local strategic planning, capacity building, project planning, targeted technological applications, project financing, collection and analysis of data for monitoring, and evaluation at national or local level.

In theme of planning, we can recall the case of the EM approach adopted by CONABIO (Mexican Biodiversity National Commission) and other public dependencies such as SAGARPA (the Mexican ministry for agriculture) in Mexico. An example can be the project implemented by GIZ on mainstreaming biodiversity into the Mexican agricultural sector. This project aims to recognize and integrate the value of biodiversity and ecosystem services into decision-making and planning instruments of public and private key actors in the Mexican agricultural sector (IKI Alliance México 2018).

With a variety of actors involved, starting from international and national organizations, sustainable land use practices are going to be tested and integrated into selected agricultural production systems and value chains for the period 2018–2021. The same practices will be embedded in public policies and in practical experiences in selected states with the purpose of up-scaling these experiences to governing and policy levels. Another major aim of the project is to promote inter-sectoral dialogue and generates concrete examples for the integration of biodiversity into agriculture with several spillovers (IKI Alliance México 2018).

In theme of capacity building, the efforts carried out for years by the Instituto Mora of Mexico City, an institution that is part of CONACYT, the Mexican national center of researches, deserve mention. Initially, a training manual was produced on EM in development cooperation projects, dedicated to practitioners, also suitable for local administrators, with numerous application examples focused on the reality of LAC (Zortea and Lucatello 2016). Subsequently a presential training event was promoted. Finally, a special capacity building path was also launched in online mode, again on the EM applied to sustainable development projects.

Many thematic areas are known and studied in LAC for which there is a significant interaction of various environmental components, both with each other and with the social and economic components. Just think of the mining sector, where "hydrocarbon processing and mining operations demand variable amounts of water and energy and can severely affect the environment and the quality of water resources" (ECLAC 2018). That interconnection has great relevance in almost all the region, but particularly in the mining countries of the Andean zone, Brazil, Mexico, Venezuela, and some of the Central American nations. For the hydrocarbon sector, the relationship can be particularly pronounced when large-scale hydraulic fracturing is located in areas where water is scarce, such as northern Mexico.

Another area very significant for these interconnections, especially in the drylands, is the withdrawal of aquifers for irrigation purposes. The growing importance of groundwater and dependence is common particularly in Central America and Mexico, where it accounts for 65% of all the water used, and in the deserts and semideserts of Argentina, Brazil, Chile, Bolivia, Mexico, and Peru. The rise in intensive exploitation and contamination of aquifers is interconnected with the three components of the Water-Energy-Food Nexus: it impacts water quantity and quality, removes land from production, and increases the energy cost for its extraction.

A relevant example of interlinkage between different sectoral policies is that of agricultural and energy policies: subsidies for energy production favor a lower energy cost and therefore an increase in the use of plants for the exploitation of aquifers for agricultural irrigation purposes. But this in turn causes over-exploitation and greater environmental damages, which—especially for arid areas—often end in the deterioration of both the ecosystems and the living conditions for resident populations (ECLAC 2018).

An Interesting Case Study from Chile

In Chile it is estimated that the desert will advance from 0.4 to 1 km per year, having as main causes climatic changes that accentuate the aridity of the territory, as well as inadequate and persistent human activities, overgrazing and erosion. Facing this scenario, it is essential to enhance the resilience of communities in drylands. Fog water harvesting represents a potential water source for communities with low annual rainfall but frequent fog events. Studies and large-scale applications indicate that the technology is viable and feasible. An interesting case of application for such technology is the fog harvesting project in an agricultural community of the Coquimbo region, in Chile, one of the most affected by desertification. That project give evidence to the value of fog harvesting projects as a community fortification instrument: fog water is a valuable contribution to drylands and to boost small rural realities. The agricultural community of Peña Blanca presents a high degraded soil, due to the extensive cultivation of wheat and the scarcity of rainfall, with prolonged droughts as a result of climate change. Because of these critical conditions the majority of the population have emigrated to the nearby center of Ovalle: nowadays the Comuneros (inhabitants of the agricultural community) that continue to live in the community are 70.

The Atrapaniebla Comuneros project is a fog harvesting project at community scale located in the Cerro Grande Ecological Reserve, 100 hectares in the middle of the desertic coastal zone of Coquimbo region. Thanks to the use of fog collectors (atrapanieblas), installed in the Cerro Grande reserve in 2006 by the Chilean foundation Un Alto en el Desierto, it was possible to develop a vegetation recovery process in the area and stimulate the relaunch of Peña Blanca community life. The success of a fog harvesting project depends on the active and long-term presence of a local organization and on the participation and training of the community. The local population of Peña Blanca recognized the "Camanchaca" (local word for fog) as a beneficial contribution to the coastal arid ecosystem, and a continuous collaboration between the community and the foundation has been in place since 2005.

The intervention of the foundation in origin was aimed at curbing this migration phenomenon and move towards a local development working with the community to develop adaptive strategies. To date, the project has three main objectives: regenerate the native vegetation of Cerro Grande to stop the advance of the desert; support the community agricultural/pastoral activities and be a water supply resource for the population in periods of extreme drought; promote the creation of local trade through the direct and indirect use of the project. It is worth mentioning the willingness of the foundation Un Alto en el Desierto to transform Cerro Grande into an open-air classroom on tackling desertification: turning the reserve into a green education site.

Desertification has environmental (climate change, soil degradation), social (need to sustain human populations in fragile territories), and economic (lack of resources to adopt environmentally friendly technologies) roots; so that desertification is both a cause and a consequence of poverty. Fog harvesting, as a source of water in arid areas, is clearly not a solution to the global problem of desertification, but can trigger positive mechanisms of local development and community resilience.

Furthermore, fog resource is complementary to rain: precipitation events occur mainly during the winter months, giving even more value to this unconventional resource. The water collected by the project (about 600,000 liters per year) serves to reforest the ecological reserve where the fog catchers are located, is a sustaining source for the inhabitants of the agricultural community during extreme droughts, and finally is used for the production of artisanal beer: the only one in the world with fog water.

Given the results, Cerro Grande is an excellent site for fog harvesting and there are all the conditions to improve: e.g., support of a local organization, community involvement. The project Atrapaniebla Comuneros proves that fog harvesting at community scale certainly succeeds in restoring the economic impulse and providing environmental and social support, without requiring unsustainable costs for the community. Finally, fog water collection is an environmentally friendly intervention that does not rely on energy consumption: fog water harvesting is a green technology. Such water collection option provides ecosystem services as well, thus increasing the climatic resilience of the context in which is used (on all this case study, see widely Dall'Osteria 2018).

Conclusions

The survival of drylands and of populations that inhabit them depends on cross-cutting and transversal factors, which often underlie socio-environmental conflicts. Scarcity creates competition and conflict, as revealed by the Nobel Prize, Wangari Maathai: "When resources are degraded, we start competing for them. [...] So one way to promote peace is to promote sustainable management and equitable distribution of resources."

The approach to sustainable development of drylands must be equally transversal, both at a strategic and operational level. In this sense, EM techniques, if applied methodically, can constitute a more complete and lasting response over time, in particular by the local government authorities. Environmental mainstreaming is a driver to improve governance in drylands because it suggests a cross-cutting work method and a road map for decision makers, also helping them analyze problems and plan solutions with a wider perspective. Furthermore, EM stimulates an inter- and trans-disciplinary approach even in non-environmental experts.

But from this perspective it is necessary to invest in specific EM training to all stakeholders, not only to policy-makers: by spreading the basic culture of EM, by training trainers at both national and local level and perhaps even by making a basic path for local administrators mandatory and following some steps who were just sketched in this article.

References

Bass S (2008) Brief review of UNDP environmental mainstreaming in relation to the UNDP-UNEP poverty-environment initiative. IIED, London

Bass S, Roe D, Smith J (2010) Look both ways: mainstreaming biodiversity and poverty reduction. IIED, London

Cherlet M, Hutchinson C, Reynolds J, Hill J, Sommer S, von Maltitz G (2018) World atlas of desertification. Publication Office of the European Union, Luxembourg

Dalal-Clayton B, Bass S (2009) The challenges of environmental mainstreaming. Experience of integrating environment into development institutions and decisions. IIED, London

Dalal-Clayton B, Bass S (2010) Environmental mainstreaming. A key lever for a green economy: challenges and approaches. IIED, London

Dalal-Clayton B, Bass S (2011) Environmental mainstreaming diagnostic. IIED, London

Dall'Osteria G (2018) Fog harvesting as a resilience tool in semi-arid environments: the case study of Peña Blanca, Chile. University of Trento—Unesco Chair in Engineering for Human and Sustainable Development, Trento

ECLAC (2018) Network for cooperation in integrated water resources management for sustainable development in Latin America and the Caribbean. The United Nations Economic Commission for Latin America and the Caribbean (ECLAC), Natural Resources and Infrastructure Division (Circular n. 49), Santiago de Chile

Gibberd J (2017) Strengthening sustainability planning: the city capability framework. Procedia Eng 198:200–211. https://doi.org/10.1016/j.proeng.2017.07.084

IKI Alliance México (2018) Mainstreaming biodiversity into the Mexican agricultural sector—IKI IBA. Ciudad de México: IKI Alliance México website (available at: http://iki-alliance.mx/en/portafolio/mainstreaming-biodiversity-into-the-mexican-agricultural-sector)

UNCED (1992) The Rio declaration on environment and development. United Nations Conference on Environment and Development (UNCED), Rio de Janeiro

UNCCD-UNDP-UNEP (2007) Proceedings of the international workshop on mainstreaming environment with a particular focus on drylands into national development frameworks (Bamako Mali 18–20 June 2007). UNDP-DDC, Nairobi

UNDP (2008) Generic guidelines for mainstreaming drylands issues into National Development Frameworks. United Nations Development Programme (UNDP), New York

UNEP (2012) GEO 5 global environment outlook. Environment for the future we want. United Nations Environment Programme (UNEP), Nairobi

UNEP (2015) Sourcebook of opportunities for enhancing cooperation among the biodiversity-related conventions at national and regional levels. Nairobi, United Nations Environment Programme (UNEP)

WCED (1987) Our common future. UNDESA, The World Commission on Environment and Development, New York

Zortea M (2013) L'integrazione ambientale nei progetti di sviluppo. FrancoAngeli, Milan

Zortea M, Lucatello S (2016) El mainstreaming ambiental en los proyectos de cooperación internacional y desarrollo. Universidad Iberoamericana and Instituto Mora, Ciudad de México

Index

© Springer Nature Switzerland AG 2020
S. Lucatello et al. (eds.), *Stewardship of Future Drylands and Climate Change in the Global South*, Springer Climate,
https://doi.org/10.1007/978-3-030-22464-6

Printed by Printforce, the Netherlands